Technical Advances
in Biomedical Physics

W0106948

NATO ASI Series

Advanced Science Institutes Series

A Series presenting the results of activities sponsored by the NATO Science Committee, which aims at the dissemination of advanced scientific and technological knowledge, with a view to strengthening links between scientific communities.

The Series is published by an international board of publishers in conjunction with the NATO Scientific Affairs Division

A	Life Sciences	Plenum Publishing Corporation
B	Physics	London and New York
C	Mathematical and Physical Sciences	D. Reidel Publishing Company Dordrecht and Boston
D	Behavioural and Social Sciences	Martinus Nijhoff Publishers The Hague/Boston/Lancaster
E	Applied Sciences	
F	Computer and Systems Sciences	Springer-Verlag Berlin/Heidelberg/New York
G	Ecological Sciences	

Series E: Applied Sciences – No. 77

Technical Advances in Biomedical Physics

edited by

P.P. Dendy
Chief Physicist
Department of Medical Physics
Addenbrooke's Hospital
Cambridge, U.K.

D.W. Ernst
Professor of Biophysics
University of Hannover
Hannover, Federal Republic of Germany

A. Şengün
Professor of Radiobiology
Department of Biology
University of Istanbul
Istanbul, Turkey

1984 **Martinus Nijhoff Publishers**
The Hague / Boston / Lancaster
Published in cooperation with NATO Scientific Affairs Division

Proceedings of the NATO Advanced Study Institute on Technical Advances in
Biomedical Physics, Istanbul, Turkey, September 14-28, 1982

Library of Congress Cataloging in Publication Data

NATO Advanced Study Institute on Technical Advances in
    Biomedical Physics (1982 : Istanbul, Turkey)
    Technical advances in biomedical physics.

    (NATO advanced science institutes series. Series E,
Applied sciences ; no. 77)
    "Proceedings of the NATO Advanced Study Institute on
Technical Advances in Biomedical Physics, Istanbul,
Turkey, September 14-28, 1982."
    "Published in cooperation with NATO Scientific Affairs
Division."
    Includes index.
    1. Biomedical engineering--Congresses. 2. Medical
physics--Congresses. I. Dendy, P. P. II. Ernst, D. W.
III. Sengün, A. (Atif) IV. North Atlantic Treaty
Organization. Scientific Affairs Division. V. Title.
VI. Series. [DNLM: 1. Biophysics--Methods--Congresses.
2. Cytological technics--Congresses. 3. Diagnosis--
Congresses. QT 34 N2796t 1982]
R856.A2N38 1982   610'.28   83-25482

ISBN-13: 978-94-009-6127-2   e-ISBN-13: 978-94-009-6125-8
DOI: 10.1007/978-94-009-6125-8

Distributors for the United States and Canada: Kluwer Boston, Inc., 190 Old Derby
Street, Hingham, MA 02043, USA

Distributors for all other countries: Kluwer Academic Publishers Group, Distribution
Center, P.O. Box 322, 3300 AH Dordrecht, The Netherlands

TABLE OF CONTENTS

PREFACE

This Advanced Study Institute was arranged to discuss in depth
the physical and technical basis of the latest developments in methods
of measurement and image analysis suitable for determining the properties of cells and tissues and for evaluating medical structures.
All topics under consideration have benefitted dramatically from an
injection of new ideas during the past 10 years, and some have
developed even more recently. The Institute brought together
lecturers and participants from 14 different countries, and the
subject matter recorded in this volume may be considered under two
general headings.

The first part of the meeting concentrated on techniques that
are most appropriate at the cellular level. One major area of development has been centered on attempts to classify cells by computerized extraction of visual features, and here, it was notable how different
techniques frequently complement each other. This part of the meeting
also examined mechanisms of damage at the cellular level caused by
different forms of radiation, and the contrasting effects of ionizing
radiation, ultraviolet light and ultrasound were highlighted.

The second part of the meeting covered a wide range of diagnostic
medical imaging methods which have been developed during the past
decade, and this volume contains informative, up-to-date reviews on
medical radiography, radionuclide and ultrasound imaging, and the
important new technique of nuclear magnetic resonance imaging. This
section of the conference also brought forth an enthusiastic and en-
lightened lecture on the clinician's expectations from these new technologies, an excellent review of the constraints operating in the
clinical environment, and finally, a brief insight into clinical hopes
for the future.

The Editors of this volume, who were also the Directors for the
Advanced Study Institute, would like to take this opportunity to
thank the Scientific Affairs Division of NATO for their financial
support of the Advanced Study Institute. Our thanks are also due
to the Faculty of Science of the University of Istanbul for their
substantial financial contribution and to the local organisational
team which, under Professor A. Sengün, worked so hard to attend to
the needs of the scientists visiting Istanbul.

P. P. Dendy

OPENING REMARKS

As Spoken at the Opening of the ASI
by Professor Ahmet Yüksel Özemre

Dean of the Faculty of Sciences
University of Istanbul
Istanbul, Turkey

En ma qualité de Doyen de la Faculté des Sciences de
l'Université d'Istanbul et de membre honoraire du Comité
d'Organisation de "l'Institut d'Etudes Supérieures de l'OTAN sur
les Progrès Techniques en Physique Biomédicale", j'ai le grand
plaisir de vous souhaiter la bienvenue et de déclarer l'ouverture
officielle des activités de l'Institut.

J'ai également l'honneur de transmettre à vous tous les
salutations de notre vénérable Recteur M le Professeur Demiroğlu
qui souhaite que cette rencontre fournisse l'occasion à de
fructueuses échanges d'idées, et qu'elle donne lieu à beaucoup
de succès.

Il m'est un devoir agréable d'exprimer aussi mes remerciements
au Conseil Scientifique de l'OTAN à notre Conseil national pour la
Recherche Scientifique et Technique et aux firmes privées qui ont
généreusement contribué, avec le concours de notre Faculté, à la
réalisation matérielle de cet Institut.

Comme tout le monde le sait, le programme de l'Institut
d'Etudes Supérieures de l'OTAN est avant tout, une activité
d'enseignement de très haut niveau avec un sujet d'étude bien
défini, et présenté dans le cadre d'un programme systématique et
cohérent. Le sujet en est toujours traité en profondeur par
d'éminents conférenciers de réputation internationale.

Un simple coup d'oeil sur le programme que vous avez tous,
permettra d'ailleurs de réaliser immédiatement que le présent
"Institut d'Etudes Supérieures de l'OTAN sur les Progrès Techniques
en Physique Biomédicale" possède toutes les chances de s'avérer un
spécimen de très haute qualité scientifique parmi ses semblables.

C'est pourquoi, je suis tout à fait confiant quant à l'issue et à l'impact de cet Institut qui n'aurait guère vu le jour s'il n'y avait eu le zèle et la diligence de ses principaux inspirateurs et organisateurs, c'est-à-dire MM les professeurs Sengün, Dendy et Ernst.

Je leur exprime franchement mon grand estime et mes félicitations.

Cet Institut n'aurait atteint son but ultime que s'il avait pu donner lieu à une collaboration ultérieure entre les scientifiques des pays membres de l'OTAN qui vont se recontrer ici.

Je crois sincèrement que les conditions propices, requises pour une telle collaboration prolongée existent potentiellement aussi bien dans le sujet qui va être traité que dans le choix judicieux des conférenciers et des autres participants.

Finalement, je souhaite à tous les participants un très agréable séjour à Istanbul et un très bon souvenir des activités aussi bien scientifiques que sociales auxquelles ils prendront part.

OPENING REMARKS

As Spoken at the Opening of the ASI
by Professor Ahmet Yüksel Özemre

Dean of the Faculty of Sciences
University of Istanbul
Istanbul, Turkey

In my capacity as Dean of the Faculty of Sciences of the University of Istanbul and honorary member of the Organization Committee of the NATO Advanced Study Institute on Technical Advances in Biomedical Physics, I have great pleasure in welcoming you and in announcing the official start of the activities of the Institute. I also have the honor of conveying to all of you the compliments of our esteemed Rector Professor Demiroğlu, who hopes that this meeting will lead to a fruitful exchange of ideas and great success.

It is also a pleasure to express thanks to the Scientific Committee of NATO, to our National Council for Scientific and Technical Research and to private companies which have generously contributed, with the assistance of our Faculty, to the realization of this Institute.

As everybody knows, the program of the NATO Advanced Study Institute is first of all a teaching activity of very high quality with a well-defined aim within the framework of a systematic and coherent program. The subject is always profoundly treated by eminent lecturers of international reputation. A simple glance at the program will cause you to realize immediately that the NATO Advanced Study Institute on Technical Advances in Biomedical Physics has every opportunity of proving to be an example of out-standing scientific quality. I am, therefore, quite confident as to the results and the impact of this Institute, which would not have been founded but for the zeal and diligence of its principal founders, Professors Şengün, Dendy and Ernst, to whom I convey my highest esteem and congratulations.

The Institute will only have reached its ultimate objective if it will lead to future collaboration between the scientists of the NATO countries who will meet here.

I sincerely believe that the favorable conditions required for such a prolonged collaboration potentially exist in the subject to be treated, as well as in the judicious choice of the lecturers and other participants.

Finally, I wish all participants a very pleasant stay in Istanbul and good memories of the scientific and social activities in which they will take part.

I. OVERVIEW LECTURES

NEW PHYSICAL METHODS IN MEDICINE AND ENVIRONMENTAL SCIENCES

H. GLUBRECHT

Institut für Biophysik, Universität Hannover, 3000
Hannover-Herrenhauser 21, Herrenhauser Strasse 2,
Hannover, G.F.R.

1. INTRODUCTION

For many years - up to the first third of this century -
biomedical research was based mainly on chemical methods. Only a
few physical techniques played a role, such as microscopy, optical
spectroscopy and the application of X-rays for direct photographic
imaging in diagnosis.

During the last 20 or 30 years, more and more physical methods
have been introduced, mainly for research in diagnosis. Nowadays
a laboratory for biological or medical research, and even a room
for clinical diagnosis, looks more like a physicist's laboratory,
full of optical, electronic and other instruments. Many, if not
the majority of the instruments now used in such laboratories are
computers or - more generally speaking - instruments for data
processing.

Actually the basis for this development in favour of physical
methods is in some measure a result of the development of Information theory. It was discovered that there were many phenomena in
physics which contain valuable information on the structural,
molecular and elementary (atomic) composition of living organisms
or parts of them. Very often this information was more hidden or
indirect than that which could be obtained by classical chemical
methods, but with the increasing availability of more and more
elaborate data processing methods, scientists succeeded in obtaining such hidden information. Eventually many of the new physical
methods proved to be not only more informative but also faster than
the classical chemical methods. There were less steps in

processing samples and less sources of error which might come from impurities or auxiliary substances used in analysis.

In summary it is a balanced mixture of knowledge and techniques from chemistry, physics, biology and mathematics which has set the scene for modern biomedical research and for the related subject of environmental sciences. The major drawback is a considerable increase in costs for such research due to the sophistication of physical instrumentation. Fortunately governments and funding institutions are normally prepared to spend large sums of money for human health.

2. SYNOPSIS OF PHYSICAL PHENOMENA USED IN BIOMEDICAL RESEARCH

The diversity of physical methods in biomedical and environmental research may be confusing for the non-physicist. Therefore Table I gives a somewhat simplified synopsis of such methods, indicating whether the information to be obtained from each method is on the macroscopic, the molecular or atomic, the electron or the nuclear level. It also attempts to indicate what kind of information this may be.

Some explanation might be reasonable. At the macroscopic level ultrasound has been introduced as it sometimes penetrates matter more easily than light and with less side-effects than X-rays. The information to be obtained by ultrasound analysis will often be complementary to the information given by electromagnetic radiation.

Of course as compared to electromagnetic radiation the resolution power of ultrasound is much lower. Corresponding to the ratio $c_{light} / c_{sound} \approx 10^6$, the wavelengths of ultrasound are considerably longer than those of electromagnetic waves. Therefore ultrasound-microscopy is much more restricted by diffraction. Nevertheless considerable progress in ultrasound analysis has been made.

Physical methods which are based on reactions with molecules and atoms are related to the outer electron shells. In the classical terminology of physics, they are optical methods and belong either to spectroscopy or to microscopy or to both. Light emission and absorption have been used to gather information for a long time. Fluorescence becomes increasingly important with the development of new fluorescent dyes and with improved methods of fluorescence microscopy.

TABLE I

PHYSICAL METHODS IN THE LIFE SCIENCES

Physical Phenomena at	Information				
	Density	Molecules	Atoms	Structure	Kinetics
Macroscopic level					
Ultrasound	+			(+)	
Molecular and atomic level					
Light emission		+	+		
Light absorption	+	+	+	(+)	
Fluorescence		+	+	(+)	
Polarisation				+	
Interference	+			+	
Scattering			+	+	
Electron level					
Electron absorption	+			+	
Electron scattering	+		+	+	
Electron spin resonance		+	+		
X-ray absorption	+		+		
X-ray scattering	+		+	+	
X-ray fluorescence	+		+		
Nuclear level					
Isotopy (stable)		+	+	+	+
Radioactive decay		+	+	+	+
Nuclear resonance absorption		+	+		
Nuclear magnetic resonance		+	+	+	+
Nuclear reactions (activation)	+		+		

Polarisation is observed not only in natural biological samples but also in the fluorescent light from stained preparations. The application of light interference and light scattering is again mainly a consequence of improved physical equipment.

All these methods, working at the molecular and atomic level, can be applied in combination with computer techniques and a full set of methods has been developed under the heading "Computer Image Analysis". Complementary to these methods is the fast evaluation of single parameters as a very large number of small particles in a very fine flow of liquid passes a beam of light in the microscope. This method, called Flow Cytometry, although it gives information on only a few parameters of the individual particles, has an extremely high statistical confidence because so many particles can be examined.

Since the discovery of X-rays and electron-radiation at the end of the 19th century it has become more and more evident that data on the inner electrons of an atom can also give valuable information. This is primarily on the nature of the atom but also sometimes on the way it is arranged in a molecular complex. Electrons hitting any piece of matter are either absorbed or scattered and information on density and structure can be obtained by analysing these processes. A well established method is Electron Spin Resonance which can be applied to free electrons as well as to unpaired electrons in paramagnetic substances. This method, simply called ESR, has been widely applied in the life sciences (for introduction see Schoffa 1964).

The interaction of X-rays with matter results in absorption, scattering and fluorescence. These processes can also give comprehensive information and two examples will be given below.

Finally nuclear physics has also contributed increasingly in opening new windows of information in biomedical and environmental research. For example stable and radioactive isotopes have both become indispensible tools in the life sciences. The ways in which they are used have become more and more refined and some of them, such as the application of activable tracers may have considerable potential, especially for solving practical problems.

Nuclear resonance absorption based on the Mössbauer Effect has been very helpful for special problems but the king of nuclear methods in recent times is nuclear magnetic resonance which has become extremely popular within a very short period.

Finally activation analysis, using various types of nuclear reactions, can be expected to give more and more information on elementary analysis in biological samples.

This list may not be complete but it gives an idea of the broad spectrum of methods now available for research in biology and medicine and environmental sciences. A few specific examples will now be discussed in a little more detail.

3. X-RAY FLUORESCENCE

X-ray fluorescence radiation is also called characteristic radiation as its energies are specific to the emitting element. This radiation can be produced by other incident X-rays or electrons or heavy particles. In the case of electrons the fluorescence radiation is accompanied by Bremsstrahlung. Heavy particles penetrating less deeply into the sample give a better yield of fluorescence radiation e.g. proton induced X-ray emission (Johansson 1982), but the accelerators used for proton production are expensive and occupy a large space. In our Institute alpha-induced X-ray emission has been developed and has proved to be a suitable tool for analysing elements with low atomic number Z (Henningsen et al 1978). We have used 10 mCi (370 M Bq) sources of Po-210 and Cm-244. The experimental system is shown in Figs. 1 and 2. It consists of:-

(i)   a pumping unit (cryosorption pump and two ion pumps with pumping rates of 8, and 20 l/s, respectively supplying an oil-free vacuum),

(ii) a vacuum chamber (Fig. 1) containing the radioactive source, beam collimator and sample holder,

(iii) a Si(Li)-detector (Kevex Corporation) area 30 mm^2, 3 mm thick with a FWHM at 5.0 keV of less than 165 eV).

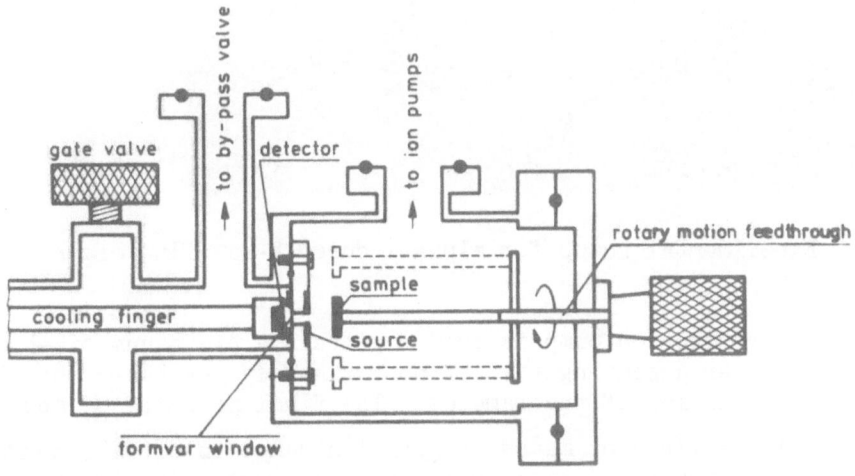

Fig. 1   Chamber for alpha-induced X-ray fluorescence

Fig. 2 Experimental setup for alpha-induced X-ray fluorescence

The cryosorption pump is able to attain a prevacuum of about 10^{-3} torr after a pumping time of approximately one hour, depending on the constitution of the samples. The final pressure of about 10^{-6} torr is realised by means of both ion pumps in circa 4 hours. During measurements the small 8 l/s pump is sustaining the high vacuum.

The cylindrical vacuum chamber (Fig. 1) is welded out of stainless steel, 15 cm in length and 10 cm in diameter. Eight aluminium shafts each supporting one sample are mounted on the axle of a rotary motion feedthrough. Therefore eight specimens can be measured successively without floating the vacuum chamber.

An aluminium disc with a hole 5 mm in diameter collimating the X-rays supports the annular source on the sample side and a thin formvar window (about 80 μg/cm^2) on the detector side. The radiation geometry combines large source area (important to prevent self absorption of alpha-particles in the source layer) with short distances between source and sample (circa 6 mm) and between sample and detector (circa 20 mm). Thus high X-ray intensities are achieved.

The thin window is needed in order to prevent precipitation of vapours from the samples onto the detector, which is cooled by liquid nitrogen and therefore acts like a cooling trap. The aluminium tube with beryllium window (25 μm) supplied originally has been removed while the detector is driven behind a gate valve by means of a spindle mechanism. The vacuum chamber has been mounted to this valve by UHV flanges. When final pressure in the vacuum chamber is reached and the detector is placed directly behind the source holder, the small volume between the formvar window and gate valve is disconnected from the pumps by a by-pass valve.

Pulse processing is performed by Kevex electronics (preamplifier 2003 with pulsed optical reset, main amplifier 4530 P, ADC 5130, MCA 5100). Data output devices are display, plotter, printer and paper tape punch for automatic interpretation of spectra by an ALGOL computer programme.

Figure 3 shows the results of X-ray fluorescence and analysis of haemoglobin in NaCl-buffer. All elements are characterised here by their K-lines. In other cases high Z elements can also be measured at the L- or M-level.

X-ray fluorescence can be used either for multi-element analysis or for measurement of one element in different substances. Figure 4 shows the measurement of oxygen in various mixtures of organic compounds. There is good proportionality between measured and calculated values.

10

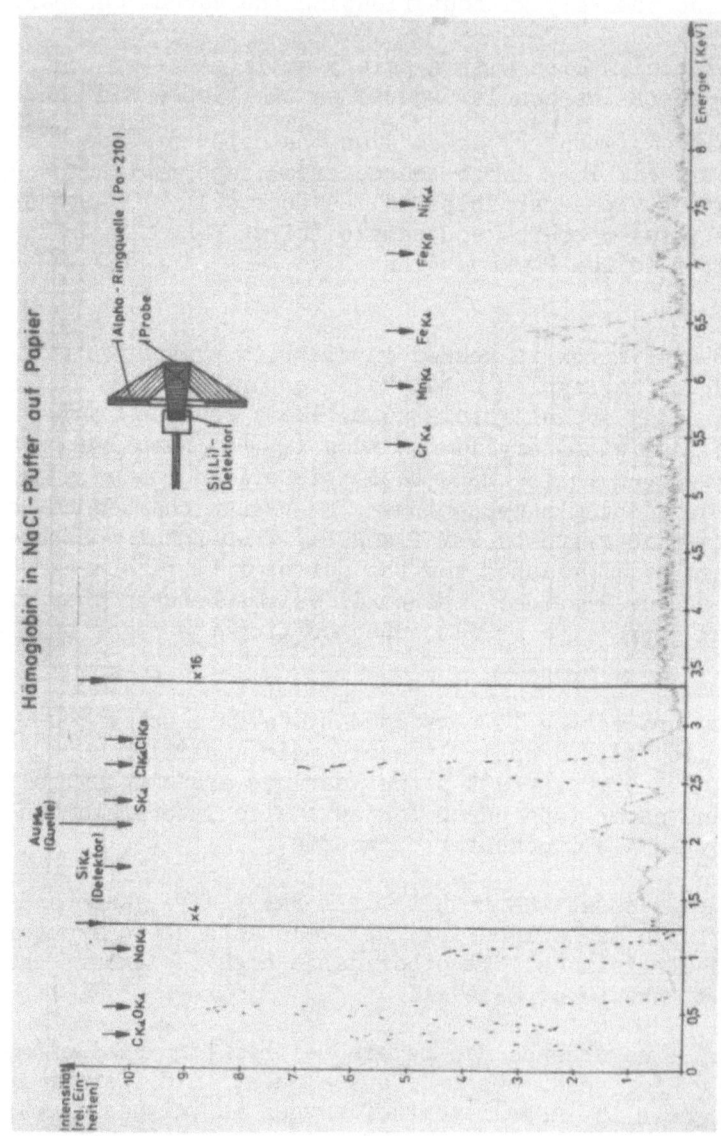

Fig. 3 X-ray fluorescence spectrum of haemoglobin in NaCl-buffer

4. MEASUREMENT OF COMPTON VERSUS RAYLEIGH SCATTERING RATIO

This method is an example of how well-established physical phenomena can be used in combination to obtain new information. The scattering of X-rays interacting with matter takes place in two different ways.

(i)   as classical or Rayleigh scattering without loss of energy (also called coherent scattering). The intensity of I_R of this radiation is roughly proportional to Z^2

(ii) as Compton scattering at electrons with some loss of energy (also called inelastic scattering). Here the intensity I_C is proportional to Z.

As the energy of the scattered radiation is different for (i) and (ii), the intensities I_R and I_C can be measured by gamma spectrometry. Their ratio is proportional to atomic number Z:

$$I_R/I_C \sim Z$$

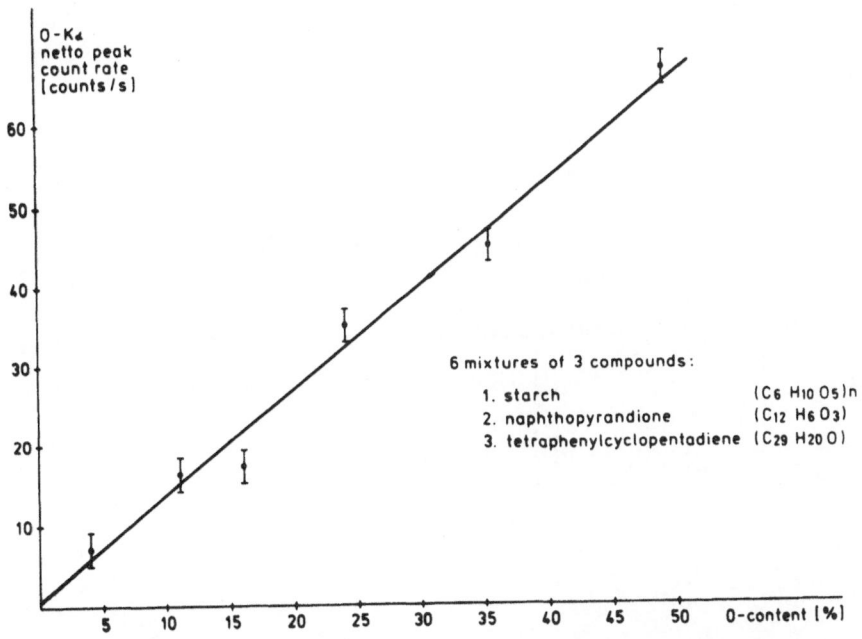

Fig. 4 X-ray fluorescence measurement of oxygen in mixtures of some organic compounds

12

If a substance containing i components of different elements with atomic weights A_i and atomic numbers Z_i is measured the effective atomic number

$$Z_{eff} = \frac{\sum_i n_i A_i Z_i}{\sum_i n_i A_i}$$

In this way one can measure the ratio of different components in a mixture of substances (Schätzler 1979, Kühn et al 1981).

Figure 5 shows the arrangement of the irradiation source, the specimen (contained in a plastic tube) and the detector. The measurement can be performed at different angles and this gives different spectra of the scattered radiation since the difference in energy between Rayleigh and Compton scattered radiation depends on the angle at which the measurement is made. Figure 6 gives an example and also the formulae for calculating the energy difference.

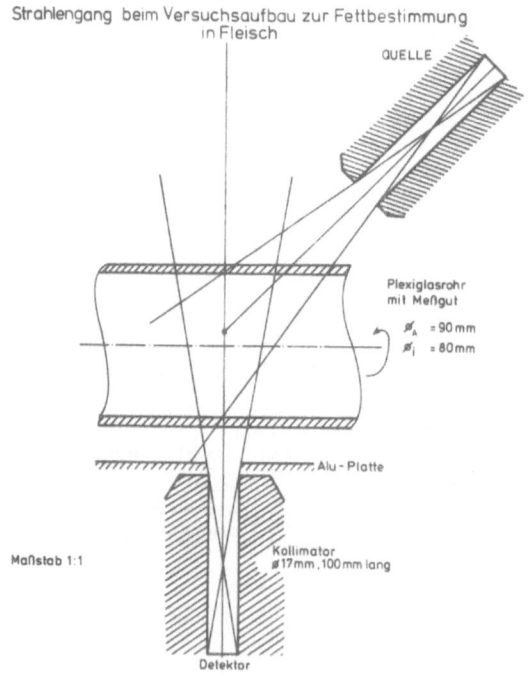

Fig. 5 Principle of measurement of Compton- and Rayleigh-scattering

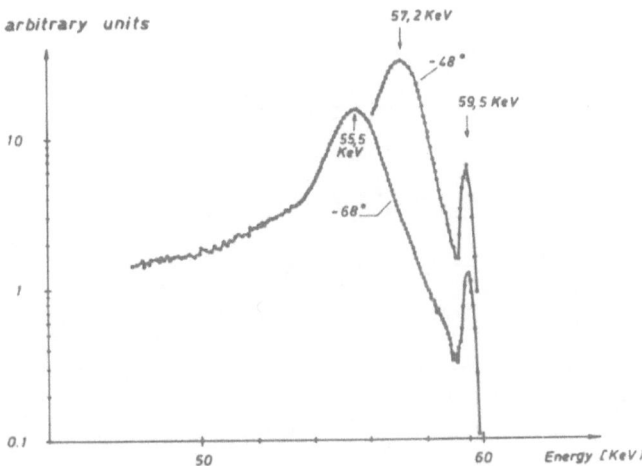

$$\Delta E = E_0 - E_C$$

$$= E_0 \ \frac{\beta(1-\cos\vartheta)}{1+\beta(1-\cos\vartheta)} \ \left(\beta = \frac{E_0}{m_0 c^2} \right)$$

E_0 , primary energy = energy of Rayleigh scattered photons in keV;
E_C , energy of Compton scattered photons in keV;
ϑ , scattering angle;
β , $E_0/m_0 c^2$
$m_0 c^2$, energy of electron at rest (511 keV).

Fig. 6 Spectra of scattered X-rays under different angles

In Fig. 7 the relationship between I_R/I_C and Z_{eff} is shown for a number of substances. From a calibration curve like this, the concentration of two components in a mixture can be evaluated as shown for meat and fat in Fig. 8.

This non-destructive method which is rather new might find wide application.

5. MOSSBAUER SPECTROSCOPY

The application of Mössbauer Spectroscopy in the life sciences is a typical example of using a nuclear effect for studying the

14

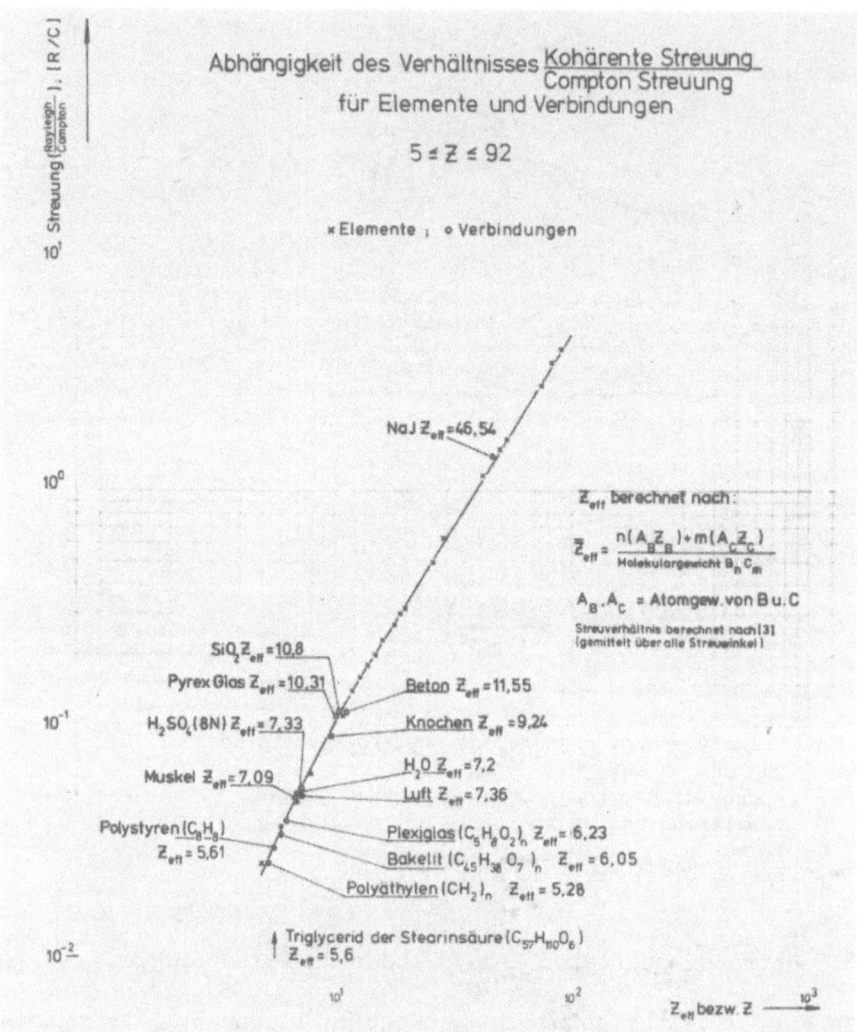

Fig. 7 Relationship between the intensities of Rayleigh- and
Compton scattered X-rays and the effective atomic number
of various substances.

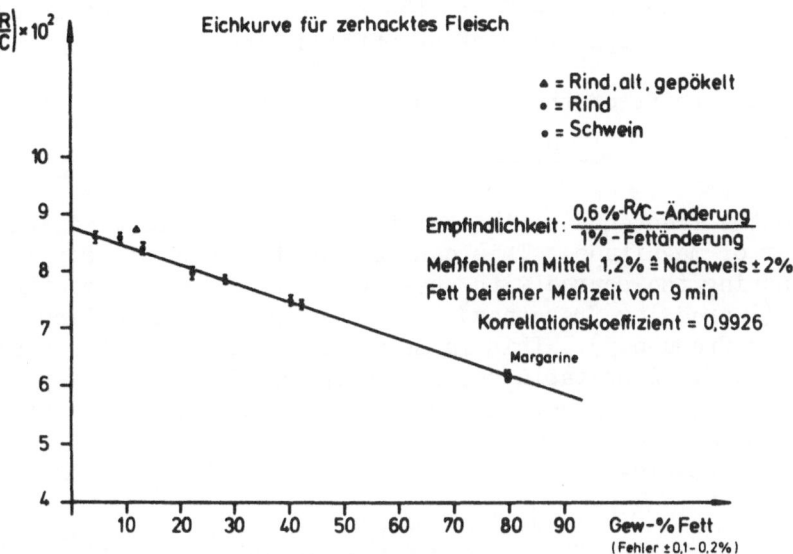

Fig. 8 Concentration of fat in minced pork, measured by Compton-
and Rayleigh-scattering of X-rays.

concentration of certain elements, especially iron, in a complex
biological sample and the molecular configuration in which it is
bound. The basis of the Mössbauer effect, nuclear resonance
absorption, is well known (see e.g. Greenwood and Gibb 1971 or
Johnson 1974). Nuclear resonance absorption is only possible if
the frequency of gamma quanta emitted by a source and absorbed by
a sample is not modified by energy losses due to recoil. This can
be achieved if the emitting nucleus is bound within a crystalline
lattice and the effect has been observed for 20 to 30 elements,
especially Fe-57.

The excited state of Fe-57 is produced by decay of Co-57
which takes place with a half-life of 72 days. The excited Fe-57
will go to the ground state within 10^{-7} seconds and gamma quanta
of 14,4 keV will be emitted. This radiation will be fully absorbed
in a piece of pure iron which normally contains 2.2% of Fe-57. If
the iron is contained in some more or less complicated chemical
compound the resonance frequency is slightly modified by various
phenomena, namely

(i)   isomeric effects due to rearrangement of the electron
      shell during chemical binding which has some influence
      on the nuclear energy levels;
(ii) the quadrupole effect (the splitting of absorption lines)
(iii) magnetic splitting of absoprtion lines.

Of course the shift of the resonance frequency ν by these factors
is very small,

$$\frac{\Delta\;\nu}{\nu}\approx10^{-11}$$

Therefore the frequency of the emitted beam has to be modified by
this order of magnitude. This can be done by using the Doppler-
effect, moving the source with speed v in the direction of the
absorber (increasing frequency) or in the opposite direction
(decreasing frequency). This is shown in principle in the upper
half of Fig. 9. A suitable speed for the required frequency shift
is in the order of mm/sec. In this way an absorption line is
obtained which is characteristic for the condition of the iron
atoms in the sample.

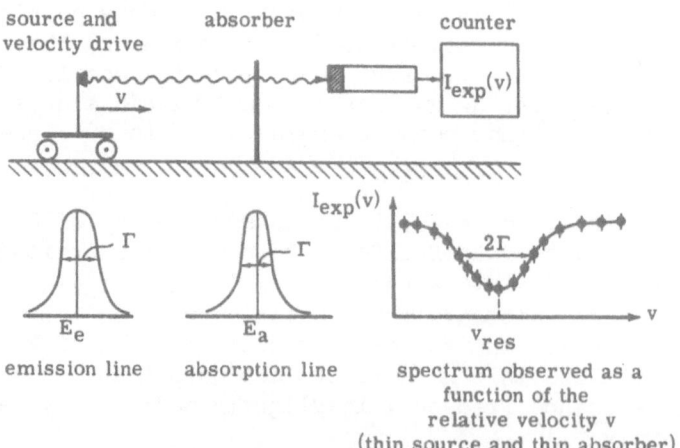

Fig. 9   Basic setup, emission and absorption lines, and velocity
spectrum in a Mössbauer transmission experiment.

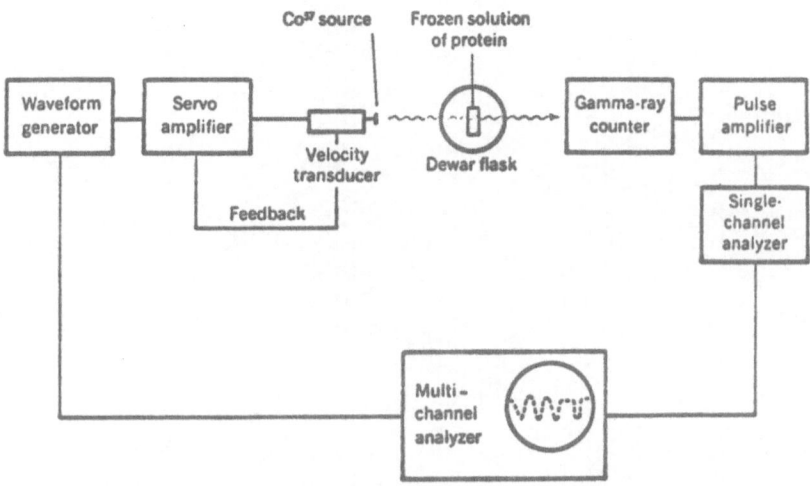

Fig. 10 Block diagram of equipment for Mössbauer Spectrometry.

This method can be applied not only to many natural biomole-
cules containing iron, but also to other molecules after an iron
atom has been incorporated. Furthermore, as mentioned above, a
number of other atoms can be used for Mössbauer studies. Figure
10 shows a block diagram of a full Mössbauer set-up where the
spectrum of gamma radiation passing through the sample is measured
by single-channel- or multi-channel-gamma radiation analysis.
Mössbauer spectrometry is already playing a big role in haematology
and in enzymology.

6. THE INDICATOR ACTIVATION METHOD

The application of radioactive and stable isotopic tracers is
a routine method today and widely used in medical and environmental
research as well as in clinical diagnosis. Sometimes this tracer
method cannot be used since radioactive isotopes will always give
a small but unavoidable radiation burden during the experiment. If
a rare stable isotope is available this can be avoided but the
sensitivity of detection is much lower in this case. In such a
situation one tries to use radionuclides with a very short half
life, but of course this limits the duration of the experiment or
the test.

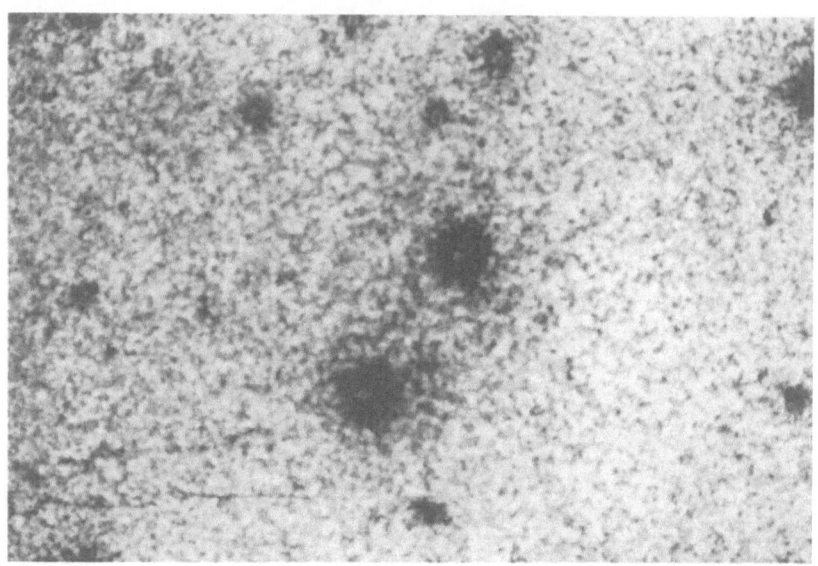

Fig. 11 Pine pollen labelled with manganese and activated by thermal
neutrons under the microscope.

Very often the chemical nature of the tracer is not important,
for example, if a special substance is to be labelled for studying
its concentration, dilution or simply localization. In all these
cases one can use isotopes of any rare element which is not normally
contained in the experimental material. If these isotopes have a
high activation cross section for either thermal neutrons or any
other activating radiation, the samples can be brought into a
radiation field after the experiment and be detected with the same
sensitivity as a normal radioactive tracer. There are many stable
elements which have high activation cross sections and can be used
for such experiments, especially rare earth elements but also e.g.
Au, In, Ag, Sb or Mn.

Many methods have been developed using this principle
(Glubrecht and Kuhn 1976). Two examples might be given here to
show how the Indicator Activation Method works. In forestry and
genetics it is important to prevent cross-pollination, that means
uptake of a foreign pollen from another species by the female
flower. Therefore it is worthwhile to investigate how far, in a
given climatic and topographic situation, pollen from a tree might
travel. It is impossible to solve this problem with radioactive
tracers as the labelled pollen grain would be a considerable

environmental danger, but by giving for example, Mn in excess to a tree, either by injection or through the soil, the pollen grain will take up a considerable amount of Mn. Therefore the grain will become highly radioactive if it is collected in the field and then brought into the thermal neutron flux of a reactor. Other activities which can also be produced may be distinguished from the characteristic lines of the Mn gamma spectrum by analytical activation analysis. In our example the pollen grains might be activated and transferred to a photographic film where the labelled grains can easily be counted (Fig. 11).

This method is in fact only a modification of the normal tracer method as shown in Fig. 12. The crucial point is that during the experiment in the field no radioactivity is used because the production of radioactive tracer through activation which is normally the first step in tracer methodology, is put at a later stage of the experiment.

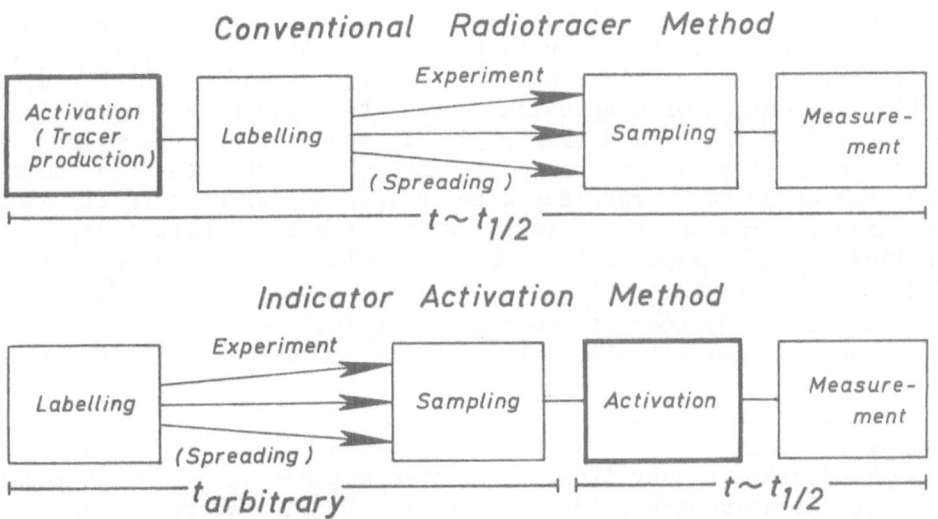

Fig. 12 Comparison of normal tracer technique and the Indicator Activation Method.

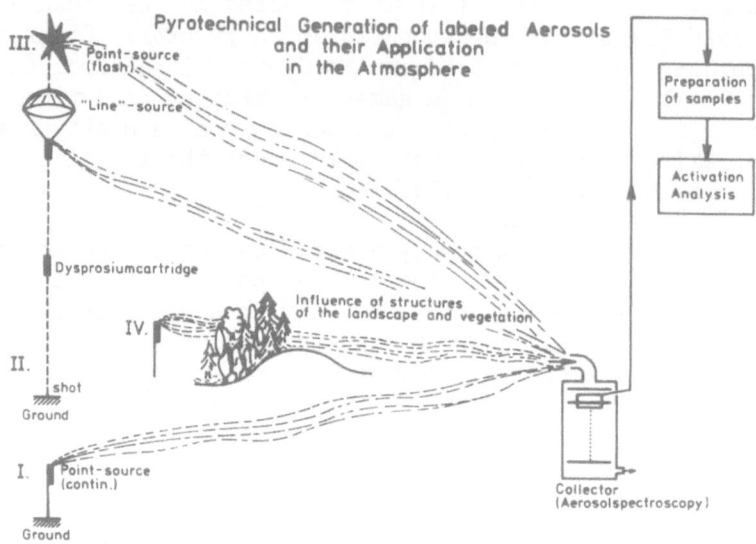

Fig. 13   Pyrotechnical generation of labelled aerosols and their
          application in the atmosphere.

        Recently the Indicator Activation Method has been used to study
the distribution of aerosols in the environment for many industrial
emitters.   A pyrotechnical product has been developed which contains
a high percentage of Dysprosium and can be exploded at any desired
distance from the ground (Fig. 13).   It can be used as a point
source as well as a line source of aerosols which cover the whole
spectrum of normal aerosols.   After collection which normally also
involves separating the different sizes of aerosol particles,
particles can be exposed to a thermal neutron flux in a reactor
and detected even after dilution by more than 1 in 10^4.   This gives
a chance to study the influence of irregularities in the topographic
or meteorological situation close to any aerosol-emitting facility.

        In medicine the Indicator Activation Method can be used e.g.
for measurement of the volume of body fluids using the principle of
tracer dilution.   Many other diagnostic methods which do not require
in vivo measurement but test samples in the laboratory can also
use activable tracers.

7. CONCLUSION

It is highly probable that the application of physical methods in biomedical and related research will increase further. This development can be very fruitful but it is a prerequisite that increasing interdisciplinary contacts must be established between the representatives of the various disciplines of modern science.

REFERENCES

Glubrecht H. and Kühn W. (1976) Die Indikator-Aktivierungs methode. Thiemig Taschenbuch
Greenwood N.N. and Gibb T.C. (1971): Mössbauer Spectroscopy. Chapman and Hall, London.
Henningsen W.P., Schätzler H.P. and Kühn W. (1978) Determination of C and O elements by alpha-induced X-ray energy spectrometry. Atomkernenergie 31, 131-134
Johansson S.A.E. (1982) PIXE, eine neue physikalische methode für chemische Multielementanalyse. Physikal. Blätter 38, No. 12, 359-364
Johnson C.E. (1974) Mössbauer Spectroscopy. Amino Acids, Peptides, Proteins 5, 215-218
Kühn W., Georgi B. and Garzke T. (1981) Zur radiometrischen Bestimmung von Stoffkomponenten durch Compton- u. Rayleigh-streuung. KfK-AFR 003
Schätzler H.P. (1979) Basic aspects on the use of elastic and inelastic scattered gamma radiation for the determination of binary systems. Int. Journ. of Appl. Rad. and Isotopes 30, 115-121
Schoffa G. (1964) Elektronenspinresonanz. G. Braun-Verlag, Karlsruhe.

PHYSICS IN MEDICINE AND BIOLOGY - 10 YEARS OF PROGRESS

P.P. DENDY

Department of Biomedical Physics and Bioengineering,
University of Aberdeen, Foresterhill, Aberdeen AB9 2ZD,
Scotland, UK.

1. INTRODUCTION

In 1967 John Mallard entitled his inaugural lecture as the first Professor of Medical Physics in Scotland "What is Medical Physics?"

During the space of one hour he was able to review virtually all the significant activities in medical physics at that time. The same title might still be appropriate today but it would have quite a different connotation. No one reading this book will be unaware that physics now makes an important contribution in medicine and biology. However, such has been the rate of progress of these contributions and their diversity, which appears to know no bounds, that few if any of us is fully aware of the current situation.

In this overview I would like to identify one aspect of health care that has developed more rapidly than most - namely the application of non-invasive physical techniques for diagnosis. I would also like to suggest two themes which should be uppermost in the minds of physical scientists during their work. The first is that development is very largely governed by technological progress. We are in a privileged position to be aware of that progress at an early stage and should be looking for every opportunity to exploit and implement it in medicine and biology. The second is the over-riding need for quantitation; where I can do no better than to remind you of the famous dictum attributed to Lord Kelvin "Anything that cannot be expressed in numbers is valueless".

2. TECHNIQUES INVOLVING IONIZING RADIATION

2.1 Nuclear Medicine

In the broadest sense, nuclear medicine means the administration of small quantities of radioactive materials for diagnosis and there are several books on the physical principles involved in this technique (see e.g. Sorensen and Phelps 1980). A major subdivision of the subject is now concerned with the use of the radiations emitted by the radioactive material to image organs or other regions of interest within the body.

It is possible to list the properties of a radioactive material that make it suitable for in vivo imaging.

(i)   It should emit monoenergetic gamma rays in the range 100 - 200 keV. Lower energy radiation is absorbed in the patient and at higher energies detection efficiency can be poor.

(ii) There should be no associated beta or alpha emission giving a high radiation dose to the patient but contributing nothing to the image.

(iii) The radionuclide should have a half life of a few hours - long enough to complete the examination but short enough to minimise the patient dose.

(iv) The daughter product of the decay should be non-radioactive.

(v)   It should be possible to bind the radionuclide firmly to a wide range of biologically important molecules.

Figure 1 illustrates very well the problems caused by technical limitations in nuclear medicine. In 1913, one of the few readily available radionuclides was radium and this was fed to a frog in its diet for several days. Delineation of specific organs is poor and furthermore, because of the nature of the radiations emitted, it was necessary to squash the animal before the radiograph could be prepared!

Any worthwhile development of nuclear medicine was out of the question until a wide range of reactor produced radionuclides started to become available about 1948 (Mallard and Trott 1979). As seen in Table I one of these, Tc-99m has excellent physical properties for imaging.

Another important technological development was still required to provide a good imaging agent. The Tc-99m must be firmly bound to a pharmaceutical which shows a high level of selective incorporation in the organ of interest (Dendy et al 1980). Indeed the major disadvantage of Tc-99m is that it is not an ideal element for this purpose.

Fig. 1   An early radiographic image of the distribution of radioactivity in a frog.

TABLE I

THE PHYSICAL PROPERTIES OF TECHNETIUM-99m

Half life	6 hr
Gamma ray energy	140 keV
Beta particle dose	Nil
Decay product	Tc-99 (half life 2×10^5 yr)

The production of high quality radiopharmaceuticals is essential to Nuclear Medicine and will be discussed in detail elsewhere in this volume. For some types of imaging, the problem has been solved. For example, high quality bone scans may now be obtained with Tc-99m labelled methylene diphosphonate and the combined effect of a good radionuclide and a good pharmaceutical on demand for the service is shown in Fig. 2.

Parallel developments in imaging equipment have of course also occurred (Rollo 1977) and in the North-East of Scotland the annual rate of diagnostic examinations using radioactive materials is now about one examination for each 200 of the population, so Nuclear Medicine has clearly become a significant diagnostic service.

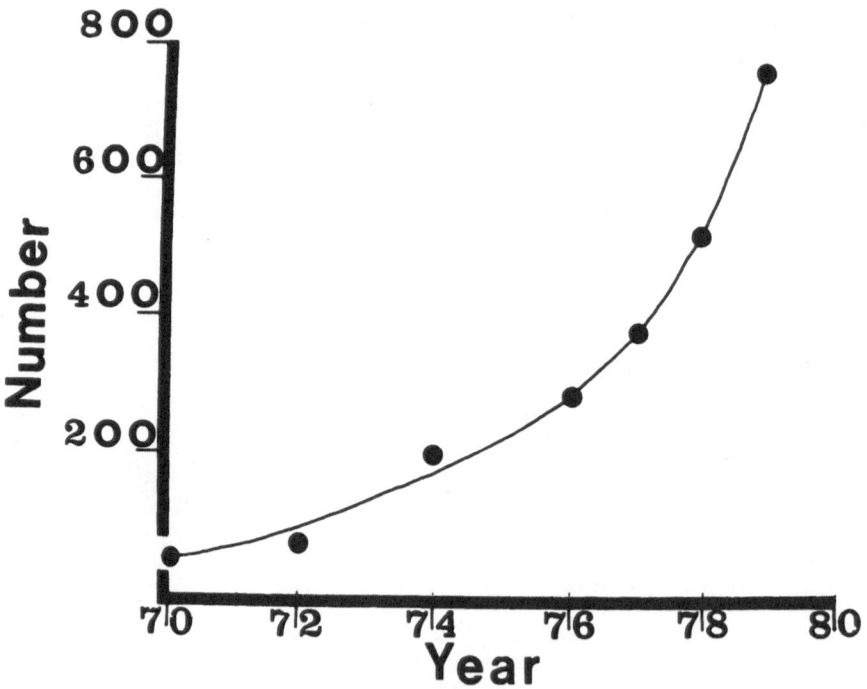

Fig. 2 Increase in the number of radionuclide bone scans carried out in the North-East of Scotland during the period 1970-1980.

2.2 Transmission Computerised Tomography

The most spectacular advance with X-rays has of course been the development of transmission computerised tomography (CT) (Hounsfield et al 1973).

Conventional radiology suffers from three major limitations:-

(i)   Superimposition - the conventional radiograph is a two-dimensional image of an inhomogeneous object with many planes collapsed into one. All depth information is lost and the confusion of overlapping planes makes detection of subtle abnormalities difficult or impossible.

(ii)   Geometrical effects - the viewer can be confused about the shape and relative positions of various structures displayed on a conventional X-ray picture and care must be taken before drawing conclusions about the spatial distribution of objects even when an orthogonal pair of radiographs is available.

(iii) Attenuation effects - when the intensity of X-rays striking the film is described by the well-known equation $I = I_o \exp(-\mu x)$, it is important to realise that beam attenuation I/I_o is dependent on both μ and x. This can lead to ambiguity in a radiograph since a given difference in attenuation can be due to a change in thickness alone (x), a change in composition alone (μ), or a change in both these factors.

CT attempts to overcome these problems by considering the object as a series of thin slices and examining each slice from a number of different angles. Figure 3 shows a simplified version of the machine with just one X-ray source and one detector. For each angular setting of the machine, transmission readings are recorded for 160 ray positions associated with each linear traverse. By stepping through 180° at 1° intervals 28,800 readings can be obtained.

If the plane of interest is divided into a large number of picture elements or pixels (a typical value might be a 160 x 160 array) and an attenuation co-efficient is assigned to each pixel, the collected data may be used in a large number of simultaneous equations to calculate the attenuation in each pixel. Since the co-efficients are additive the total effective attenuation coefficient along a line is equal to the sum of the attenuation coefficients of all the pixels along that line.

The number of pixels, N, represents N unknown values of μ and thus N independent equations, i.e. data from N independent X-ray

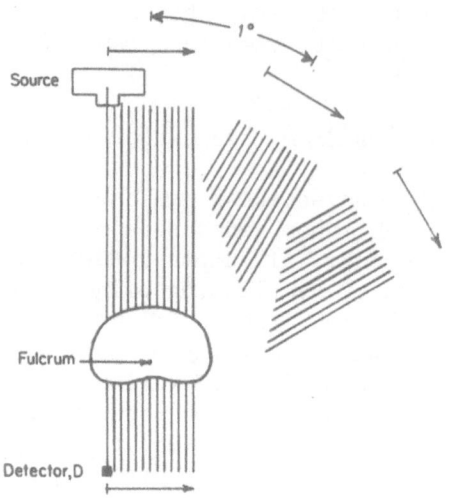

Source

Fulcrum

Detector,D

Fig. 3 The basis of computerised tomographic scanning using a single detector. A single linear scanning sequence is performed; the source detector system is rotated slightly, the scan repeated and so on (reproduced with permission from Craig and Lowry 1980).

beams, are required for solution. When attempting to reconstruct an image from a matrix of 160 x 160 elements, typical of that used in a brain scan, more than 25,000 readings are required and a similar number of simultaneous equations must be solved. Now the importance of computer technology to this development becomes apparent since correlating all this information is beyond the capability of the human brain but is an ideal problem for the computer, especially since it is a highly repetitive numerical exercise.

Second and third generation machines have largely concentrated on faster scanning times, first by using several pencil beams and an array of detectors and more recently by using a fan-shaped beam wide enough to cover the whole body section. Several hundred detectors are used and machine movement is reduced to continuous rotation only.

Modern machines can collect all the data in 5 sec, sometimes even less, and provide images in which the pixel width is typically 1 - 2 mm and in which changes in soft tissue density of less than 0.5 per cent can be clearly delineated. Figure 4 shows convincingly the effect of ten years progress on X-ray CT.

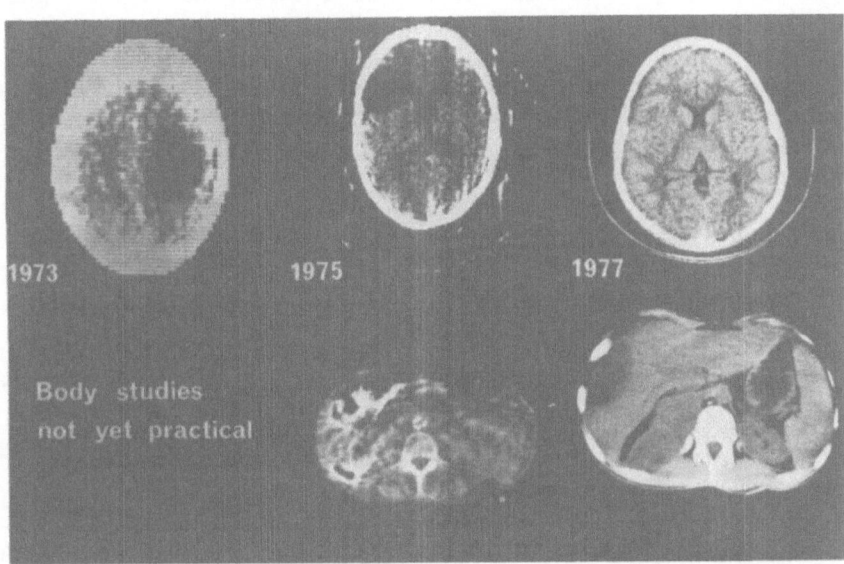

Fig. 4 A set of images showing the rapid development of the
quality of X-ray CT images during the 1970s - upper row
head scans, lower row whole body scans. Reproduced by permission International General Electric Company of New York Ltd.

This is an appropriate point at which to refer to costs. The
X-ray CT scanner was by no means the first very expensive piece of
high technology equipment introduced to hospitals - high energy
Co-60 machines and linear accelerators for radiotherapy preceded
CT scanners by several years. However, CT scanners seemed to
signal something of a watershed and the start of ever growing
demans to buy increasingly more costly equipment. A whole body CT
scanner costs in the region of $800,000 but when revenue coneequences
e.g. building, maintenance, staff for say ten years, are added the
total bill may reach $2,000,000. Even in the relatively well-off
countries this is a large sum of money and we are approaching a
situation where levels of health care will be determined by cost
considerations rather than by limitations on our medical knowledge.

2.3 Digital Radiology

Conventional radiology has made relatively little progress for upwards of 30 years, perhaps because there has been a strong emphasis on qualitative visual interpretation of X-ray photographs. Progress would possibly have been quicker if the recording medium had been more suitable but the highly non-linear dose-response curve of photographic film is itself a strong deterrent against quantitative work.

Encouraged by developments in CT and in nuclear medicine, many groups are now taking a more quantitative approach to radiology and the sub-specialty of digital radiology is developing rapidly (Pullan 1981).

If an array of high quantum efficiency detectors is used, each measuring say 1 mm x 1 mm then it can be shown that for an exposure of 1 roentgen at about 60 keV, 10^8 photons will be incident on

each 1 mm x 1 mm element of the patient and about 10^6 photons will emerge unscattered from an object equivalent to 20 cm of water.

The fluctuation in this signal due to Poisson statistics is N or 10^3 which is only 0.1% of the signal. It follows that changes of this order should be detectable.

A central problem in digital radiology is that for visual display the eye can only resolve about 20 separate levels and this dynamic range has to be matched to the very small changes in signal that are known to be statistically significant. One way to do this is by only displaying edges, but a more promising approach is by using a subtraction technique on digital radiographs obtained under slightly different conditions. One possibility might be to use X-ray beams of different energies or an alternative would be to obtain images before and after the injection of small quantities of contrast medium. By digital subtraction techniques, it may be possible to visualise 1 mm diameter vessels with an exposure of one roentgen if a differential concentration of 1.5 mg/ml of contrast medium can be achieved inside and outside the vessel.

On the clinical side most progress seems to have been made in applying digital radiology to non-invasive cardiovascular measurements (see e.g. Harrington et al 1982) but there are a number of other applications and it is encouraging to see this new quantitative approach to a subject that has remained static for so long.

3. NON-IONIZING RADIATION TECHNIQUES

All the techniques we have discussed so far require the patient to be exposed to ionizing radiation. In most instances the radiation dose is small - for example about 1 rad (10 m Gy) in a radionuclide bone scan or about 2 rad (20 m Gy) in CT imaging localised to the scanned slice. However, current thinking in radiological protection is that there is no safe level of ionizing radiation so even small doses carry a small but finite risk.

Largely for this reason, there has been great interest during the past 10 - 15 years in non-invasive diagnostic techniques with non-ionizing radiations.

3. 1 Ultrasound

The earliest of these to make a significant impact was ultra-sound and the publications of Wells (1977a, 1977b) are valuable reference works on the subject.

Two main techniques have been applied. In the pulse-echo method the ultrasonic probe emits a short-duration stress wave into the medium. When the stress wave meets a boundary, part of it is reflected and returns to the probe where it generates another signal on the CRO screen. Information can be obtained from both the time interval between the transmitted and returning signals which depends on the distance to the boundary and the speed of sound in the medium $V = \sqrt{K_a/\rho}$ where K_a is the adiabatic bulk modulues and also from the relative sizes of the signals which depends on the difference in acoustic impedance $Z = \sqrt{K_a \rho}$ and attenuating properties of the medium.

Two basic methods of displaying the data have been adopted. The principle outlined above provides what is essentially an "A-scan" with the display showing the amplitude of the reflected signal. Nowadays the A-scan has only limited usefulness, for example in echo encephalography or at higher frequencies sometimes in ophthalmology. The alternative display is a B-scan in which the returning signal is simply represented as a bright spot on the display. By linking the direction of the CRO time base to that of the ultrasonic beam across the patient, a two-dimensional image can now be synthesised.

There have been a number of important developments in ultra-sound diagnosis in the past 10 years - for example development of grey scale images. The original B-scan equipment had bistable displays with the positions of all echoes marked by points of the same intensity on the display screen. The advent of grey scale displays where the amplitude of the echoes is represented as a level of brightness was of major importance. This allowed the radiologist much greater discrimination and greatly improved the diagnostic usefulness of ultrasound.

A further development has been in real time scanning. The simple B-scan required the operator to move the transducer head to build up an image. This meant that the ultrasound could not be used to image movement within the body. In real time scanning the movement of the ultrasound beam is controlled mechanically or electrically at a rate high enough to image internal organs in real time. This is of great importance in cardiology and in obstetrics where movement is of diagnostic importance.

A third important development in recent years adopts a quite different principle, the Doppler effect, whereby the frequency of ultrasound waves will change if they are transmitted through, or reflected from a moving target. Figure 5 illustrates how a pair of transmitter-receivers may be used to obtain simultaneous readings of upstream and downstream flow rates. The main application of this technique is in the investigation of the vascular system where changes in blood flow characteristics can be closely monitored. It is also used to listen to the foetal heart and is being invest- igated as a system for cancer detection in the breast where abnormal blood supply is a common occurrence.

Since the rationale for introducing ultrasound is that ionizing radiation has a potential for causing long-term harmful effects, there has been an intensive study of ultrasound bioeffects (Fry 1979 Lele 1979). There is well-documented evidence for thermal and mech- anical damage at high intensities but the evidence for genetic damage has been somewhat conflicting. We have recently studied the

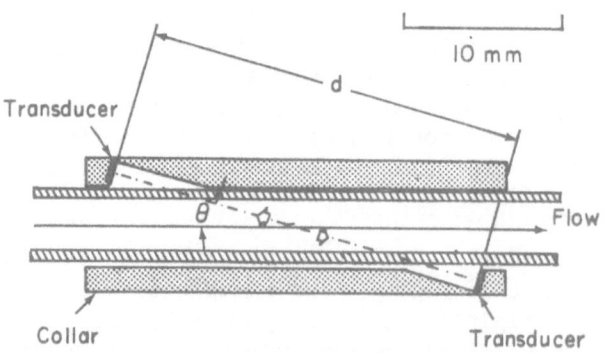

Fig. 5 A diagram showing the positioning of a pair of ultrasound transmitter-receivers to make measurements of flow rate using the Doppler-effect principle (Reproduced with permission from Wells 1977a)

clonogenic behaviour of HeLa cells following exposure to therapeutic intensities of ultrasound with the cells firmly embedded in
agar but could find no clear evidence to suggest disturbance in
growth pattern with a continuous exposure at 15W total power and
a peak intensity of 24W cm^{-2} (Law 1982).

3.2 Tele-diaphanography

Whilst developing this theme of the increasing application of
non-ionizing radiations for diagnosis, I would like to mention
briefly a very old technique that has recently been revitalised,
that of tele-diaphanography.

There is much interest in the early detection of breast cancer
since cure rates are very dependent upon the stage of the disease
at the time of detection. Whereas palpation coupled with X-ray
mammography probably offers the most reliable diagnosis at present,
the mid-plane dose of between 0.5 rad (5m Gy) and 0.05 rad (0.5 m Gy)
is probably unacceptable for mass screening of well women, many of
whom would be screened on many occasions over the course of several
years.

Thermography has been somewhat disappointing as a screening
procedure when used in isolation due to a rather high false positive
rate. This arises because superficial blood vessels do not always
indicate an underlying tumour, non-malignant diseases may also
increase the temperature, the contraceptive pill causes abnormal
patterns and with bilateral disease there is no normal breast for
comparison.

In the tele-diaphanography system shown in Fig. 6 the power
supply PS feeds a water cooled bright light tungsten lamp source
and the light is then guided through a light-pipe to the underside
of the breast. Peak transmission through soft tissue occurs at a
wavelength of about 1 μ m and either a transmission colour photograph is taken with infra-red sensitive film or as shown here in a
more recent development, a colour TV system and display may be
used for immediate interpretation.

In general, carcinomata are seen as dark shadows with some
evidence that this is partly due to the tumour itself and partly
due to increased absorption of light by the blood in the venous
dilation associated with the tumour. Pilot studies with this
technique have given results which compare favourably with mammography. Lesions as small as 5 mm have been detected in a study in
Sweden but even smaller lesions will need to be detectable if
diaphanography is to become a really effective screening method.

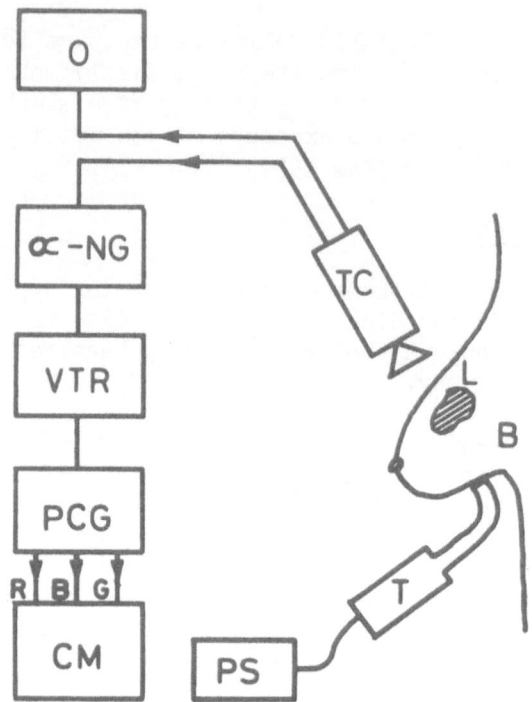

Fig. 6   Schematic diagram of tele-diaphanography equipment T = torch,
         PS = power supply, B = breast, L = lesion, TC = infra-red
         sensitive television camera, VTR = video tape recorder,
         O = oscilloscope, α-NG = alpha numeric generator, PCG =
         pseudo colour generator, RBG = red, blue, green inputs,
         CM = colour monitor (from Watmough 1983 with permission)

3.3 Nuclear Magnetic Resonance

    It would not be possible to conclude this overview without
mentioning the newest and most exciting development, nuclear
magnetic resonance imaging (Lauterbur 1973; Kauffman et al 1981).
This is discussed in detail elsewhere in the book. For the purpose
of this overview, it will be sufficient to remind you that, just as
a spinning top will precess about a vertical gravitational field,
so the proton in a hydrogen atom is spinning and will precess about
an applied magnetic field. Furthermore, if the material is irradiated with electromagnetic radiation of exactly the same frequency
as the frequency of precession, energy is absorbed from the field
and alignment of the proton spin directions changes relative to the
applied magnetic field.

    The change in spin direction is proportional to the duration
of exposure to the magnetic field and two specific changes are of
particular importance. If the spin vectors of the nuclei rotate

through 90°, the nuclei radiate the surplus energy to their surroundings at the same resonant frequency. The strength of the signal gives information about the proton density.

If the spin vectors rotate through 180°, the excess energy is dissipated by exchange with neighbouring molecules and the rate of exchange is a measure of the degree of ordering of the structure. Since most of the protons are present as water in biological materials, this behaviour essentially investigates the way in which water is bound to macro-molecules. The value of T_1 may be calculated by applying a 90° pulse at a later time when the residual energy not yet dissipated by exchange mechanisms will be re-irradiated. Thus the method is capable of measuring either proton density or the T_1 relaxation time.

Quite apart from the intense interest generated by any new approach, there are two major reasons why NMR imaging should receive so much attention. The first is the magnitude of the differences between T_1 values of different tissues (Table II). Whereas with X-ray CT it is often necessary to attempt to identify differences in X-ray opacity of 0.5% or less, T_1 values between different normal and diseased tissues vary by 200% or more. This must surely make identification easier. The second point is that the proton

TABLE II

Some typical spin-lattice relaxation times of normal and pathological human liver tissue measured from in vivo NMR images at 1.7 MHz (Adapted, with permission from Foster 1983)

	T_1 (milliseconds)
Normal liver	140 - 170
Chronic active hepatitis	170 - 180
Cirrhosis	180 - 300
Secondary liver tumour	280 - 450
Hepatoma	300 - 450
Ascites fluid	700 - 1000
Simple serous cyst	800 - 1000

density and T_1 relaxation times reflect biological properties that no other technique has the capability to investigate. Although at the present time we still do not know precisely what T_1 values mean at the molecular level, we now have the means to study many diseases that depend on subtle changes in body water balance, a subject where we know very little at present, and this must also be important because such a high proportion of the body is water.

4. CONCLUSION

Space has not permitted discussion of many other developments of physics in medicine and biology in the past ten years - computerised treatment planning systems in radiotherapy, applications of very short half life cyclotron produced radionuclides (Kellershohn 1981), ultrasound and microwave hyperthermia, and a wide range of applications in electro-physiology and biomechanics. However, as if to emphasise that the subject seems to know no bounds, consider once again the subject of computerised tomography. Table III lists many different forms of CT. For each technique, five crosses have been placed to indicate the present position on the long road from scientific speculation to routine clinical use. X-ray CT is well established, and we see NMR CT moving rapidly to the left on this chart. Single photon emission CT and positron emission CT have also found limited clinical usefulness. Some of the other techniques have not yet reached the hospital but who can tell what this chart will look like in 1993 when perhaps someone else will be attempting an overview of "Physics in Medicine and Biology - 10 years of progress."

REFERENCES

Craig D. and Lowry W.S. (1980) The CAT scanner in "Physical Techniques in Medicine Vol. 2" (Ed. J.T. McMillan) John Wyley & Sons Ltd. pp. 2-22

Dendy P.P., Sharp P.F., Keyes W.I. and Mallard J.R. (1980) Radionuclide emission imaging. Single photon techniques including radiopharmaceutical developments. Brit. Med. Bull. 36, 223-230

Foster M.A. (1983) in "Magnetic Resonance in Biological Systems" Pergammon Press pg. 267

Fry F.J. (1979) Biological effects of ultrasound - a review. Proc. IEEE 67, 604-19

Harrington D.P., Boxt L.M., and Murray P.D. (1982) Digital Subtraction Angiography. Overview of technical principles. Am. J. Roentgenol. 139, 781-786

Hounsfield G.N., Ambrose J., Perry J. et al (1973) Computerised transverse axial scanning (tomography) Parts 1, 2, 3. Brit. J. Radiol. 46, 1016-1051

TABLE III

A schematic representation of the current status of various cross-section imaging techniques based on computerised tomography (CT) (adapted from Pfeiler 1981)

Status	Routine Clinical Use	Experimental Stage	Scientific Idea
X-ray CT	XXXXX		
Dynamic heart X-ray CT			XXXXX
Single photon emission CT		XXXXX	
Positron emission CT		XXXXX	
Nuclear magnetic resonance CT		XXXXX	
Ultrasonic CT			XXXXX
Proton CT			XXXXX
Impedance CT			XXXXX
Microwave CT			XXXXX

Kauffman L., Crooks L.E. and Margulis A.R. (Eds) Nuclear Magnetic Resonance Imaging in Medicine (1981) Igoku-Shoin, New York and Tokyo

Kellershohn C. (1981) Positron emitting radionuclide in the study of metabolic and physiopathologic mechanisms. Br. J. Radiol. 54, 91-102

Lauterbur P. (1973) Image formation by induced local interactions, examples employing nuclear magnetic resonance. Nature 242, 190

Law A.N.R. (1982) The effects of therapeutic ultrasound on the clonogenic survival of cells in liquid and semi-solid media. Ph.D thesis, University of Aberdeen

Lele P.P. (1979) Safety and potential hazards in current applications of ultrasound in obstetrics and gynaecology. Ultrasound Med. Biol. 5, 307-320

Mallard J.R. (1967) Medical Physics - what is it? Aberdeen University Review Vol. XLII 1, 12-29

Mallard J.R. (1981) The noes have it! Do they? Sylvanus Thomson
    Memorial Lecture February 18, 1981. Br. J. Radiol. 54, 831-849
Mallard J.R. and Trott N.G. (1979) Some aspects of the history of
    nuclear medicine in the United Kingdom. Semin. Nucl. Med. 8,
    283-298
Pfeiler M. (1981) CT techniques in medical imaging in "Lecture
    Notes in Medical Informatics 15 Digital Image Processing in
    Medicine" (Ed K.H. Hohne) Springer Verlag, Berlin, Heidelberg,
    New York pp. 42-92
Pullan B.R. (1981) Digital radiology in "Physical Aspects of Medical
    Imaging"Eds B.M. Moores, R.P. Parker and B.R. Pullan. John Wyley
    and Sons Ltd. pp. 275-287
Rollo F.D. (1977) (Ed)"Nuclear medicine physics, instrumentation
    and agents" C.V. Mosby & Co., St Louis
Sorenson J.A. and Phelps M.E. (1980)"Physics in Nuclear Medicine"
    Grune and Stratton, New York
Watmough D.J. (1983) Transillumination of breast tissues: factors
    governing optimal imaging of lesions. Radiology (in press)
Wells P.N.T. (1977a) "Biomedical Ultrasonics" Academic Press,
    London
Wells P.N.T. Ed. (1977b) "Ultrasonics in Clinical Diagnosis (2nd
    Edit)" Churchill Livingston, Edinburgh

CLINICIANS' EXPECTATIONS FROM THE NEW TECHNOLOGY FOR IN VIVO
DIAGNOSTIC EXAMINATIONS

C. CONSTANTINIDES

Director of the Laboratory of Nuclear Imaging, Alexandra
University Hospital, Vas. Sofias Ave., 80, Athens 611,
Greece.

1. INTRODUCTION

Medical technology refers to the wide range of equipment,
devices, drugs and procedures employed in the care of patients. It
includes the capital and human investments that establish the
capability for medical practices.

In medical technology we have to answer not only the question:
"When should a clinician order a test?" but also "When should a
hospital purchase a new instrument that can perform the test?" In
order to answer these questions the clinician must be fully
informed on what to expect from the new technology because this
new technology has recently increased geometrically for the benefit
of the patient and of society.

2. SOME NEW MEDICAL TECHNOLOGY

First, let me introduce some of the most advanced pieces of
new medical technology.

In the field of nuclear medicine, the gamma-camera in combination with a mini-computer offers the possibility to evaluate the
function of the circulation. This led to the creation of a new
sub-specialty of nuclear medicine, nuclear cardiology. (Serafini et
al 1976).

Positron emission tomography (PET), represents today the new
glory of nuclear medicine. PET visualizes the chemical activity of
the brain (Sokoloff et al 1977) and the heart (Phelps et al 1978).
The level of F-18 deoxyglucose metabolism makes it possible to

predict the histological grading of brain tumours. Thus, the
biochemical classification of brain lesions by means of positron
tomography is an idea whose time has come.

The CAT scan (Hounsfield 1973) and NMR (Damadian 1971) provide
for the clinician the opportunity to see inside the human body
from outside and to make photographic slices of the organs to be
examined as if they were on a tray in the pathology laboratory.
These two devices can reduce the cost of health care because their
use minimizes the number of other laboratory procedures.

The Dynamic Spatial Reconstruction (Crummy et al 1980) promises
much for medical diagnosis in the near future. Endoscopic colour
photo-documentation has also attained a high standard of sophistication through pioneering new technical developments and advances.
Nearly all the cavities in the human body can be explored endoscop-
ically by expeditiously combining appropriate equipment (Hirschowitz
et al 1958).

Finally, microprocessors and microcomputers are assuming an
increasingly important role in clinical and biomedical research.

All these new achievements have brightened the medical profess-
ion and have made the clinician proud and confident.

3. BENEFITS FROM NEW MEDICAL TECHNOLOGY

It is well known that the most recent technological advances
have helped the medical sciences to make enormous leaps in the
diagnostic sector, to reduce almost unbelievably the time required
for diagnosis and, most important of all, to guide the medical
scientist into new avenues of research with the result that today,
medicine is a high level science and easily keeps pace - whether
or not in first place - with its allied sciences.

In the field of nuclear medicine which is very dependent on
new medical technology, almost 50% of the diagnostic problems of a
General Hospital may be solved through nuclear medicine procedures.
Also, for the first time, nuclear medicine offers methods for
taking pictures of the brain which show metabolic activity in
regions concerned with hearing, vision and thoughts.

The wide use of the CAT scanner has reduced the cost of medical
care because other diagnostic tests have been avoided. CAT scanning
is seen as an expensive investigation because its cost is readily
visible, but the costs that the scanner saves are difficult to see,
detect and understand.

In the area of medical research medical technology has helped
in a way that is hard to quantify but impossible to over-estimate.

Very few of todays research projects have no contribution from
medical technology.

Finally, in everyday medical practice, medical technology has
contributed in shortening the hospitalization period of the patients.

4. FACTORS AFFECTING THE USE OF MEDICAL TECHNOLOGY

Many considerations might affect the choice and success of
technological applications (White et al 1972).

The first is that health care is not a manufacturing industry,
but a service industry in which personal aspects are valued, and
health care organizations are social systems. Technological methods
and innovations should be applied to health services systems, not
because these methods are available and feasible but because they
materially assist in meeting the needs and demands of people for
health care. Technological innovations that do not serve socially
determined ends are unlikely to contribute much of practical utility
and can add substantially to the costs of medical care.

As a corollary to this point, the needs and demands of patients
must be specified before technological methods can be successfully
employed. Technology cannot solve problems that require other
measures such as the creation of formal health care organizations
with defined objectives. Many potential applications of technology
must await the development of explicit national policies and
standards for health services and the evolution of centres for
decision-making in what is now a fragmented industry.

The third consideration is that the problems we tackle initially
by technology should affect large numbers of persons and their
health problems and large numbers of health personnel and institutions rather than events affecting few persons and providers. There
are a number of reasons for using this criterion: costs per
episode of illness, health problem, or person cared for can be
minimized when economies of scale and aggregation of markets exist;
skills required to achieve established standards of quality can be
developed and maintained when large batches of services or products
are processed by technological means; changes in attitudes and
behaviour that condition the applications of technology are most
likely to occur when "critical masses" of both consumers and
providers exist; and acceptance of technological innovations is
most likely when accompanied by increased satisfaction for large
numbers of providers and consumers.

Fourth, technology can and should be used to enhance efficiency
whenever possible. This may take the form of economies in inputs
of labour, physical capital or, perhaps most important, expensive
human skills. However, the nature and extent of the proposed

savings can prove illusory. Large computer systems, for example, require highly skilled personnel to operate and maintain the systems, back-up systems in case of failure, and capital funds for periodic modernization.

Fifth, there is little hard scientific evidence that many of our medical treatments, regimens and procedures are efficacious. We should avoid heavy investment in technology that automates or otherwise attempts to increase the efficiency with which we provide services of dubious efficacy and concentrate in areas where the underlying methods of treatment have been demonstrated to prevent, cure and ameliorate disease.

Sixth, technological applications should be advocated on the basis of evidence that they are both "worthwhile" and "practice ready" if costly failures are to be avoided. They should be preceded by studies of clinical and administrative efficacy and cost effectiveness, taking into consideration operating and service reliability, safety hazards, performance standards, calibration problems, and redundancy requirements. Scientific and technological work leading to invention and development of prototypes should be followed by related studies on marketability and acceptability before innovation is advocated.

Finally, controlled field testing, evaluations and large-scale applications of technology require adequate numbers of suitably sized settings each with the requisite sources to support the effort, trained manpower for managing, operating and servicing it and the interest and means to evaluate it.

5. PROBLEMS CONCERNING THE APPLICATION OF TECHNOLOGY IN MEDICINE

Undoubtedly the development of Medical Sciences in recent years and the immense impact which it has made on the progress of human society is due to a great extent to the contribution of technology which during this time has undergone tremendous advancement in all fields.

In spite, however, of all this much lauded contribution of technology to medicine, many problems have been created which increase rapidly with the passage of time and pervade all areas of medicine.

The basic problem which we face, especially in small countries, is the difficulty of acquiring high technology equipment in a reasonable time. Problems arise both from a lack of availability of financial funding necessary for the acquisition of such machines, and also from technical problems associated with importation and the beaurocracy associated with transporting, installing and commissioning such machines.

On the other hand, there is also a lack of specialised personnel to work in centres which use such high technology equipment. It is suggested that special seminars should be organised to train existing personnel both before the equipment is acquired and also during its use. Suitable conditions need to be created to attract personnel. For this, it is necessary to provide special stimuli to increase the interest of those who are capable of working in these fields.

The enormous problem of maintenance of machines also needs to be faced. It should be stressed that many such medical machines lie unattended and unused in dark store rooms either from lack of operators or lack of specialised personnel to maintain the machine in good working order. At this point it should also be mentioned that with regard to maintenance company representatives show almost complete indifference once they have sold their product.

The same companies invent many different methods for the favourable projection of their own products, with the result that some confusion exists over the true effectiveness and usefulness of each machine and thus its ultimate value. In other words, there is a danger that medicine may become involved in various cliques that start to play the sales game. With this in mind we must make financial incentives neutral with regard to technology, consider carefully the criteria we use to select the machinery most suitable for our purposes, and not be drawn by various attractive additions or alterations without careful consideration of the uses to which they could be put. At this stage correct advice is needed from experienced and impartial scientists to help in the solution of this particular problem.

By degrees, laboratories with high technology medicine must be developed in the provincial hospitals. Again the problem arises of selection criteria as to where these centres should be placed. Centres near provincial universities should be selected initially because the necessary scientific personnel are more easily attracted to an academic centre and technical support is more readily available.

Advances in new technology for diagnosis do not necessarily translate into increased survival. For example, George and Wagner examined the effects of radionuclide scanning for brain tumours at Johns Hopkins from 1962 to 1972 (George and Wagner 1975). The number of brain scans increased ten-fold in that decade and for patients with tumours the average interval between onset of symptoms and operation fell from nearly four years to less than one. Yet survival after operation remained unchanged and less than half the patients operated on each year throughout the decade were alive two years later.

The phrenetic dissemination of technology especially in the field of computers has led scientists and others who work in this area to the most complicated problem of data overload. Smith from the Mayo clinic stated last year in Naples that the dynamic spatial reconstructor produces data which correspond to 130 volumes of Tolstoy's book "War and Peace" every second (Smith 1981). This is an international problem. Progress is too fast and the feast of technological developments may have led to indigestion. Historically technology has led to a geometrically increasing multiplication of new devices and as technology divides up its tasks, the clinician becomes isolated and alienated and loses sight of the complete picture of the sick human being.

The revolution in new technology has created other problems, such as increasing the cost of medical care (now more than 9% of the gross national product in the USA), speeding up the medical practice of participating in iatrogenic illness and accelerating the anxiety of clinicians who now publish books such as that by John Knowles (1977) entitled "Doing better and feeling worse".

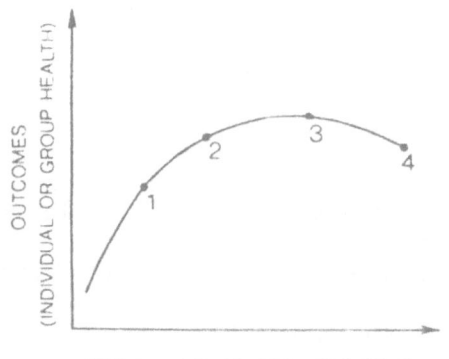

Fig. 1. The relation between medical inputs, measured as services or dollars, and outcomes, measured as individual or group health. The points illustrate theoretical situations with different marginal returns to health from increasing medical inputs.
(From Schroeder, 1981)

The phenomenon of doing better and feeling worse, whilst spending more is illustrated in Fig. 1, which shows four hypothetical points on a curve relating costs or inputs to outcome for either population groups or individual patients. Point 1 is a very underserved area such as a southeast Asian refugee camp where a small amount of medical care can make a big difference in outcome. Point 2 is an underserved area such as might be found in some rural areas. Point 3 is what John Bunker (see Schroeder 1981) called "the flat of the curve" where no further improvement in outcome can result no matter how much more medical care is given. Finally, Point 4 illustrates clinical iatrogenesis where further medical care is, on balance, more injurious than helpful. As a reaction to this, the patient sometimes longs for the good old days, the days of nostalgia and the stethoscope.

It is, at first glance, very curious that dissatisfaction with medicine, in recent years noted all over the world, is at its most vociferous just at a time when doctors have at their disposal the most powerful medical technology the world has yet seen. The "old fashioned" general practitioner, with few drugs in his bag, is for some reason held in higher esteem. Modern diagnosis provides an effective range of instruments and procedures that are highly diagnostic for a long list of diseases. How, then, can we explain the paradox that more effective medicine has resulted in "nostalgia" for a medically primitive past?

Some complaints concentrate on rising new technology because it is more expensive or because it requires more skilled technical personnel and more experience on the part of the clinicians. Clinicians always wish to have the last word not because they are super scientists, but because they have to deal with the most inestimable good, which is the health. The clinician as member of society faces a phobia for the unimaginable production of new instruments especially in diagnosis. The clinician stands in the centre of a technological revolution which he should accept, learn and apply.

The whole problem of medical technological revolution is deep and complicated and the narrow limits of this presentation do not permit us to examine it more widely.

6. ASSESSMENT OF MEDICAL TECHNOLOGY

Progress in medical technology assessment and consequent improvements in medical practice depend upon many factors (Fineberg and Hiatt 1979): development of a strategy for identifying targets of evaluation; improved methods of assessment and of promoting changes in practice; training and involvement of analysts from a variety of disciplines; dedicated government agencies, third-party payers and medical institutions; adequate financial support; and, above all, committed physicians and medical societies.

The number of potential subjects for technological assessment is enormous. We need practical criteria for deciding whether and to what extent a practice should be evaluated. The degree of uncertainty about benefits and safety and the expected costs of a practice can help determine the magnitude of its evaluation. We also need clearer grounds for differentiating practices whose value has not been proven from those that are clinically establish- ed. A gross distinction between "experimental" and "proved" does not account for the possibilities that the safety and technical components of a practice may be known, although its clinical effectiveness is not, that a practice may be proven effective for one class of patients and not for another, or that it may be proven in one setting but not in others.

Because of the complexity of technological assessment, experts from disciplines in addition to clinical medicine must be enlisted in these efforts. Data specialists, epidemiologists and statist- icians can help us clarify the natural history of disease and the effects of interventions, develop uniform methods for collecting, analyzing and storing medical data, and improve designs for clin- ical studies. Behavioural scientists can work with physicians to develop better techniques for assessing the quality of life and other tangible benefits of medical care. Since assessment of medical practices includes comparing costs and benefits, the contributions of economists are essential. In addition, some forms of assessment require the perspectives and knowledge of the lawyer, the ethicist and the political scientist.

Interdisciplinary research requires not the sequential advice of a series of experts, but the interaction of experts who under- stand each other's languages, points of view and value systems, and who can jointly reach balanced judgments. This implies a broadened education for physicians, with greater emphasis on the quantitative and social sciences. Teaching hospitals should encourage assessment of medical practices, just as they have promoted studies of the biological aspects of disease and treatment.

7. THE FUTURE OF MEDICAL TECHNOLOGY

It is clear that the issue is not whether medical technology is good or bad but rather how to foster its appropriate use. How much knowledge, beyond the medical years which he has already spent and experience he has already gained, should the clinician accumulate to reach the point where he will be able to predict the future of new technology? Clinicians ask for better sensitiv- ity resolution and quality for their diagnostic examinations. They ask for simple diagnostic instruments, not a complexity that will drive them to the position where they may say "Does anyone remember where I put the patient?" They ask of technologists more basic research for better production of diagnostic instruments. They do not like technologists to participate in programmes which proliferate complicated and sometimes useless instruments. They ask of physicists to be able to deal more effectively in combining clinical and technical data.

As far as technology is concerned, medicine is not in the forefront of industrial interest because of the limited market by comparison with the larger markets of other sciences. On the other hand, this market becomes more restricted in periods of financial crisis when first priority becomes expenditure for health.

There are distinct examples of a considerable delay in the exploitation of a new discovery before it is applied in medicine. The experimental foundations of NMR spectroscopy were laid by F. Block and M. Purcell 30 years before NMR became available to medicine. Better co-operation can shorten this time lag. This implies a broadened education for clinicians with greater emphasis on the basic sciences. A further very important issue might be the development in every hospital of a workshop where technologists could convert medical ideas into technological practice. This area has been the dream of clinicians for many years but until now there has been little progress towards achieving it except in large medical centres.

Physicists and technologists who collaborate with clinicians should inform them continuously of new achievements in their own area. This collaboration can become a continuous stimulus for clinicians to approach problems in the technological sphere. Here, an international "gap" exists in collaboration which must be bridged as soon as possible. Further, co-operation and mutual respect are essential.

Ways must be found to prevent inadequacies in instruments which require very expensive development. Development evolves secretly not only by competing companies but also by nations.

An ideal collaboration will lead both clinicians and technologists to better investigations of medical needs and to the creation of simpler and more useful technology. Improved certainty as to diagnosis, better quantification and better medical decision-making are worthwhile objectives.

Clinicians see for the future production of diagnostic instruments which will be able to give better information from the area of the pancreas for example and the paraaortic space. Better radiopharmaceuticals than thallium (Tl-201) for more precise localization of the site and estimation of the size of a myocardial infarction and better radiopharmaceuticals for the study of metabolic activity of the brain and the heart with improved image displays and simpler computer processing are required. In digital subtraction angiography, exact registration is required and patient sensitivity to iodinated contrast media remains a problem.

The unique features of NMR should provide better information in the future on the chemistry of tissues. Proton images from NMR should be comparable to those from transmission computerised tomography. PET scanning of the brain and other organs can interact with NMR imaging in two mutually helpful ways - NMR images can help identify which regions of the brain are the sites of biochemical measurements made by positron tomography; PET measurements can help determine the biochemical significance of NMR images, such as those of relaxation times.

The development of PET will lead to an increase in the number of hospital cyclotrons, although financing remains a problem. With PET imaging and more cyclotron produced positron radiopharmaceuticals, perhaps one day physicians will be able to use molecular data assessed by quantitative imaging in the way the pathologist now uses the microscope to study autopsy specimens.

Finally, radiolabelled monoclonal antibodies to tumour-associated antigens will be a valuable approach to the in vivo diagnosis of cancer.

All these efforts should be carefully directed to one single target, the optimization of human health. Therefore, progress in medical technology must be under continuous evaluation and assessment. Some relevant factors are: development of a strategy for identifying targets of evaluation; improved methods of assessment and of promoting changes in technology; training and involvement of analysts from a variety of disciplines; dedicated government agencies and medical institutions; adequate financial support; and above all, committed physicians and medical societies. At any time, the body of knowledge that forms "the technology", especially that dedicated to diagnosis, is a curious mixture of highly effective research interspersed with dogma, empiricism and at times, superstition. With exponential growth of technology this situation can no longer be tolerated. All efforts which lead to the progress of technology should be very carefully validated by very skilled personnel who will thereby contribute to diminishing the diagnostic error.

8. CONCLUSION

Technology must give clinicians the chance to master it and not to serve at its altar. Only in this way can clinicians return to their real role as physicians. They may have to abandon some of their metallic self-satisfaction, or the glittering instruments piled high in their offices and laboratories, but by doing so will have the opportunity to participate in a kind of medical practice that should satisfy both themselves and all of society as well as their patients.

It should not escape our attention that all the problems generated by the application of technology in the medical sciences are concerned with the saving, the repair, and the maintenance of that most valuable human possession, health, before which neither money, politics, nor any other interests should take precedence.

REFERENCES

Crummy A.B., Strother C.M., Sackett J.F. et al (1980). Computerized fluoroscopy: digital subtraction for i.v. angiocardiography and arteriography. Am. J. Roentgenol. 135, 1131

Damadian R. (1971). Tumour detection by nuclear magnetic resonance. Science, 171, 1151

Fineberg V.H. and Hiatt H.H. (1979) Evaluation of medical practices. The case of technology assessment. New Eng. J. of Med.15, 1086.

George R.O. and Wagner H.N. (1975). Ten years of brain tumor scanning at Johns Hopkins: 1962-1972, Noninvasive brain imaging: computed tomography and radionuclides. Ed. H.J. deBlanc Jr., J.A. Sorenson. New York Soc. of Nucl. Med. pp 3-16.

Hirschowitz B.I., Curtiss L.E. and Peters C.W.(1958). Demonstration of a new gastroscope, the "fiberscope". Gastroenterology. 35, 50

Hounsfield G.N. (1973) Computed transverse axial scanning (tomography) Part I: Description of system. Br. J. Radiol. 46, 1016

Knowles H.J. (1977) Doing better and feeling worse. Health in the United States. W.W. Norton and Co., Inc. New York.

Phelps M.E. Hoffman E.J., Selin C. et al (1978) Investigation of F-18 deoxyglucose for the measure of myocardial glucose metabolism. J. Nucl. Med. 19, 1311

Schroeder A.S. (1981) Medical technology and academic medicine. The doctor-producers' dilemma. J. Med. Educ. 56, 634.

Serafini N.A., Gilson J.A. and Smoak M.W. (1976) Nuclear Cardiology Plenum Medical Book Company, New York and London

Smith H. (1981) Dynamic spatial reconstruction. Lecture, Naples Computerized Medicine, Naples Sept. 19-23.

Sokoloff L., Reivich M., Kennedy C. et al (1977) The C-14 deoxyglucose method for the measurement of local cerebral glucose utilization. Theory, procedure and normal values in conscious and anesthetized rats. J. Neurochem. 28, 897

White L.K., Murnaghan H.J. and Gaus R.C. (1972) Technology and health care. Special article. N. Eng. J. of Med. 287, 1223

RECENT DEVELOPMENTS FOR IN VIVO DIAGNOSTIC EXAMINATIONS USING PHYSICAL METHODS - THE CLINICIAN'S VIEWPOINT

I. URGANCIOĞLU

Cerrahpasa Medical Faculty, Centre of Nuclear Medicine, Istanbul, Turkey.

Medical technology has progressed greatly in recent years. It has made a major contribution to the improvement of the diagnostic methodology. Among the new techniques, X-ray computed tomographic scanning, single photon emission computed tomography and nuclear magnetic resonance (NMR) imaging can be mentioned. Development of these methods has closely paralleled recent rapid development in physics and computer sciences and their application to medical methodology.

These techniques have many common problems:

(i)   They are provided by a team comprising physicians, physicists, radiopharmacists and well trained technicians. Careful organisation is required if they are to work in collaboration.

(ii)   Are there enough well trained physicians for sufficiently good interpretation?

(iii) What measures can be taken to decrease observer variability in the interpretation of diagnostic images?

(iv)   It is preferable to apply non-invasive diagnostic investigative methods since risks and complications are less.

In all investigative methods, especially in imaging, it is important to decrease observer variability, to have repeatable results and good statistics. Observer variability has been explained from two basically different points of view. The first regards variations as the result of potentially correctable "mistakes" made by the observer for a variety of reasons, for example, incomplete knowledge, faulty visual search or occasional lapses of attention (Turner 1978). The other approach to the problem of observer variability is to regard the interpretation of a diagnostic image as a statistical decision made under conditions of uncertainty. A diagnostic image is a complex structure in which abnormalities indicative of the presence of disease are seen against a background of confusing noise, such as normal variations, artefacts, or anatomical structures that obscure or simulate the abnormalities. This condition creates uncertainty for the observer. The sensitivity of the test (correct identification of true positives) and its specificity (correct identification of true negatives) then become important.

Now let us consider the advantages of non-invasive techniques both for patients and for physicians. The aim of any medical investigation is to obtain maximal information without hazard to the patient, or at least to obtain the same information with the least hazard to the patient. For this purpose several methods have been suggested recently. Here, by way of illustration, I wish to discuss briefly, emission computer tomography (ECT), radionuclide cardiac investigative methods and NMR in medicine from the clinicians point of view.

First, what is radionuclide emission tomography? It is a method similar to computed transmission (CT) tomography. The principle of CT scanning is to produce an exact three-dimensional reconstruction of a tomographic section which can be accomplished in the transverse or longitudinal direction or at any angle between these two extremes. Important physical factors which must be taken into account in ECT, according to Phelps (1977) are:

(i)   The requirement of uniform or near uniform detector resolution and response (efficiency) with depth,
(ii) Accurate correction for photon attenuation
(iii) Removal or significant reduction of scattered radiation to ensure that the detector response represents the linear sum of the activity viewed and to provide high contrast and sensitivity
(iv) Accurate detector positioning and sampling to provide optimum reconstructed image quality
(v)   A detection system with high efficiency to satisfy the stringent statistical requirement.

ECT systems can be divided into two main categories:

(i)   Systems which employ single photon counting such as scanners and cameras for Tc99m, I-131, I-123, etc.
(ii) Systems which employ annihilation coincidence detection of positron emitting radionuclides.

The question of whether ECT can perform dynamic studies, for example in a manner similar to the conventional approach to dynamic studies such as blood flow with rapid time sampling, has arisen frequently. The answer is probably negative because of statistical limitations (Jaszczak et al 1977).

Now it seems that computed tomography can be further divided into:

(i)   Morphological tomography with transmission CT

(ii) Physiological and functional tomography with ECT.

The success of the ECT approach is very dependent on the availability of several features, which play a role in determining the success of the ECT and the selection of either single photon emission tomography or annihilation coincidence detection. ECT is being used to approach physiological or functional problems by using radionuclides for the measurement of metabolism and physiological function (Jaszczak et al 1977, Oppenheim 1980, Watson et al 1980).

The second method I wish to discuss is nuclear cardiology. Non-invasive methods employing radiopharmaceuticals have been used in cardiology for a long time - the first diagnostic use of radio-nuclides was in the study of the heart by Blumgart and Weiss in 1927. The evolution and growth of nuclear cardiology have been very rapid and many cardiac investigations can now be done by non-invasive methods.

Differential diagnosis of chest pain, cause of dypsnoea and fatique can easily be investigated by these methods. The pathology and physiology of the cardiovascular system can also be studied easily. In addition, nuclear cardiology has permitted the visual display of function. Common procedures in nuclear cardiology are assessment of myocardial function, including ejection fraction and wall motion studies, investigation of the heart muscle to demonst-rate acute myocardial infarction and radionuclide angiograms (Holman et al 1979, Thrall et al 1980, Soin and Brooks 1980, Bradley and Tosteson 1981).

The third subject I wish to mention briefly is NMR. Although this will be discussed in detail elsewhere, it is important to emphasise at this stage the importance of this new technique which shows great promise both physiologically and tomographically.

NMR is based on the idea that certain atomic nuclei behave as small magnets. According to the strength of the applied magnetic field, these nuclei will orientate themselves in the direction of the field. In this configuration, the nuclei will rotate about the field of direction (precess) at a frequency depending on the strength of magnetic field. If these aligned and precessing nuclei are exposed to an alternating magnetic field (a radiofrequency or RF field), some of them will be forced from their equilibrium state to a higher energy level. When the interrogating RF is removed, the excited nuclei emit the added energy and return to their former state. The analysis of certain parameters of the emitted energy can be used to create images.

Relaxation time is a parameter describing the exponential return to equilibrium of the nuclear magnetism of the selected nucleus. Two main relaxation times are noticed, T1 (longitudinal relaxation time) -"Spin lattice relaxation time" and T2 (transverse relaxation time) - "Spin-spin relaxation time". The longitudinal magnetic relaxation time, T1, measures the time it takes the sample initially to become polarised in the fixed external magnetic field. It is also the time constant of return to equilibrium in the fixed magnetic field after a radiofrequency pulse. T1 is shorter in liquids (seconds) than in solids (minutes to hours).

T2 is a measure of how long the resonant nuclei hold the temporary transverse magnetisation. It indicates the relationship between the strength of the external field and the strength of local internal fields.

Since different nuclei have different magnetic moments and spins, they will undergo resonance at different frequencies when placed in the same magnetic field. NMR images provide different information from that obtained by conventional radiography, computed tomography, nuclear medicine, ECT or ultrasound images. In NMR imaging, the signal is proportional to the concentration of resonating nuclei in the resonant volume or plane and may also be proportional to the relaxation times. The potential of this technique includes evaluating cellular chemical properties or proton density (Fullerton 1982) and it is hoped that much information will be obtained with the NMR technique for greater understanding of diseases.

REFERENCES

Bradley W. and Tosteson H. (1981) Basic physics of NMR. Nuclear Magnetic Resonance Imaging in Medicine. IGATU-Shoin
Fullerton H.D. (1982) Basic concepts for nuclear magnetic resonance imaging. Magnetic Resonance Imaging, 1, 39
Holman B.L., Abrams H.L., and Zeitler E. (1979) Cardiac Nuclear Medicine, Springer-Verlag.
Jaszczak R.J., Murphy P.H., Huard D. and Bardine J.A. (1977) Radionuclide emission computed tomography of the head with Tc-99m and a scintillation camera. J. Nuc. Med. 18, 373

Oppenheim B.E. (1980) Computer assisted emission imaging. J. Nuc. Med. $\underline{21}$, 286

Phelps M.E. (1977) What is the purpose of emission computed tomography in nuclear medicine. J. Nuc. Med. $\underline{18}$, 399

Soin J.S. and Brooks H.L. (1980) Nuclear cardiology for clinicians. Futura publishing company.

Thrall J.H., Pitt B. and Brady T.J. (1980) Radionuclide wall motion study and ejection fraction in clinical practice. $\underline{64}$, 99

Turner D.A. (1978) Observer variability; what to do until perfect diagnostic tests are invented. J. Nuc. Med. $\underline{19}$, 435

Watson N.E., Cowan R.J., Ball M.R., Moody D.M. et al (1980) A comparison of brain imaging with gamma camera, single photon emission computed tomography and transmission computed tomography Concise Communication, J. Nuc. Med. $\underline{21}$, 507

II. TECHNICAL ADVANCES APPLICABLE MAINLY IN VITRO

III. CLINICAL ADVANCES APPLICABLE MAINLY IN VITRO

CELL IMAGE ANALYSIS IN QUANTITATIVE CYTOLOGY

Peter H. BARTELS, M. BIBBO, G. OLSON and G.L. WIED

10625 East Speedway, Tucson, Arizona 85710, U.S.A. and International Academy of Cytology, 5841 South Maryland Avenue, Chicago, Illinois 60637, U.S.A.

1. INTRODUCTION

The goal of computer analysis of high-resolution digitized images of cells and tissues is to improve diagnostic and prognostic differentiation. Consequently, the emergence of clinical cytology and histopathology as a quantitative science is a recent development that follows advances in image recording technology and computer design. This article reviews the potential of this new methodology, presents the underlying rationale, and provides selected examples from recent results.

There are several major research efforts under way. The first is automation of clinical laboratory procedures. Examples are automated reading of differential white blood cell count (Preston 1962, Ingram et al 1968, Green 1979, Landeweerd 1981) an automated search for chromosome metaphase spreads (Bishop and Young 1977) automated recording of DNA histograms for cells from benign and malignant lesions (Wied et al 1982) and automated prescreening for cervical and bladder cancer (Bartels & Wied 1975, Koss & Bartels 1980). All of these tasks, which until now were performed by trained personnel, demand full attention; yet they are tedious and tiring, and may, during periods of diminished alertness, cause the personnel to make errors. Automation offers untiring, consistent performance at a known and controlled level of reliability. It usually allows the examination of a larger sample size, and thus may add increased reliability. In general, though, the automated device is expected to perform only those tasks that a well-trained human specialist can perform.

The second research effort concerns the potential of machine performance in tasks where humans either perform poorly or may not be capable of providing an answer. Quantitation has been shown to improve markedly diagnostic and prognostic procedures that are based on image information not readily perceived by human visual assessment, and thus never before used for diagnostic purposes (Bartels & Wied 1981). Computation has been shown to be capable of providing image information based on higher order dependence schemes in the digitized imagery to which the human visual sense is not sensitive. Such computed image properties may provide valuable diagnostic and prognostic clues. We see here a direct expansion of our diagnostic ability to perceive, based on the analytical power of computation. Furthermore, automated data processing of a clinical sample permits a comprehensive and usually more exhaustive utilization of all the collected information than is the case in human evaluation. The human diagnostician frequently bases a diagnosis on the "worst instance" criterion, i.e. patient diagnosis is usually based on the appearance of the most abnormal cells. However, when frank tumour cells are not detected in a sample, a diagnosis of "malignancy present" would not normally be made. Cell image analysis has clearly established that a conclusive diagnosis of malignancy could be made on the basis of clues computed from the non-malignant cells in a clinical sample (Wied et al 1980, Wied et al 1982). Statistical assessment of a clinical sample allows diagnostic decisions to be made with known power of the test and known sensitivity, particularly in instances when diagnostic clues are subtle. This capability is central to the development of early detection schemes, avoidance of the need for curative treatment, pre-screening of well-populations, monitoring of workers exposed to hazardous substances at the work place, and biological test and monitoring systems set up to control environmental pollution.

It is not true that diagnostic improvements remain irrevocably tied to the use of computer-analytical procedures. Instead, experience has shown that a definite feedback to human diagnostic judgment is evolving. Computer assessment may point toward certain diagnostic clues, and frequently experienced cytologists then learn to recognize and utilize those clues in their visual assessment (Koss et al 1977, Bartels et al 1978, Bibbo et al 1981).

An exciting development is the use of computer graphics as an exploratory tool for human diagnostic decision making (Dytch et al 1983). Computer graphics may be used to recall from a data bank cell images ordered according to selected cell image properties; for example, the nuclear-cytoplasmic ratio. Human visual judgement often fails to recognize a difference when it is presented as an isolated instance without a reference standard. When, however, an ordered sequence of instances is presented, trends frequently become evident.

Computer graphics may also be used to project a cell image together with a two-dimensional graphic display formed by two suitably chosen image properties as axes (Wied et al 1983). One can then see how a given cell relates to other cells of the same or other diagnostic classification, or how a given cell image relates to cells of the same cell diagnostic classification, but from patients of different disease categories. Finally, computer graphics may generate simulated images of cells and tissues. Such imagery helps us to define the diagnostic clues that experienced diagnosticians are using. It aids us in determining how such clues have to occur in context with other clues. Simulation helps to define the bounds of diagnostic categories as they exist in the minds of experienced human diagnosticians. A mere verbal description delineates very inadequately what constitutes a cell type, but by formalising a description and generating random instances on the computer, one can establish whether the description is complete. A number of specific questions may then be answered. (i) Will the system ever generate images that a human expert would not put into that class? (ii) is the description too limiting so that the system cannot generate random images which a human expert would consider extreme examples of a particular calssification? (iii) are all the mutual dependencies of image properties in the simulated imagery as one would expect them to occur in real images? (iv) does the system draw decision boundaries between one simulated class and all others where a human would draw the distinction? There really is no other way to explore the validity and implied delineation of a description of a cell type than by computer graphic simulation. This is true for cell images and is even more critical for the complex pictorial scenes seen in tissue sections. Computer graphics allow us to verify whether the concept of a class of cells or sections has been fully "understood" by the decision logic.

Once such definitive descriptions are established, it is possible to determine the statistical sampling requirements for the detection of minute and early changes, as required for the development of test procedures in environmental pathology, for example. Understandably, most work in quantitative cell image analysis has been in areas of potential clinical application, but cell image analysis will also play an important role in our understanding of basic cell biological processes.

The key to all the above-mentioned capabilities is in the representation of microscopic imagery in numerical form.

2. NUMERICAL REPRESENTATION OF IMAGE INFORMATION

The optical light microscope has been one of the most important tools in diagnostic medicine for more than a century. This is not only because diagnostically significant changes in tissues and cells can be visualized at image magnifications accessible to the light

microscope, but also because reliance on the optical light microscope as a diagnostic tool rests on the predominance of vision among
the human senses. Diagnostic assessment, detection, and discrimination in microscopic preparations have traditionally relied on the
human eye/brain complex, with its superb ability to understand
complex visual scenes.

Less than 20 years ago, the notion that a mere machine not
only might be able to perform some recognition tasks in diagnostic
microscopy, but also surpass human perceptual acumen was considered
preposterous. Yet, this is exactly what is being accomplished now.
This exciting potential for expansion of our diagnostic and
perceptual capabilities lies at the heart of what might be called
"digital microscopy". Technological advances have made it possible
to digitize optical images at high spatial resolution and at rates
of up to 100 MHz, with 6 to 8 bits of photometric resolution per
pixel. The image may be digitized for several spectral bands
simultaneously. (Analog Devices Computer Labs, AD 5010KD, 100 MHz,
6 bit monolithic ADC's, Greensboro, NC 27409, 1982. TRW LSI
Products, La Jolla, California, "Flash A/D converted printed circuit
assembly model TDC 1025 EIC, eight bit, 75 MSPS, 1982).

In comparison, one may roughly and generously estimate the
upper limit for a steady rate of image intake by humans to be about
7 MHz, which is at the flicker-fusion frequency. This estimate
neglects the fact that a human observer does not take in every
pixel in the grey value of an image every 30th of a second on a
video screen, and that in complex pictorial scenes grey values at
different locations are highly correlated.

The digital microscope is presently being redesigned as an
optical input peripheral to a fast computer, with the recorded
images in numerical form, rather than as an instrument to facilitate
visual observation (Shack et al 1979, Shoemaker et al 1982). The
implications of this reach far beyond mere technical accomplishment
and will change our very approach to the use of information from
clinical samples.

The representation of histological and cytological images in
numerical form offers several distinct advantages - (i) the images
are quantitative, so they can be reproduced, operated on, or
compared, (ii) they can be standardized and defined in terms of
their descriptive statistics, (iii) measures for defined image
properties or "features" can be estimated by their mean values and
their mutual dependencies by their variance/covariance matrix. This
holds true both for individual images of single cells and for
sample fields in histological sections.

Quantitation, however, can be extended to characterise all the
material from a given patient. The composition of the clinical

sample of cells of different cell types, their average feature values, and their indices of typicality may be computed and compared to data from patients in the same diagnostic condition. Measures of distance, of atypicality, and of direction of change from normal can be defined and computed for cells of all types to provide a detailed assessment of a patient's state of health (Bartels et al 1978). In tissue sections the same approach applies, with additional information being available in descriptors and measures of the nuclear placement pattern in the tissue.

The next higher level of quantitation provides data on clinical samples from patients with different diseases, different diagnostic categories, and how these relate to each other.

Numerical representation allows data comparison and application of powerful multivariate statistical tests to detect change. Again, comparison, discrimination, and detection of change enter at two different levels. There is the capability to distinguish a given cell from another with known reliability. Statistically, this capability is limited by the overlap of tolerance regions in the feature space for the two multivariate distributions from which the two cells were samples. Then, there is the capacity to detect small changes between cell populations, and here the limitations are set only by the sample size, the power of the multivariate test statistics, and skillful experimental designs. Statistically, the detection of differences between cell populations is limited by the overlap of confidence regions for the means of the multivariate distributions representing them.

In all of these applications there are clearly instances where human diagnostic assessment is highly developed, and may be used as baseline for the "truth" as established, for example, by the clinical course of disease.

Human visual perception excels in distinguishing different tissue components, in following boundaries, and in substituting, from context, information that should be there but that the particular section shows either only incompletely or not at all. Human visual perception is superb at seeing a pictorial scene as a whole, where different components form an arrangement or a placement pattern. The characteristics of a particular placement pattern can be judged with great reliability whether or not it constitutes what is expected in a certain diagnostic situation.

A human observer is good at detecting relative differences, such as a change in the nuclear/cytoplasmic ratio. We have a superb capability for understanding complex pictorial scenes, something which has not yet been established in an automated system. We are only now beginning to understand how human visual perception works, how diagnostic information is stored, recalled, and used, and how associations are formed and maintained.

However, in all cytodiagnostic tasks there are also instances where human diagnostic assessment is not fully effective and where computed image information is clearly superior (Bartels et al 1980).

There is an abundance of image information that the human visual sense simply does not perceive but that can be retrieved by computation. Textural properties of chromatin or the placement pattern of nuclei are determined by mutual dependencies. If a dense granule is located at a certain place, then the occurrence of another granule of the same or different density at a certain distance is given with a known probability. Such probabilities control the co-occurrence not only of two granules but also of all of the chromatin distributed in the nuclear area. Mathematically, a mutual dependence scheme may be described by a set of parameters - transition probabilities, eigenvectors and eigenvalues of matrices, and similar multivariate statistics. Human beings do not distinguish between textures that differ only in their higher order dependence schemes. This is best demonstrated by an example.

Figure 1 shows a five-variate random process: the grey-value at any point depends on the values at five other locations. To the human observer, the texture field looks homogeneous in its gross structure. But to a computer this field consists of four quadrants with four distinctly different textures. The mutual dependence schemes in the four quadrants are sharply different and the probability that a computer would mistake one for the other is in this instance less than one in a million. Human observers simply do not perceive such high-order dependencies for which there may have been no evolutionary advantage. We would immediately have noticed a difference in the first-order statistic if the average grey-level in the four quadrants had been different. However, and this is the salient point, we now have the capability of extending our power of perception by using computation.

The human eye also is not particularly good at perceiving distributional statistics. Figures 2 and 3 show two fields into which randomly positioned circles of different size have been placed, modelling, for example, cells with nuclei of different diameters. To the observer, the two fields look quite similar. Figure 4 shows though that the distribution of radii in Fig. 2 is unimodal, whereas in Fig. 3 a bimodal distribution prevails.

The human eye is not capable of recognising small differences. In Figs. 5 and 6, two distributions of circles are shown with the same random placement pattern. Although they look identical, the diameter of every circle in Fig. 6 is 3% less than its corresponding counterpart in Fig. 5.

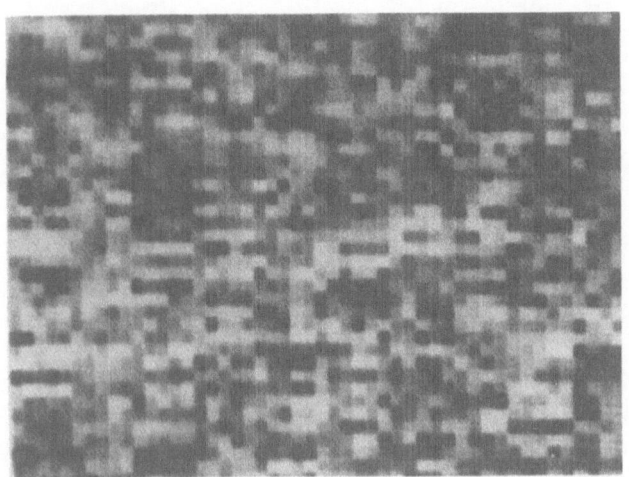

Fig. 1. Textured field in which grey values follow a fifth-order
mutual dependence scheme. While the area appears to be of homogeneous texture to the human eye, to a computer the field clearly falls
into four quadrants with statistically highly significant difference
in texture, as in fact is the case.

Fig. 2. Model for assessment of nuclear diameter distributions,
         first sample field.

Fig. 3. Same as in Fig. 2, second sample field

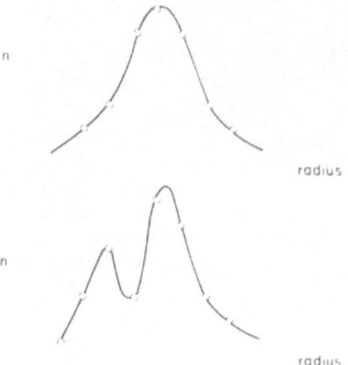

Fig. 4. Unimodal and bimodal distributions of radii from first and second sample fields shown in Figs. 2 and 3.

Fig. 5. Model for assessment of nuclear areas, first sample field.

Fig. 6. Same as Fig. 5, second sample field

Each of these different aspects of computed image information plays a role in practical applications. One of the surprising results of cell image analysis has been the consistency and reproducibility of chromatin distribution patterns in cell nuclei.

3. CHROMATIN DISTRIBUTION PATTERNS

The chromatin distribution pattern in a cell nucleus is determined by (i) the number and size distribution of the chromatin granules, (ii) their staining density, and (iii) their spatial arrangement. The chromatin may be described as coarse or fine in granularity as well as uniformly distributed or aggregated. In addition, one may assess the tendency of the chromatin to concentrate near the nuclear membrane or the centre of the nucleus. For cells of a given type or in a given physiological state, the chromatin distribution pattern is surprisingly consistent and provides highly reproducible features. Features derived from the chromatin distribution pattern enable one to detect very early responses of a cell to (i) changes in physiological conditions or (ii) external influences such as ionizing radiation, virus infections, immunologically active agents, and pharmaceutical and toxic substances.

Research efforts towards objective characterisation of nuclear chromatin distribution patterns have addressed two major problems. The first is the methodology of introducing mensuration to what has been a strictly subjective procedure - visual microscopic examination of cells by a cytologist. How can one characterise such a chromatin distribution pattern, and what measures or features should be defined? How can one compute and evaluate these features? How may one apply them for objective classification or description of a response or a trend? The second problem refers to the biological meaning of these measures. How can one interpret the measures themselves or the observed changes in terms of biological processes?

These two problems may be addressed independently, but it is clear that they are closely related. One can establish a highly successful automated classification procedure for different cell types; for example, normal and malignant cells, using features that consistently render excellent discrimination. To accomplish this, one does not require any insight into the biological reasons as to why these features assume certain values in malignant cells and markedly different values in normal cells. On the other hand, a more satisfactory situation exists whenever a known biological process can be closely correlated with the value distribution for a given feature and a biological interpretation is possible. Under these conditions one can measure a biological response directly.

It is well known that the degree of condensation of nuclear chromatin as heterochromatin or euchromatin reflects different levels of functional activity, (Grundmann and Stein 1961) hetero- chromatin being associated with a lower level of genetic activity. The relationship between heterochromatin and euchromatin and the functional state of cells was explored in several studies by Kiefer and Sandritter (Sandritter et al 1967; Sandritter and Kiefer 1970; Kiefer et al 1974). Increasing differentiation or maturation leads to larger proportions of the nuclear chromatin assuming a condensed form. The ratio of heterochromatin to euchromatin is well defined and varies only within narrow tolerances in cells of a given cell type and in a given physiological state. Sandritter and Kiefer (1970) determined the proportion of heterochromatin in peripheral lymphocytes as 80% whereas lymphoblasts contain 40% to 60% hetero- chromatin. Dedifferentiating cells are characterized by a predom- inance of euchromatic material. The increase in heterochromatin and the loss of genetic information has been described as a patho- genic principle by Sandritter and others (Harbers et al 1968). Marked changes in the ratio of heterochromatin to euchromatin were observed in transforming lymphocytes (Bartels et al 1969, Rowinski et al 1972).

4. PRACTICAL APPLICATIONS

The methodologies of image recording, data reduction, feature selection and evaluation, the stategies for automated classificat- ion self-learning computer programmes employed for this purpose have been extensively described during the past decade. We refer the reader to the historical review by Preston (1981),to a detailed description of practical procedures by Bartels and Olson (1980) and Simon et al (1975) to the proceedings of conferences on clinical cytology automation (Wied et al 1976, Pressman and Wied 1979) and to a review of the state of the art (Bartels and Wied 1975, Bahr et al 1979, Bartels et al 1980, Wied et al 1981).

The sensitivity of image analytical methods is very high. However, one must distinguish between two aspects of this sensitiv- ity. First, there is sensitivity for detection of change in observed image features, given by the analytical power of the employed statistical procedures. Since sample size is usually not a problem in cell image analysis, and since almost routinely at least half a dozen independent variables can be identified, it can confidently be said that existing statistical methods provide more than adequate sensitivity. Second, there is the sensitivity of chromatin organization to changes in the cells' environment. We can detect and substantiate the occurrence of subtle changes in the chromatin, but how small a biological stimulus will produce a change?

Two examples may serve to demonstrate this. In a study by
Lockart et al (submitted for publication) Daudi cells were incubated
with interferon. Such incubation may, in its initial step, activate
a very small number of four to six additional genes. Even such a
gentle stimulus leads to statistically significant changes in the
chromatin condensation, as can be seen in Fig. 7. Two image properties are plotted: one a discriminant function, that is a compos-
ite feature for the discrimination between controls and incubated
Daudi cells, the other a measure of the amount of noncondensed
chromatin. Shown are the confidence ellipses for the estimates of
the bivariate means, control cells, and interferon-treated cells;
the 50% tolerance ellipses; and the Bayesian decision boundary for
the classification of a cell as a control cell or as an affected
cell.

The second example is taken from an application in environmental pathology; the detection of exposure to subtoxic doses of
a toxicant. The tremendous potential for computer analysis of cells
and tissues in environmental pathology is demonstrated in a study
by Nair et al (1980). Experimental rats were exposed to doses of
chlordane several orders of magnitude below the toxic level. Cytophotometry revealed, as expected, a statistically significant
increase in the number of liver cells with doubled DNA contents,
indicating the occurrence of repair processes (Fig. 8 and 9) More
interesting though was the finding that in the liver cells of
treated animals, even the cells with regular (8N) DNA content
showed consistent, measurable changes in their chromatin pattern.

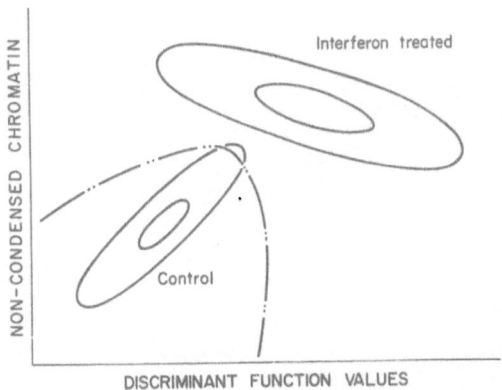

Fig. 7. Changes in the chromatin condensation in Daudi cells
         incubated with interferon.

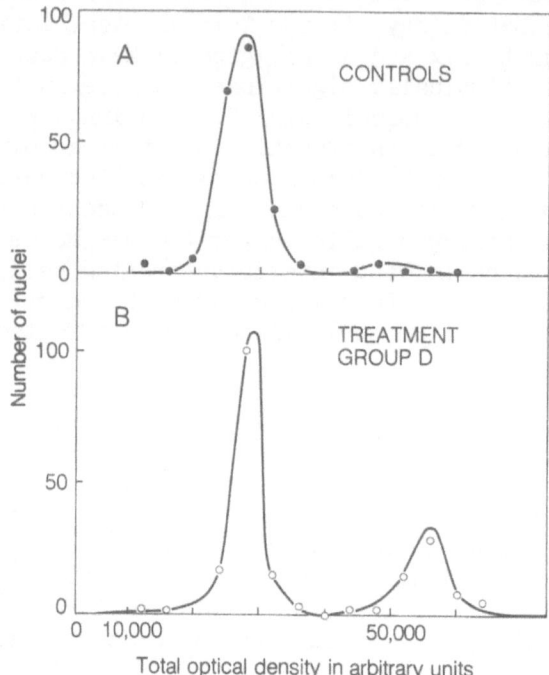

Fig. 8. Distribution of hepatocytes as a function of total optical
density from control rats(A) and rats administered feedings
of CCl$_4$ and chlordane (B).

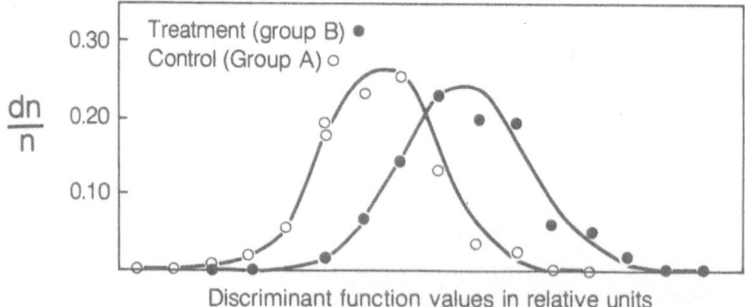

Fig. 9. Distribution of linear discrimination function values for
hepatocytes from control rats and animals fed chlordane.

There can be no doubt that the ability to measure even more subtle changes will require a new determination of what constitutes toxic exposure. It may be that observed changes of this nature are not lasting, or that they will remain asymptomatic. It may also be that mere visual examination of stained tissue sections may no longer be adequate in the determination of thresholds of toxicity.

There are numerous examples where the sensitivity of image analytical techniques for the detection of change is clearly demonstrated. It is also likely that the limits for detection have not been approached. On the other hand, relatively few studies have been directed at probing the specificity of changes in the chromatin patterns. The few studies that have been completed indicate that there appears to be a surprising specificity to changes in chromatin distribution and staining.

Lymphocytes exposed to ionizing radiation (Anderson et al 1975 a, b) to cytotoxic drugs (McKee 1975) and to a virus infection (Olson and Bartels 1981) all exhibit measurable changes in their chromatin pattern, even though these changes generally remain below the threshold for visual detection (Olson and Bartels 1982). This is demonstrated in Fig. 10 where Feulgen-stained lymphocytes exposed in vivo to X-ray irradiation, to cyclophosphamide, and to a Friend virus infection are shown.

Figure 11 shows that in the two-dimensional feature space, each exposed population deviates from the untreated cells to a different extent and in a different direction. Table 1 shows classification results.

One of the most captivating examples of the use of non-visual diagnostic clues is the marker features for the presence of dysplastic and malignant disease expressed in normal-appearing intermediate cells from the ectocervix (Wied et al 1980, Bibbo et al 1981, Wied et al 1982). This discovery has two important implications. First, there must be a "field effect" causing subtle changes in normal intermediate cells in patients who have malignant cervical disease. Second, if this finding is consistent for a large sample of patients, it will have a major impact on the strategy for cervical cancer prescreening. No longer would there be a need to search for possibly rare tumour cells among large samples of 50,000 to 200,000 cells in a clinical preparation, nor would there be any need for substantial computer power to do the image processing. Instead, a modest sample of readily detected intermediate cells could be examined, and a powerful statistical test could be applied to see whether a patient is likely to require further attention.

72

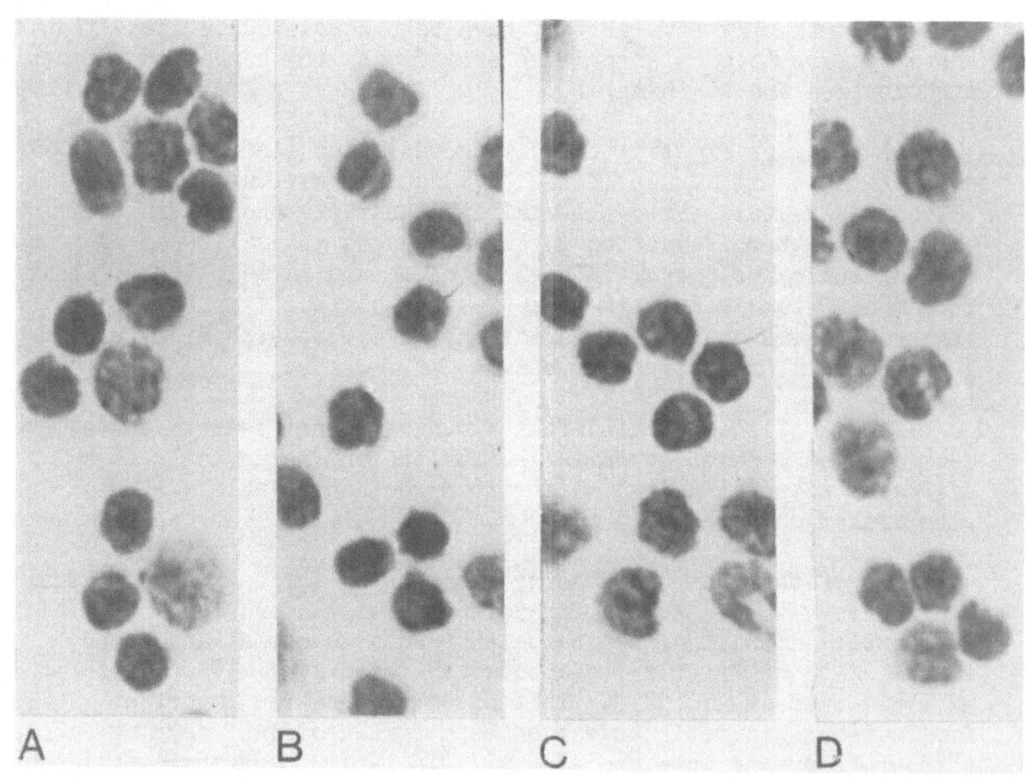

Fig. 10. Appearances of Feulgen-stained lymphocytes after
exposure in vivo to (A) no treatment (B) X-ray
irradiation (C) cyclophosphamide (D) Friend virus
infection.

73

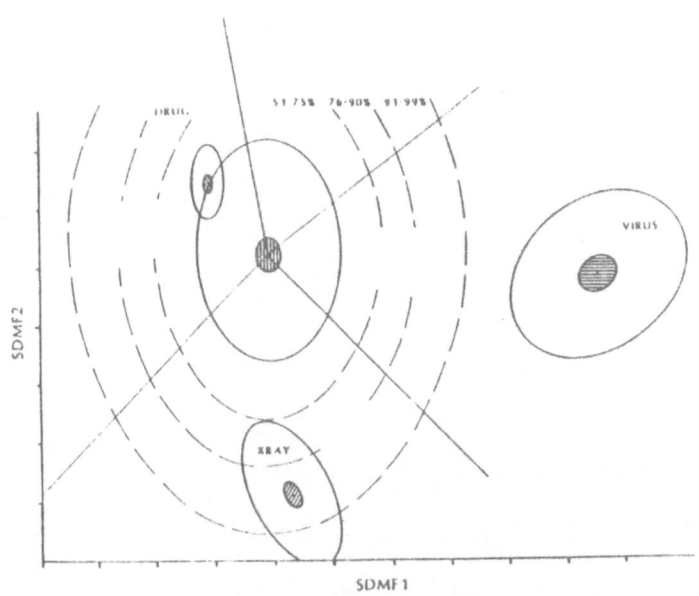

Fig. 11. Results of computer analysis - expressed in two-
dimensional feature space - for the nuclei shown
in Fig. 10.

CLASSIFICATION OF LYMPHOCYTES BY PROGAMME PROBE

ACTUAL GROUP MEMBERSHIP	PREDICTED GROUP MEMBERSHIP (%)			
	CONTROL	VIRUS	DRUG	X-RAY
CONTROL	82.4	0	10.9	6.7
VIRUS	6.9	93.1	0	0
DRUG	4.3	0	95.7	0
X-RAY	0.6	0.6	1.7	97.1

% GROUPED CASES CORRECTLY CLASSIFIED 92.56%

TABLE 1

74

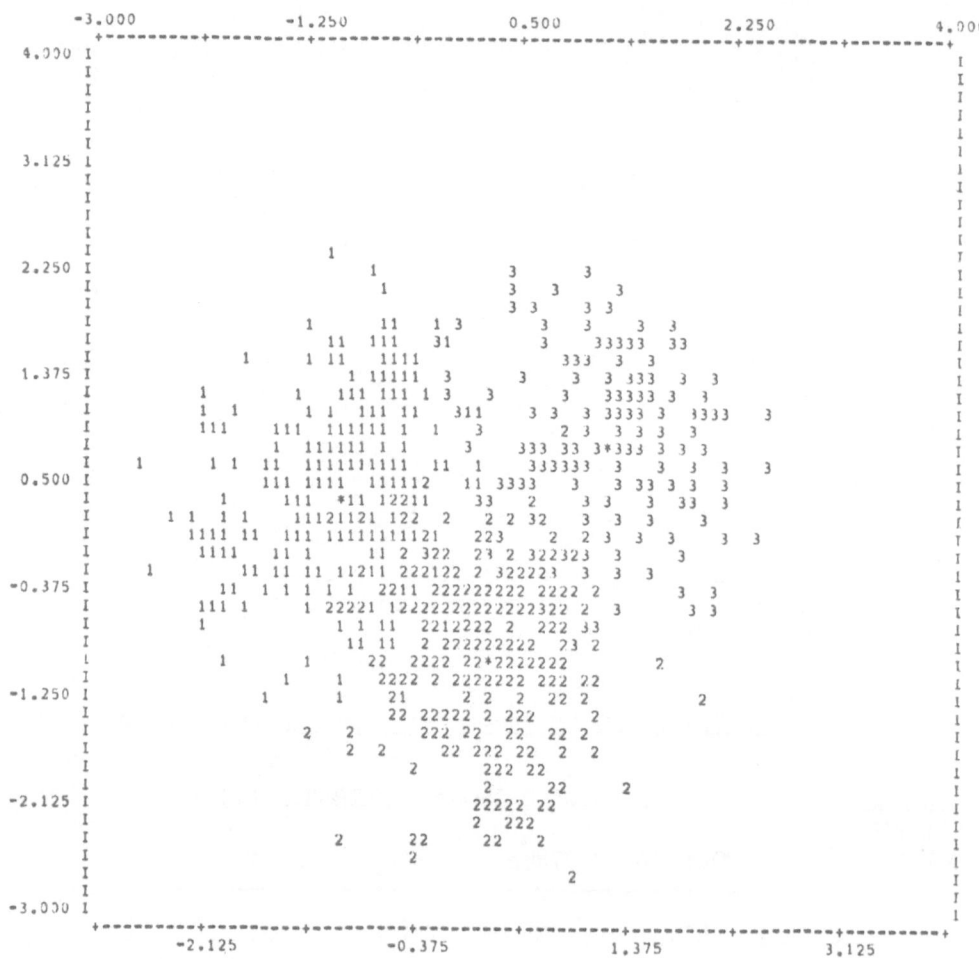

Fig. 12. A plot of two discriminant functions for cell data from patients with normal cytology (code 1), patients with severe dysplasia/carcinoma in situ (code 2), and patients with invasive cancer (code 3).

Computer assessment has discriminated remarkably well between
normal intermediate cells from the ectocervix of patients with
normal cytology and normal-appearing intermediate cells from
patients with malignant disease. Figure 12 indicates the chromatin
changes in intermediate cells from patients with normal cytology,
carcinoma in situ, and invasive cancer, respectively. The extent
of change is about the same for both categories of patients with
abnormal cytology; however, the direction of change is different
and specific. Figure 13 shows the corresponding confidence regions,
tolerance ellipses, and Bayesian boundaries for the discrimination.

How should the assessment of a given patient sample by conveyed
to a cytologist? One option is to draw a curvilinear regression
line through the centroids for the three groups shown in Fig. 13
and project data for all of the patient's cells down onto this
line. Characteristic patterns that can be easily assimilated by a
cytologist are shown in Figs. 14a, 14b, and 14c for a patient with
normal cytology, a patient with carcinoma in situ, and a patient
with invasive cancer (Bartels et al 1982).

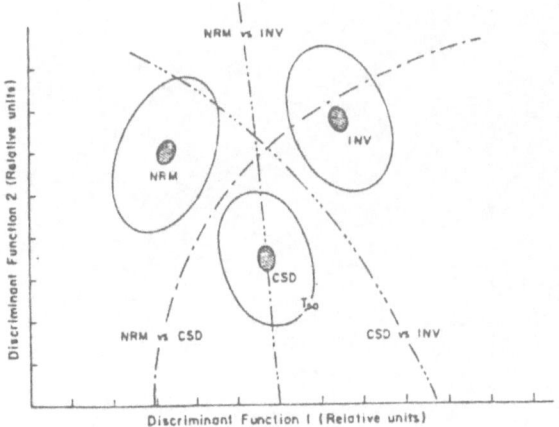

Fig. 13.   Confidence regions, tolerance ellipses and Bayesian
boundaries for the data shown in Fig. 12.

76

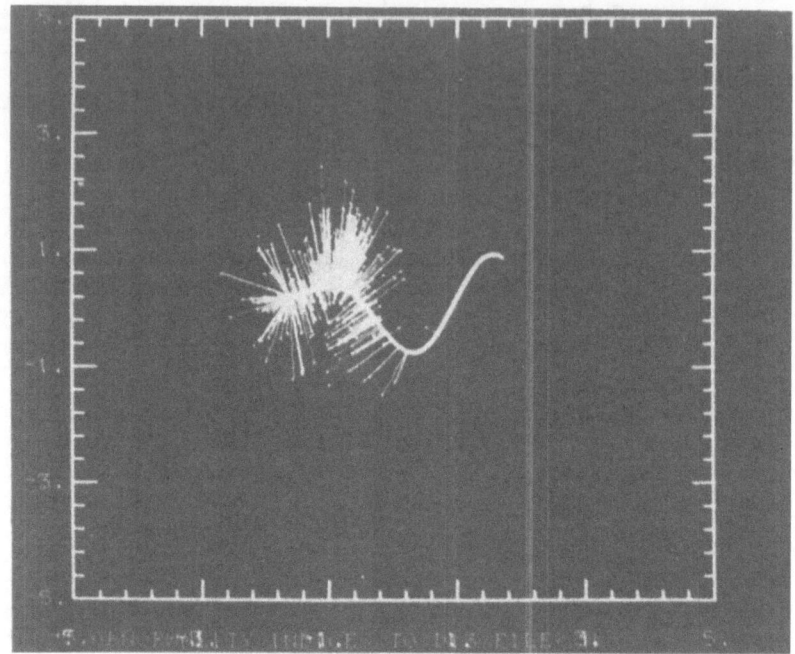

Fig. 14a

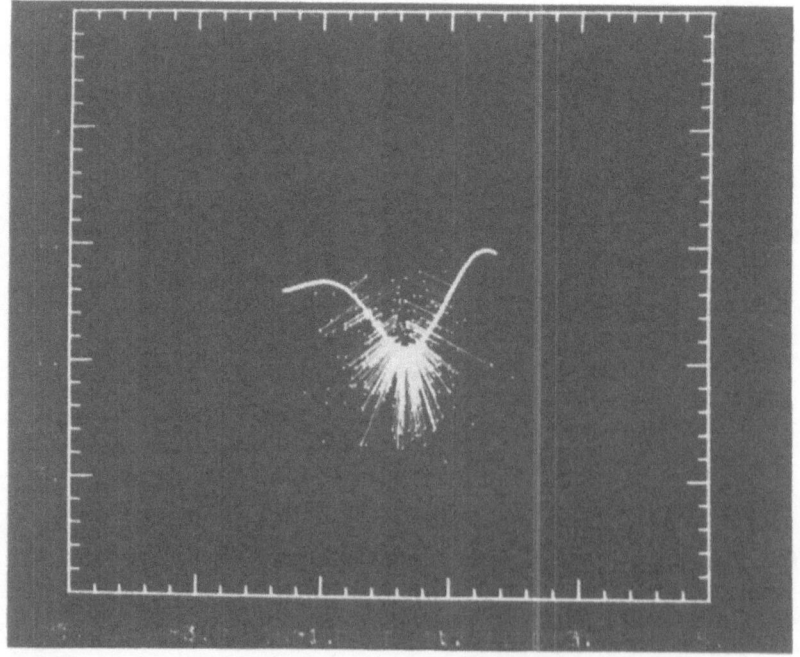

Fig. 14b

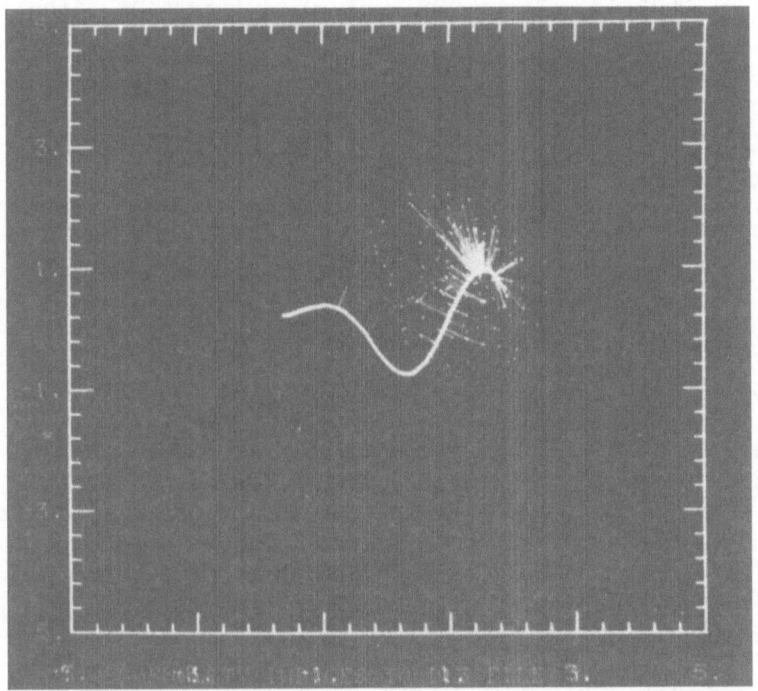

Fig. 14c

Data reduction patterns obtained for a) normal cytology
b) carcinoma in situ c) invasive cancer. For details of the
procedure, see text.

It is essential in studies of this kind to establish the
statistical significance of differences between categories over
differences between individual patients. For this, nested designs
in analysis of variance are suitable (Bartels et al 1983).

For the blue intermediate cells from patients with normal
cytology (INMT NRM) with moderate dysplasia (INMT MOD), and with
severe dysplasia/carcinoma in situ (INMT CSD), a 3 X 10 X 30, two-
level nested design was used with three diagnostic categories, 10
patients each, and 30 cells measured on each slide. This provides
2 degrees of freedom for the mean square of the diagnostic categories, 3 X 9 = 27 degrees of freedom for the patient to patient mean
square, and 3 X 10 X 29 = 870 degrees of freedom for the cell-to-
cell variability.

Table II gives the results for the marker feature "red/green"
contrast of nuclear chromatin (feature #106). The mean values for
the three categories were NRM 0.0822; MOD 0.125, and CSD 0.112.

	Sum of Squares	df	ms	% total ms	F-value	alpha
Diagnostic Category	0.205143	2	0.102571	90.5	9.773	< 0.01
Patients	0.283354	27	0.00496	9.2	36.534	< 0.01
Cells	0.249909	870	0.000287	0.25		

Table II

Of the total mean square, 90.5% is attributable to difference between categories, 9.25% to differences between patients, and only 0.25% is due to differences between cells. If one relates the cell-to-cell mean square to the measured mean value for feature #106 over all observations, a coefficient of variation of 15.9% is obtained. The patient-to-patient mean square reflects changes that have a coefficient of variation of 95%, or roughly a factor of two. The diagnostic category variation by comparison changes by a factor of three.

Expression of marker features in intermediate cells from the ectocervix presents an excellent example of image information that has escaped the attention of cytologists, but that can be measured with consistency to provide clues as to the diagnostic condition of the patient. However, when cells are arranged in the order of a discriminant function score, experienced observers can see the trend in staining properties. This is demonstrated in Fig. 15. A discriminant function was computed that separated intermediate cells from normal patients and intermediate cells from patients with malignant disease. Several cell images were selected and arranged in order of their scores. In this instance positive values for the score indicate cells with strong clues to the presence of malignancy in the patient; negative values indicate patients with normal cytology. This is also an excellent example of the creative use of computer graphics in quantitative and diagnostic cytology.

Numerical representation of cytological image data allows image manipulation for training purposes. Images of cells of known cell type can be offered in the form of high-resolution, full-colour graphic displays. Trainees may examine the contents of a very large data base. Images may be recalled in a selected order; for example, a trainee may ask the file system to produce images ordered according to increasing abnormality.

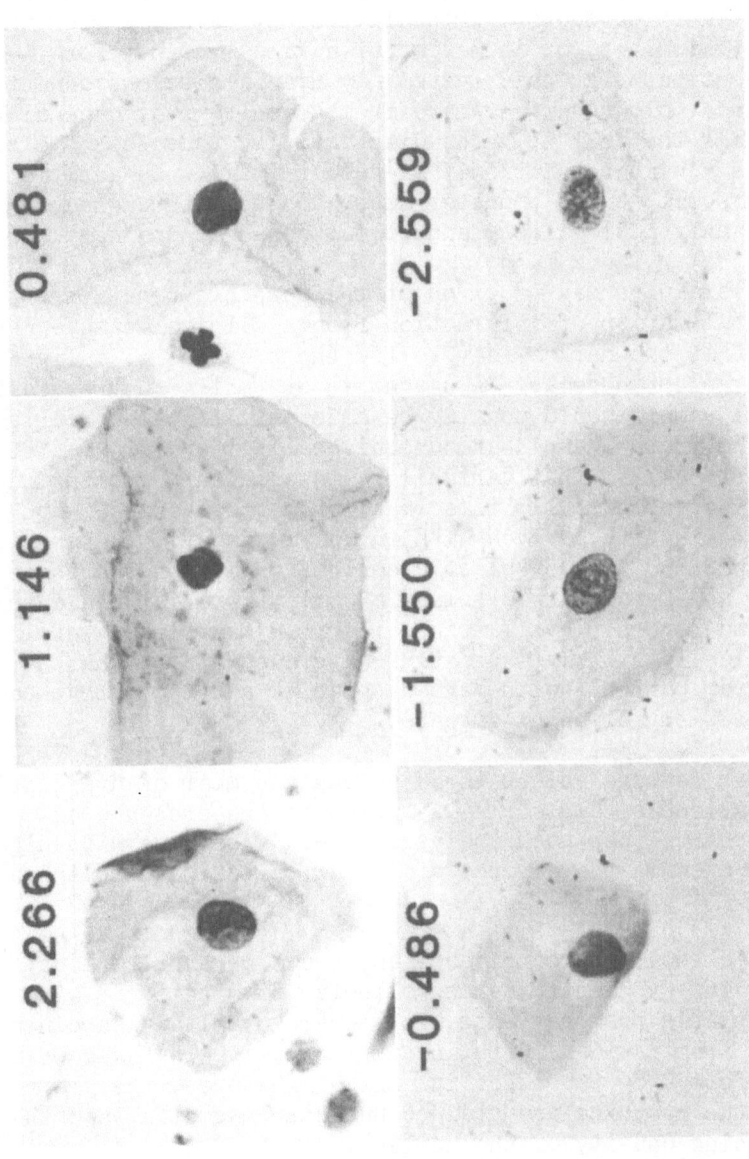

Fig. 15. Intermediate cells ranked along the axis DF_2. The first three are from patients with carcinoma in situ; the second three are from normal patients.

Numerical representation allows us to find effective ways of conveying the meaning of non-visual, computed diagnostic clues to a clinician by the use of computer graphics. Histologists and cytologists are superbly capable of evaluating visual clues. The computer diagnostic system may present its assessment of a cell in two forms. It may either provide a collection of numbers such as discriminant function scores, probabilities of being this or that, or atypicality indices, or it may offer a visual clue. For example, the computer graphic system may draw a circle around the cell image in a colour hue anywhere from green to red, depending on how "normal" the cell is. The diagnostician thus sees the cell as always, as a high-resolution full-colour image. In addition, he also sees the cell as the computer does, assessed through a statistical/analytical filter, and colour coded accordingly.

Several examples may be given of the possible benefits of exhaustive utilization of information from a clinical sample. A case in point is the marker feature for the presence of malignant disease in dysplastic cells. A diagnosis of "malignancy present" in a cervical sample would usually be made only if tumour cells had been found in the sample and unequivocally identified. Yet it is well established that variability in clinical sampling may fail to provide tumour cells in a small proportion of patients with carcinoma in situ or even with invasive cancer. It has recently been established that dysplastic cells from any of the three cell types - non-keratizing dysplastic cells, keratinizing dysplastic cells, and severely dysplastic cells from metaplasia - fall into statistically clearly separated sub-groups: those that originated from patients with mere dysplasia, and those that came from patients with malignant disease.

If marker feature values in an adequate number of dysplastic cells in a patient's sample clearly point to the presence of malignant disease, then such a diagnosis could be made with high reliability, even in the absence of tumour cells in the clinical sample.

Figure 16 shows the confidence and the tolerance regions for non-keratinizing dysplastic cells collected from patients with moderate dysplasia from patients with severe dysplasia/carcinoma in situ, and from patients with invasive cancer. The axes of the display are formed by two effective marker features, the green/red contrast of the Papanicolaou-stained nuclear chromatin, and the average staining density of the nucleus.

Typical examples based on the discriminant function scores for non-keratinizing dysplastic cells and severely dysplastic cells from metaplasia are shown in Figs. 17 and 18.

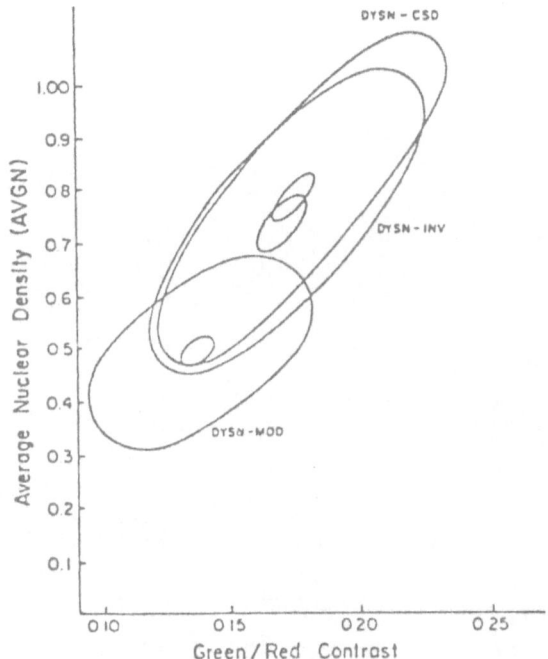

Fig. 16. Confidence and tolerance regions for non-keratinizing
dysplastic cells collected from patients with i) moderate
dysplasia (DYS N-MOD), ii) severe dysplasia/carcinoma in
situ (DYSN-CSD) iii) invasive cancer (DYSN-INV)

One may assign dysplastic cells to three categories: patients
with moderate dysplasia, patients with severe dysplasia/carcinoma
in situ, and patients with invasive cancer. The majority of cells
from a given patient should be classified into the disease category
that gave rise to them. Figure 19 shows such patient profiles.

It is only fair to say that most of the new diagnostic
capabilities are still being tested and applied in the context of
research projects. One notable exception is the use of ploidy
assessment to provide prognostic information. It has been known
for a long time that cell nuclei from most malignant lesions
contain increased amounts of DNA and are aneuploid e.g. Mellors et
al 1952, Sandritter et al 1966, Antandilov 1976).

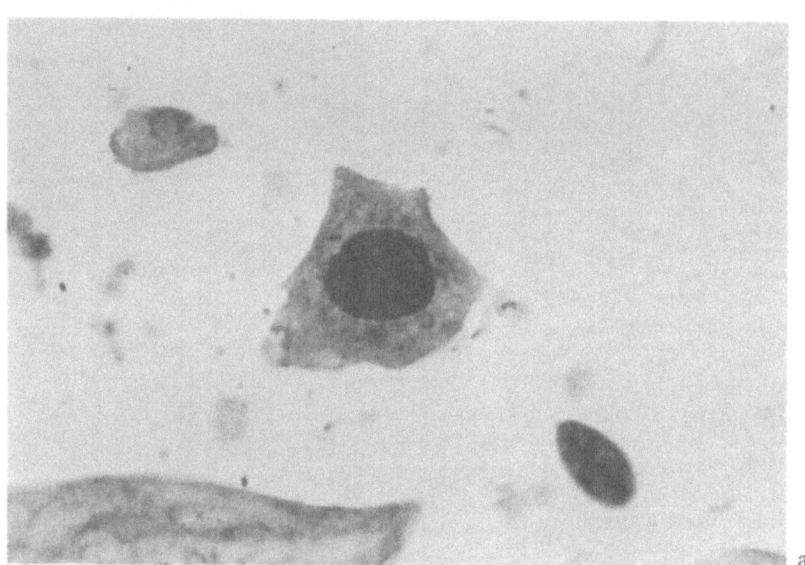

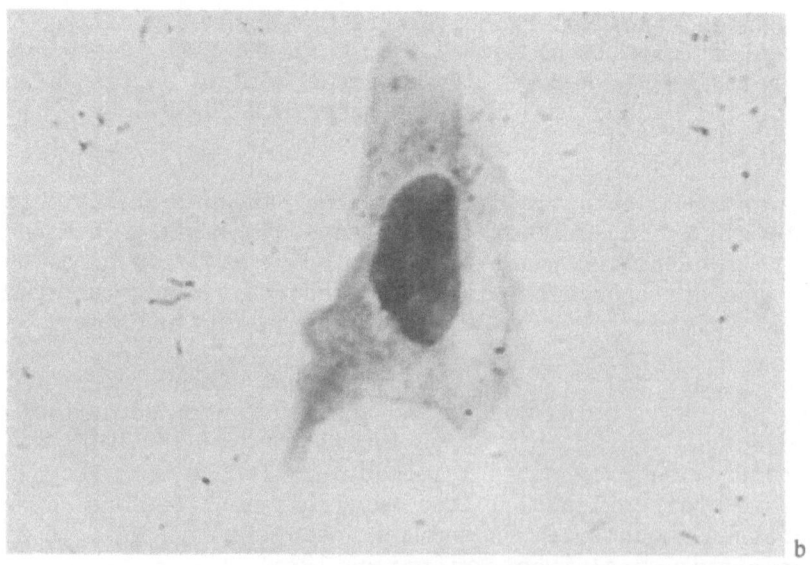

Fig. 17 Typical examples based on discriminant function scores for
non-keratinising dysplastic cells a) patient with moderate
dysplasia b) patient with carcinoma in situ.

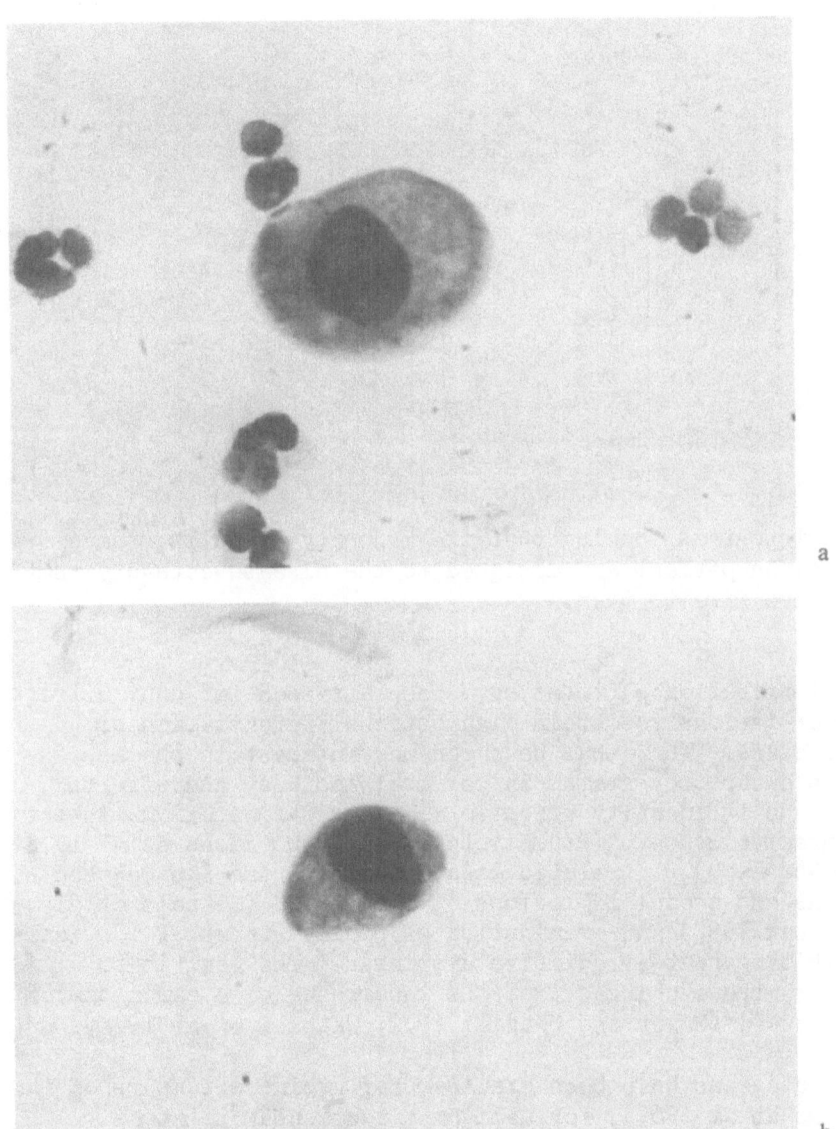

a

b

Fig. 18 Typical examples based on discriminant function scores for
severely dysplastic cells from metaplasia a) patient with
moderate dysplasia b) patient with carcinoma in situ.

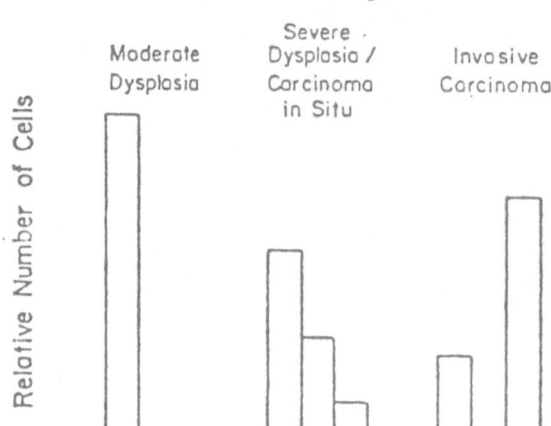

Fig. 19. Histograms showing that the majority of cells from a
given patient are assigned to the disease category that
gave rise to them.

DNA distribution patterns have been assessed for many different types of lesions to obtain diagnostic differentiation or
prognostic clues. There has been renewed interest in the assessment of the aneuploidy status in cervical lesions, and efforts
have been made to identify precursor lesions for carcinoma in situ
and for invasive cancer. (Hrushovetz et al 1970, Sachs et al 1972,
Fu et al 1982 a, b). A related area is differentiation between
condylomatas and aneuploid lesions of the cervix (Meisels et al
1979, Shevchuk 1982) and examination of ploidy status of cervical
cells exhibiting post-irradiative dysplasia. The goal is to
differentiate these changes from the recurrence of a carcinoma of
the cervix uteri (Meyer and Okagaki 1972, Okagaki et al 1974).

DNA histograms have been examined for mucinous tumours of the
ovary (Weiss et al 1969), for malignant testicular teratoma
(Lederer 1976) and have been extensively studied in fibrocystic
disease and lobular carcinoma of the breat (Boquoi et al 1975,
Sprenger et al 1979, Fossa et al 1982). There are clinical tests
now for pre-operative differentiation of thyroid aspirates to
ascertain whether one is dealing with a follicular adenoma or a

carcinoma (Sprenger et al 1977). There are numerous applications
in the assessment of lymphoid cell populations (Braylan et al 1978)
of astroglial tumours (Reuck et al 1979) and of bronchial cell
populations (Nasiell et al 1978). Laboratory tests have been
developed for prognostic differentiation between chondromas that
are diploid and chondrosarcomas that may be diploid to hyperploid,
with the diploid instances having a more favourable prognosis
(Cuvelier and Roels 1979). Such ploidy determinations in chondrosarcomas are considered to provide better prognostic information
even than conventional histopathologic grading.

Other areas in which DNA histograms are valuable are differentiation of urinary bladder papillomas versus carcinoma in situ
of the bladder, an otherwise difficult diagnostic task.

DISCUSSION AND CONCLUSIONS

Computer assessment of high-resolution images of cells and
tissues has undergone a remarkable development over the past 20
years. In the early 1960s, microscope photometers with scanning
stages first became commercially available. The rate of data
acquisition at that time was typically 10 to 50 measurements per
second. The first computer operated on-line to a microscope photo-
meter (Wied et al 1968) was a DEC LINC 8, with 4 k of memory,
capable of 10^3 to 10^4 pixel operations per second. Today, ultra-
fast scanners capable of data acquisition rates of up to 128 MHz,
and computers capable of up to 10^{12} pixel operations per second
are becoming available (Preston 1979, Maenner et al 1980). In fact,
high-resolution cell image analysis is approaching the data
acquisition rates that have made flow cytometry such an attractive
methodology. We have thus seen an advance in instrumentation
capability by factors of from 10,000,000 to 100,000,000 at remarkably no increase in price.

Quantitative methodologies have quickly progressed from initial
efforts to achieve discrimination between cells of types well
recognised by human experts to more challenging and dramatically
novel applications. These applications include environmental
pathology to detect sub-threshold levels of toxicity, use of cells
from the lymphoid system to monitor a patient's immune surveillance
capability, use of lymphocytes as biological dosimeters, research
into marker features as early indicators for the development of
chronic and possibly malignant disease, and novel kinds of diagnostic clues from individual cells indicating directly the state of
health of the patient. Thus mensuration has led not only to
quantitation, but also to basic questions about disease development. The much more difficult problem of analysis of tissue
sections is now beginning to be addressed (Bartels et al 1976,
Simon et al 1980, Wenzelides et al 1981, Preston 1981, Castillo et

al 1982). Because it is now clear that cytophotometric image
analysis has had a fundamental impact on our views of clinical
cytology, it is equally clear that a similar and possibly ever
farther reaching change will occur in histopathology. Clinical
practice follows conservative rules. It is only now that the DNA-
histograms, first researched over 20 years ago, are finding entry
into actual clinical practice. It may be a long time before many
of the results of the advanced analytical methods described here
will become part of daily clinical practice. Computer graphics
will be a major aid in bridging the gap of comprehension and
intuitive understanding between visual expertise and mathematical/
analytical assessment.

REFERENCES

Anderson R.E., Olson G.B., Shank C., et al (1975) Computer analysis
    of defined populations of lymphocytes irradiated in vitro. I
    evaluation of murine thoracic duct lymphocytes. Acta Cytol. 12,
    126-135
Anderson R.E., Olson G.B., Howarth J.L. et al (1975) Computer
    analysis of defined populations of lymphocytes irradiated in
    vitro. II Analysis of thymus dependent vs bone marrow derived
    cells. Am. J. Path. 80, 21-32
Avtandilov G.G. (1976) Perspectives on the use of microspectrophoto-
    metry in the diagnosis of pretumorous processes and malignant
    tumors. Cancer 38, 254-258
Bahr G.F., Bartels P.H., Wied G.L. and Koss L.G. (1979) Automated
    cytology, Chap. 31, pp.1123-1164 in Diagnostic Cytology.
    J.R. Lippincott, Philadelphia, 3rd edition.
Bartels P.H., Bahr G.F., Griep J. et al (1969) Computer analyses on
    lymphocytes in transformation. A methodologic study. Acta Cytol.
    13, 557-568
Bartels P.H. and Wied G.L.(1975) Extraction and evaluation of
    information from digitized cell images, pp. 15-28 in Mammalian
    Cells: Probes and Problems, Proceedings of the First Los Alamos
    Life Sciences Symposium, Los Alamos, New Mexico Oct. 17-19, 1973
    C.R. Richmond, D.F. Petersen, P.F. Mullaney, and E.C.Anderson eds
    ERDA
Bartels P.H. and Wied G.L. (1975) High resolution prescreening
    systems for cervical cancer. pp. 144-184 in Proceedings of the
    International Conference on Automation of Uterine Cancer, Chicago,
    Illinois. Tutorials of Cytology, Chicago
Bartels P.H., Marluce Bibbo, and Wied George L.(1976) Modelling of
    histologic images by computer. Acta Cytol. 29, 62-67
Bartels P.H., Olson G.B., Chen Y.P. et al (1978) Discrimination
    between human T and B lymphocytes by computer analysis of digit-
    ized data from scanning microphotometry. II Discrimination and
    automatic classification. Acta Cytol. 22, 530-537

Bartels P.H., Bibbo M., Richards D., et al (1978) Patient classification based on cytologic sample profiles. I Basic measures for
profile construction. Acta Cytol. 22, 253-260

Bartels P.H., Olson G.B. (1980) Computer Analysis of lymphocyte
images in Methods of Cell Separation Vol. 3 pp.1 - 99.
N. Catsimpoolas Ed. Plenum Press, New York

Bartels P.H., Koss L.G., and Wied G.L. (1980) Automated cell
diagnosis in clinical cytology pp.314 - 342 in Advances in
Clinical Cytology. G. Koss and D.V. Coleman Eds. Butterworths &
Co. London

Bartels P.H., Olson G.B., Lockart R. and Wied G.L. (1980) Cytophotometric studies of cell populations. Cell Biophys. 2, 339-351

Bartels P.H. and Wied G.L. (1981) Automated image analysis in
clinical pathology. Am. J. Clin. Pathology 75, 489 - 493

Bartels P.H., Bibbo M., Dytch H. et al (1982) Diagnostic marker
displays for intermediate cells from the uterine cervix. Acta
Cytol. 26, 29 - 34

Bartels P.H., Bibbo M., Dytch H.E. et al. Marker features for
malignancy in ectocervical cells: statistical evaluation. Cell
Biophysics (In press 1983)

Bibbo M., Bartels P.H., Sychra J.J. and Wied G.L. (1981) Chromatin
appearances in intermediate cells from patients with uterine
cancer. Acta Cytol. 25, 23 - 28

Bishop R.P. and Young I.T. (1977) The automated classification of
mitotic phase for human chromosome spreads. J. Histochem.
Cytochem. 25, 730 - 740

Boquoi E., Krebs S. and Kreuzer G. (1975) Feulgen DNA cytophotometry on mammary tumor cells from aspiration biopsy smears. Acta
Cytol. 19, 326 - 329

Braylan R.C. et al (1978) Cell volumes and DNA distributions of
normal and neoplastic human lymphoid cells. Cancer 41, 201 - 209

Castillo X., Yorkgitis D. and Preston K. (1982) A study of multi-
dimensional multicolor images. IEEE Trns Biomed. Eng. BME-29,
111 - 121

Cuvelier C.A. and Roels H. (1979) Cytophotometric studies of the
nuclear DNA content in cartilaginous tumors. Cancer 44, 1 1361 -
1374

De Reuck J., Roels H. and Vander-Eecken H. (1979) Cytophotometric
DNA determination in human astroglial tumours. Histopathology 3,
107 - 115

Dytch H.E., Bartels P.H., Bibbo M. et al.Computer graphics in
cytodiagnosis. J. Analytical Quantitative Cytol. (In press 1983)

Fossa S.D. et al (1982) Nuclear Feulgen DNA content and nuclear
size in human breast carcinoma. Human Pathology 13, 626 - 630

Fu Y.S., Reagan J.W., Hsiu J.G. et al (1982) Adenocarcinoma and
mixed carcinoma of the uterine cervix. I. A clinicopathologic
study. Cancer 49, 2650 - 2570

Fu Y.S., Reagan J.W., Fu A.S. and Janiga K.E. (1982) Adenocarcinoma
and mixed carcinoma of the uterine cervix II. Prognostic value of
nuclear DNA analysis. Cancer 49, 2571 - 2577

Green J.E. (1979) Rapid analysis of haematology image data. The
ADC-500 Preprocessor. J. Histochem & Cytochem. 27, 174 - 179

Grundmann E. and Stein P. (1961) Untersuchungen ueber die Kernstruk-
tur in normalen Geweben und im Carcinom. Beitr. z. Patholog. Anat.
und zur allgem. Pathologie 125, 54 - 76

Harbers E., Lederer B., Sandritter W. and Spaar U. (1968) Untersuch-
ungen an Nukleohistonen. IV. Heterochromatisierung in der
Rattenleber waehrend der carcinogenese. Virchows Arch. B. Cell
Pathology 1, 98

Hrushpvetz S.B. and Lauchlan S.C. (1970) Comparative DNA content of
cells in the intermediate and parabasal layers of cervical intra-
epithelial neoplasia studied by two-wavelength Feulgen cytophotometry. Acta Cytologica 14, 68-77

Ingram M., Norgren P.E. and Preston K. (1968) Automatic different-
iation of white blood cells. pp. 97-117 in Image Processing in
Biological Science, D.M.Ramsey Ed. University of California Press,
Berkley and Los Angeles, California.

Kiefer J.G., Kiefer R., Moore G.W. et al (1974) Nuclear images of
cells in different functional states. J. Histochem. Cytochem. 22,
569 - 576

Koss L.G., Bartels P.H., Bibbo M. et al (1977) Computer analysis of
atypical urothelial cells, I. Classification by supervised
learning algorithms. Acta Cytol. 21, 247-260

Koss L.G. and Bartels P.H. (1980) Urinary cytology. Device
capabilities and requirements. J. Analytical Quantitative Cytol.
2, 59-65

Landeweerd G.H. (1981) Pattern recognition of white blood cells.
Ph.D. Thesis, Universiteit te Amsterdam

Lederer B. (1976) Statistical analysis of cytophotometric DNA
measurements demonstrated on malignant testicular teratoma.
Acta Cytol. 20, 5-6

Lockart R., Bartels P.H. and Olson G.B. Detection of Daudi cells
which had been incubated with human interferon, by computer
assisted cell image analysis, submitted for publication

Maenner R., Schneider W., DeLuigi B. and Posner K. (1980) A general
purpose multi-micro-system with high fault tolerance and unlimited
system capacity. Euromicro J. 6, 388-390

McKee P.H., (1975) Computer analyses of lymphocyte alterations
induced by immunosuppressive chemical agents. Ph.D. dissertation,
Department of Microbiology, University of Arizona.

Meisels A., Roy M., Fortier M. and Morin C. (1979) Condylomatous
lesions of the cervix. Am. J. Diag. Gynecol. Obstet. 1, 109-116

Mellors R.C., Keane J.F. and Papanicolaou G.N. (1952) Nucleic acid
content of the squamous cancer cell. Science 116, 265-269

Meyer A.A. and Okagaki T. (1972) Microspectrophotometric study of
post irradiation dysplasia in vaginal smears. Cancer 30, 964-971

Nair K.K., Bartels P.H., Mahon D.C. et al (1980) Image analysis of
  Feulgen stained rat hepatocytes: effects of chlordane. J. Analyt.
  Quantitative Cytol. 2, 285-289
Nasiell M. et al (1978) Cytomorphological grading and Feulgen DNA
  analysis of metaplastic and neoplastic bronchial cells. Cancer
  41, 1511-1521
Okagaki T., Meyer A.A. and Sciarra J.J. (1974) Prognosis of irradiated carcinoma of cervix uteri and nuclear DNA in cytologic
  post-irradiation dysplasia. Cancer 33, 647-652
Olson G.B. and Bartels P.H. (1981) Differentiation of splenocytes
  and peripheral blood lymphocytes from mice infected with Friend
  murine leukemia virus. Pattern Recognition 13, 57
Olson G.B. and Bartels P.H. (1982) Assessment of environmental,
  induced insults upon lymphoid cells by computerized morphometric
  analysis. Proceedings Biologic Dosimetry Conference, Munich-
  Neuhergerg, Oct. 1982.
Pressman N. and Wied G.L. Eds. The Automation of Cancer Cytology
  and Cell Image Analysis. (Tutorials of Cytology, Chicago,
  Illinois, 1979)
Preston K. Jr., (1962) Machine techniques for automatic leukocyte
  pattern analysis. Ann. N.Y. Acad. Sci. 97, 482-490
Preston K. Computer hardware for biomedical pattern recognition.
  pp. 213-232 in Biomedical Pattern Recognition and Image Process-
  ing. K.S. Fu and T. Pavlidis Eds. Dahlem-Konferenzen, Berlin
  Verlag Chemie Weinheim 1979.
Preston K. (1981) Tissue section analysis: feature selection and
  image processing. pp. 17-36 in Pattern Recognition 13,
Preston K. (1981) Digital image analysis in cytology, in Digital
  Image Analysis, Azriel Rosenfeld Ed. (Springer Verlag, New York)
Rowinski J. Pienkowski M. and Abramszuk J. (1972) Area represent-
  ation of optical density of chromatin in resting and stimulated
  lymphocytes as measured by means of Quantimet. Histochemie 32, 75
Sachs H., Bahnsen J. and Stegner H.E. (1972) Scanning-mikroskopisch-
  cytophotometrische Untersuchungen an Dysplasien des collum uteri.
  Arch. Gynaek 212, 401-412
Sandritter W., Carl M. and Ritter W. (1966) Cytophotometric measure-
  ments of the DNA contents of human malignant tumors by means of
  the Feulgen reaction. Acta Cytol. 10, 26-30
Sandritter W., Kiefer G. Schlueter G. and Moore W. (1967) Eine
  cytophotometrische Methode zur Objektivierung der Morphologie von
  Zellkernen. Histochemie 10, 341-352
Sandritter W. and Kiefer G. (1970) Objectivization of chromatin
  patterns using the fast-scanning stage of the UMSP1, p. 177 in
  Automated Cell Identification and Cell Sorting. G.L.Wied and
  G.F. Bahr Eds. (Academic Press, New York 1970)
Shack R., Baker R., Buchroeder R., Hillman D. et al (1979) Ultrafast
  laser scanner microscope. J. Histochem. Cytochem. 27, 153-159

90

Shevchuk M.M. and Richart R.M. (1982) DNA content of condyloma acuminatum. Cancer 49, 489-492

Shoemaker R.L., Bartels P.H., Hillman D.W. et al (1982) An ultra-fast laser scanner microscope for digital image analysis. IEEE Trans. Biomed. Eng. BME-29, 82-91

Simon H., Kunze K.D. Voss K. and Herrmann W.R. (1975) pp. 155-158 in Automatische Bildverarbeitung in Medizin und Biologie (Theodor Steinkopff Verlag, Dresden)

Simon H., Kranz D., Voss K. and Wenzelides K. (1980) Automated data sampling in sections from selected liver diseases. Path. Res. Pract. 170, 388-401

Sprenger E., Lowhagen T. and Vogt-Schaden M. (1977) Differential diagnosis between follicular adenoma and follicular carcinoma of the thyroid by nuclear DNA determination. Acta Cytologica 21, 528-530

Sprenger E., Ulrich H. and Schondorf H. (1979) The diagnostic value of cell nuclear DNA determination in aspiration cytology of benign and malignant lesions of the breast. Analytical Quantitative Cytol. 1, 29-36

Weiss R.R., Richart R.M. and Ikagaki T. (1969) DNA content of mucinous tumors of the ovary. Amer. J. Obstetr. Gynecology 103, 409-424

Wenzelides K., Kranz D., Simon H., and Voss K. (1981) Anwendung der automatischen Mikroskopbild Analyse in der Praxis am Beispiel verschiedener Lebererkrankungen. Zbl. allg. Path. u. patholog. Anat. 125, 532-538

Wied G.L., Bartels P.H., Bahr G.F. and Oldfield D.G. (1968) Taxonomic intracellular analytic system (TICAS) for cell identification Acta Cytol. 12, 177-210

Wied G.L., Bahr G.F. and Bartels P.H. Eds. The Automation of Uterine Cancer Cytology. (Tutorials of Cytology, Chicago, Illinois 1976)

Wied G.L., Bartels P.H., Bibbo M. and Sychra J. (1980) Cytomorpho-metric markers for uterine cancer in intermediate cells. J. Analytical Quantitative Cytology 2, 257-263

Wied G.L., Bibbo M. and Bartels P.H. (1981) Computer analysis of microscopic images: applications in cytopathology. pp.367-409 in Pathology Annual, P.P. Rosen Ed. (Appleton-Century-Crofts, New York)

Wied G.L., Bartels P.H., Dytch H.E. et al (1982) Diagnostic marker features in dysplastic cells from the uterine cervix. Acta Cytol. 26, 475-483

Wied G.L., Bartels P.H., Bibbo M. and Dytch H. Rapid DNA cytometry in clinical cytologic diagnosis. J. Analytical Quantitative Cytol. in press 1982

Wied G.L., Bartels P.H., Bibbo M. et al. Rapid high resolution cytometry. J. Analytical Quantitative Cytol. in press 1983

A CELL BIOLOGIST'S VIEW OF CURRENT TECHNOLOGY - ITS ACHIEVEMENTS
AND LIMITATIONS

Atif ŞENGÜN

Radybiyoloji Kürsüsü, Fen Fakultesi, Istanbul University,
Vezneciler, Istanbul, Turkey.

1. INTRODUCTION

The number of physical and chemical methods used in cytology
has increased rapidly in recent years. Forty years ago, the light
microscope and its associated preparative methods were sufficient
for cytological research. Nowadays, in any modern cytology textbook,
one can find more than 25 important physical methods used in cytological research. The price of some of the instruments is very
high and the understanding of their operation and the evaluation
of their measurements has sometimes been quite difficult even for
the less experienced physicist.

A comprehensive review of modern physical methods used in
cytology and the new horizons which they have opened up for cytology as a whole would be almost impossible. Therefore I shall try
to summarise what has been established with the light microscope
and leave others to discuss progress made with newer methods. In
this way a comparison of old and new methods can be established.

Early scientists, using light microscopes which were very
simple when compared with those of today, observed that most of the
organs in an organism were composed of innumerable minute bodies
which were termed "cells" by Hooke in 1765. In the years which
followed, many researchers used the light microscope to obtain,
bit by bit, the information necessary for the cell theory which
was to be put forward by Schleiden and Schwann almost 80 years
later. Even though a further 140 years have elapsed since then,
the cell theory has not yet reached its final formulation. The
main reason for this may be that the cell size and structure, and

its components vary in different organisms and various organs of a multicellular organism. Furthermore, one cell does not always form a functional unit.

2. RESULTS OBTAINED WITH THE LIGHT MICROSCOPE

During the early development years of cytology, the light microscope was used almost exclusively. In order to have a better view of the cell or its different parts, various physical and chemical methods, such as sectioning and staining were developed. From Willson's book, "The cell in development and heredity" issued in 1929, one can cite the observations concerning the structure and the functional units of cells made by using the light microscope as follows: cell wall; plasma membrane; cortical layer with hyaloplasm inside; plastids; chondriosomes; vacuoles; passive metaplasmic bodies; central bodies; Golgi bodies; nucleus; nucleolus; basichromatin; oxychromatin; karyosome or chromatin nucleolus; cytoplasmic granules (microsomes, chromidia, metachromatic granules, secretory granules, storage granules); fibrilli (myofibrils, neurofibrils).

In multicellular organisms the nature of the interconnections of cells was being investigated. It had also been established that these structures forming the cell showed variations in different organisms and different tissues of multicellular organisms with respect to size, form and structure. It was known that the cell itself showed variations in form, size and internal structure.

With these primitive microscopes, the functions of the cell and of the elements forming the cell's structure were investigated. Some examples taken from Wilson's book include

(i)   Differentiation of cells during development
(ii)   Alteration of the appearance of the cytoplasm and its organelles during function e.g. appearance of secretion granules
(iii) Mitosis with particular reference to variations occurring in the chromosomes, formation of centromere, astrosphere and of spindle apparatus
(iv)   Meiosis and related changes in the cytoplasm and in chromosomes
(v)   Spermato and organo-genesis, fertilisation and development, especially formation of a multicellular organism from one egg cell
(vi)   Cytoplasmic movement
(vii) Localisation of the genetic material in the chromosomes
(viii) Mutational variations of chromosomal structure
(ix)   Variation of chromosome number in different species, even in the same organism

(x)   Endomitosis
(xi)   Some features of chromosome structure including the
       centromere, telomeres, heterochromatic and euchromatic
       regions of chromosomes and their variations, single and
       paired spiral structures of chromosomes
(xii) Division of the chloroplasts
(xiii) Budding of the nucleus and nucleolus in some cancer cells
       and in cells from the amniotic membrane
(xiv) Nucleo-cytoplasmic relations

It was also observed the size, shape and internal structure of
a cell differentiated depending on heredity and external factors.
The similarites between a single cell and the cells of a multicellular organism were being discussed in 1929 just as they are
nowadays.

The light microscope permits us to observe not only cell
structure, but also the behaviour of the organelles forming the
cell's structure and the alterations they undergo during their
function. For instance, when the salivary glands of Chironomus
larvae are examined in a drop of paraffin oil or some other solvent,
alteration of the structure of the giant chromosomes can be observed
depending on the time of examination. A giant chromosome is a
bundle of chromonemata and exhibits bands, interbands and hetero-
chromatic regions. Under the light microscope it can be observed
that at certain regions the number of bands increases or decreases.
Their thickness and general appearance is also changing during
observation (Fig. 1).

If we examine the homologous chromosomes in different organs
of Chironomus larva again, it will readily be observed that they
differ from each other, even though they cannot be compared with
each other (Fig. 2). However, when homologous chromosomes are
investigated in different cells of the same tissue, it can easily be
seen that they are almost similar in size and structure (Fig. 3).
These observations indicate that the structure of giant chromosomes
changes depending on their function and on the chemical and physical
factors of the environment.

There are limits to the use of the light microscope which has
been a great help in the development of cytology. It is well known
that these limits depend on the minimum object size and the minimum
distance between two small objects that can be resolved. Structures
smaller than these limits are unobservable with the light microscope.
For this reason those units smaller than 0.2μm or two objects
having a smaller separation than 0.2μm cannot be seen. Unfortunately, the size of some viruses is 10 nm and many of the cell
components are between 200 and 250 nm. Therefore, in order to
see these small structures, we need new techniques.

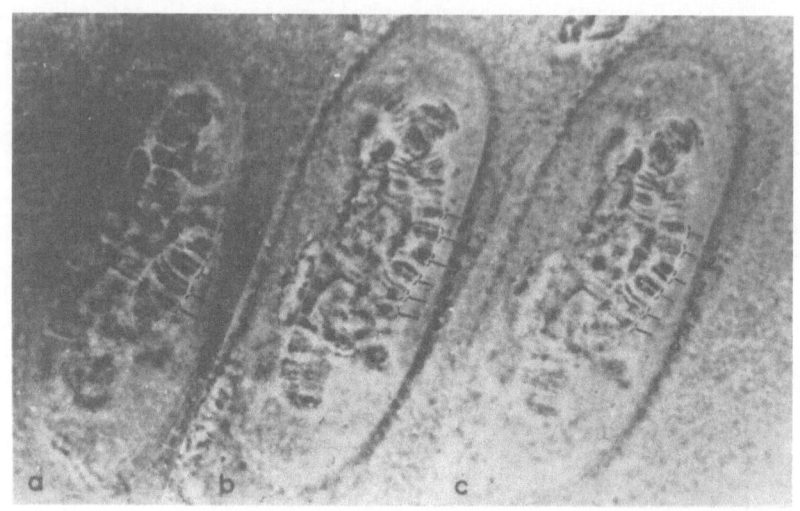

Fig. 1 Alteration of chromosomal structure observed under a Zeiss
phase contrast microscope. Some bands of the chromosomes
are photographed with an interval of half an hour: a at
13 hr 30, b at 14 hr and c at 14 hr 30. If the corresponding regions are compared with each other it can be seen:

No	at 13 hr 30 (a)	at 14 hr (b)	at 14 hr 30 (c)
1	One thick band	Two bands become distinguished	Two bands are visible
2	Two bands	Three bands and some dark substances	Four bands embedded in dark substance
3	Three bands	Three bands, one becomes unclear	Two clearly visible bands
4	Two clear bands	Two bands	Two thick bands
5	Two clearly vis- ible bands and one unclear band	Two bands	Two bands
6	Two bands	One band	Two unclear bands

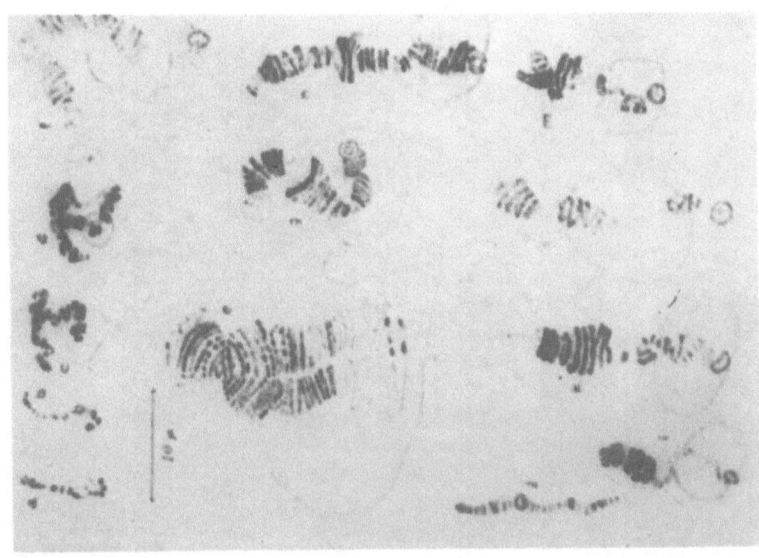

Fig. 2 Hand-drawn pictures of the smallest chromosomes from
different tissues of old 24 mm long Chironomus larva. a - b
from Oesophagua, c-d from proventriculus, e: from salivary
gland, f: from midgut, g-h from Malpighian tubules, i - j
from vorder rectum, k: from hind rectum, l: from blood gills
m: from anal region. (Şengün, 1951, Rev. Fac. Univ. Istanbul. Ser. B, 16: 1-44.

In addition to the structures of some cell components, sometimes their functions can be investigated much better outside the
cell. Thus physical methods are needed for the extraction of very
small cell components, and their collection with minimal damage to
them during these procedures.

The cell and its components are dynamic structures. Their
structures and function change under the conditions in which they
exist and many of these changes cannot be observed under the light
microscope. For example, one cannot follow the DNA synthetic
process with the light microscope, but autoradiography enables us
to find out the complicated and time-dependent mechanism of this
process. In addition, there are specific relationships among the
various parts of the cell. For instance, the ciliary movement in
Paramaecium cytopharynx is controlled by the initial region of the
cytopharynx. When this region is destroyed by UV-microbeam irradiation, the cilia do not move. In this process a physical method
has been used to destroy the critical region of the cytopharynx.

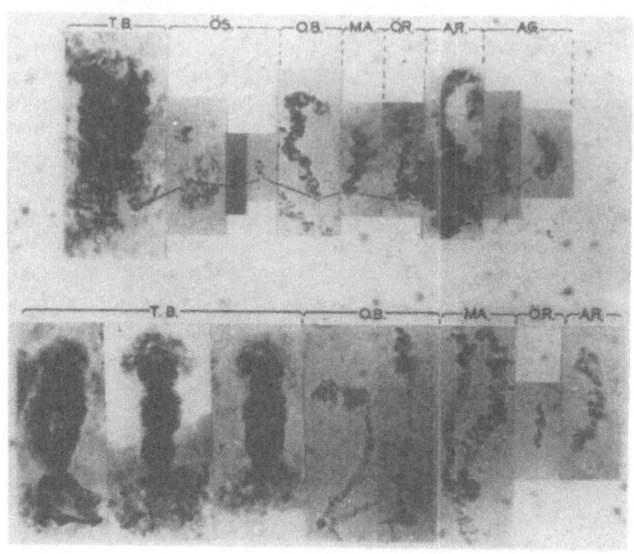

Fig.3 The smallest chromosome from different tissues of two different larvae of Chironomus plumosus. The upper pictures are
from one, the others from another larva. Mag. X 3000. TB from
salivary gland, OS from oesophagus, OB from mid-gut, MA from
Malpighian tubules, OR from vorder rectum, AR from hind
rectum, AG from anal region (Şengün A. 1963, Rev. Fac. Sci.
Univ. Istanbul, Ser. B, 28: 41-49)

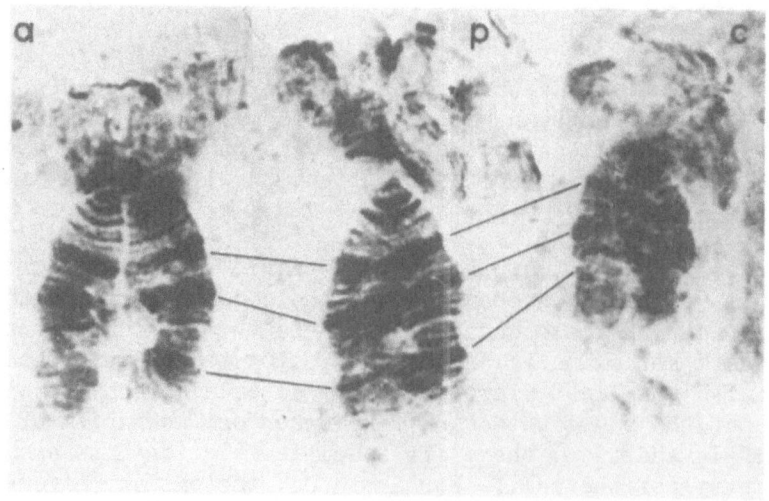

Fig. 4 Homologous chromosomes from different cells of the same
tissue (Şengün A.1964, Rev.Fac.Sci.Univ Istanbul Ser B 29:
73-79)

In experiments on the role of the nucleus, nucleolus, and even of chromosomes, transplantation from cell to cell is required. This process can only be realised by a new physical method. These few examples show that physical instruments besides the light microscope are required during the investigations related with cell biology.

When the history of cytology is examined, it is seen that great developments in this scientific field follow those of physics or chemistry. The application of the results or methods obtained in physics and chemistry to cytology, creates a stimulus. The physical methods used in cytology during recent years, according to our standards, are either very expensive or the evaluation of the results obtained from them needs specific specialisation. From this point of view, research with these new methods in a laboratory creates financial and personnel problems. For this reason, the number of laboratories using all of the modern physical methods is quite small in financially under-developed countries. Mostly, only one or two of these instruments are present in a laboratory. Therefore in countries having financial problems like Turkey, the assembly of a range of expensive physical instruments used in cytology and in other branches of biology in one specific laboratory would be a good idea. On the other hand, there are still many topics which can be investigated using the light microscope which is much simpler to use and much less expensive than the new methods. That is to say, the light microscope is still of great importance in cytological investigations.

DEVELOPMENT OF OPTICAL METHODS FOR QUANTITATIVE CYTOCHEMISTRY

Lore ZECH

Department of Medical Cell Genetics, Karolinska Institutet, Medical Nobel Institute, Box 60400, S-104, 01, Stockholm, Sweden

1. INTRODUCTION

Cytochemical techniques, based on optical methods, have been of considerable value during the development of modern cell biology. The main interest of these studies has been focussed on quantitative determination of nucleic acid and protein content of the cell. By far the greatest part of the applied work to date has concerned growth, differentiation and function of normal cells, but more recently interest in applying biophysical techniques to a study of tumour material has been growing rapidly.

Different procedures and instrumentation for cytochemical studies have been developed and tested over the years. From the historical point of view microspectrophotometry is the oldest of the methods which will be described. It originated from ultra-violet (UV) microscopy when around the turn of the century August Köhler, the father of the principle of illumination in the light microscope, built the first UV-microscope at Carl Zeiss in Germany. In his first publication on UV-microscopy, Köhler reported the remarkable observation that non-stained biological specimens showed specific absorption of UV-light (Köhler 1904). There was almost no interest in this extraordinary finding during the first decades of the century. However, around 1930 development of UV microspectrophotometry and other cytochemical methods commenced.

2. SPECTROPHOTOMETRY

2.1 UV-microspectrophotometry

The first attempts to use the UV-microscope as a tool for studies of the cytochemical composition of the cell were made by T. Caspersson who developed the first UV-microspectrophotometer at the Karolinska Institutet in Stockholm. He applied the principle of spectrophotometry of solutions to individual cells, using the specific absorption of nucleic acids at 265 nm according to the formula

$$E = \log I_o/I$$

where E is the extinction, I_o the intensity of the incident light and I the intensity after transmission through the object. The very first measurements of nucleic acids in individual cells were performed with an UV-microscope combined with equipment for photography. For wavelengths below 270 nm a spark gap between rotating electrodes was used as light source. The transmission of UV-light was measured on photographic plates (Caspersson 1936).

During the following decades, technical improvements and automation of instrumentation for absorption measurements led to high resolution instruments that operate with great accuracy (Caspersson 1950, 1979; Caspersson and Kudynowski 1980). Among the many technical approaches, special mention should be made of the scanning system, because it reduced one of the most serious errors in spectrophotometry caused by inhomogeneous distribution of the biological material. Different types of microscope stage for scanning were developed allowing faster and more exact measurements than before (Zeiss, 1964; Carlson 1970a).

In the 1950s Carl Zeiss in Germany built the first commercial instrument for UV-microspectrophotometry, the UMSP I, which was based on the experience and early tests of Caspersson. The Zeiss instrument which is still used in different laboratories has the great advantage that it is easy to handle by biologists and does not require special technical training (Zeiss 1964). As light source the Zeiss microspectrophotometer uses a Xenon lamp which has a continuous spectrum and is fairly stable. The monochromator can select wavelengths in the UV as well as in the visible region of the spectrum. The objectives are neofluars with high numerical apertures, especially developed for this instrument (although now widely used for other purposes), allowing the measurement of very tiny particles. The Zeiss UMSP uses a two beam method for photometric measurements. The monochromatic light is split into a measuring beam and a reference beam by two synchronously rotating mirrors. Each beam passes through a complete UV-microscope and thus the two beams are as similar as possible. The light from both

beams is directed in rapid alternation onto a single detector and
from this a recorder, connected via a photomultiplier, records the
relationship between measuring and reference beams. Evaluation
of the measurements is made automatically and the integrated values
are registered on curves (Trapp 1966, Zeiss 1964).

The Zeiss UMSP as well as other microspectrophotometers
constructed during the first two decades of development of cytochemistry, was built mainly for studies inside individual cells.
Therefore, these instruments operate slowly with the registration
of measured values being the speed limiting factor.

For cell population studies, mainly in tumour research,
these methods are too slow and there is a demand for fast instruments. Therefore, Caspersson and his group constructed spectro-
photometers of the non-recording scanning type. Omitting the
registration, combined with high automation, increases the speed
of the measuring procedure considerably but the scanned area is
still set by the observer under visual control This means that the
automated integrating measurement is still dealing with individual
cells (Caspersson 1979; Caspersson and Kudynowski 1980). Therefore,
this type of analysis is in contrast to flow measurements where
the individual cells can no longer be distinguished.
hed.

2.2 Sources of error

There are a number of potentially serious problems in UV-
microspectrophotometry most of which are caused by the biological
object. One of the questions most extensively discussed is
unspecific light-loss resulting from absorption of heterocyclic
and aromatic amino acids and from scatter and reflection (Caspersson
1950). The contribution of amino acids to nucleic acid absorption
has been determined in model experiments, e.g. mouse fibroblasts,
by the following procedure. First the cells were measured at 260nm.
Then they were extracted by RNase, DNase and finally trichloracetic
acid. After this treatment acridine orange staining showed that no
nucleic acids were left in the cell. They were now measured again
at 265 nm and these experiments showed that the contribution of
nucleic acids to the 265 nm absorption was about 77%, the contribution of proteins to the absorption at 265 nm about 23% (Killander
1966). From this it is evident that the contribution of proteins
to absorption at 265 nm is relatively small, in spite of the fact
that proteins are quantitatively the most prominent compounds in
the cell type studied (mouse fibroblasts). The nucleic acid/
protein ratio is more or less constant in exponentially growing
cell cultures, as indicated by determination of the correlation
coefficient 0.75 - 0.85 (nucleic acid/protein) (Killander 1966).

Light scatter and reflection are greatly dependent on pre-
treatment of the biological material, mainly fixation and mounting
medium. Unfortunately it is impossible to exclude this type of
light loss totally even if all precautions are taken. In a solution
which is more or less homogeneous and contains particles of uniform
size the unspecific absorption can be calculated. In biological
objects, however, the situation is different, because the particles
are of different sizes, some being very small and behaving according to Rayleigh's formula and other bigger particles and conglomer-
ates for which the laws of goemetrical optics are more appropriate.
The problem has been discussed by several authors, and it is
generally assumed that light loss resulting from refraction is
independent of the wavelength of the light, whereas light loss
caused by scatter is dependent on the wavelength. This means that
we are unable to determine light loss caused by scatter and
reflection exactly since we cannot distinguish between the different absorbing particles (Caspersson 1950).

For the biologist who wishes to obtain information on exact
amounts of substance in the cell, the question arises how to treat
the non-specific extinction. A general rule is to measure first
at the wavelength where the substance to be investigated has its
specific absorption maximum and then at a wavelength where nucleic
acids and proteins do not absorb, e.g. at 315 nm. Minimum and
maximum values can be calculated for the required absorbance. The
minimum value is based on the assumption that the non-specific
extinction is independent of wavelength, i.e. it is always equal to
the recorded total extinction at 315 nm. In this case

$$E_{265} - E_{315} = \text{specific extinction at 265 nm}$$

The maximum value is based on the assumption that the unspecific extinction obeys Rayleigh's law for light scattering of
infinitely small particles. Roughly speaking this means that at
265 nm the non-specific absorption lies somewhere between the
extreme values of

$$E_{315} \text{ and } E_{315} \text{ X } (E_{265}/E_{315})^4$$

2.3 Extinction coefficient

In ordinary spectrophotometry the following formula is used

$$E = k \, c \, d$$

where k is the extinction coefficient (which is the specific
absorbance), c the concentration of absorbing substances and d the
cuvette thickness. This formula is "the general absorption law"
and is a combined expression for Bouger-Lambert's law ($E = k_1 \, d$)

and Beer's law $E = k_2 c$). The same formula may be used in micro-spectrophotometry. It may then conventionally be written

$$E = k\ m$$

where m = cd and is the mass per unit area of absorbing substance at the measuring point.

We must realise, however, that there is a great difference between ordinary spectrophotometry and microspectrophotometry. Ordinary spectrophotometry generally deals with solutions, obeying the formula

$$E = k\ c\ d,$$

which means, that k has the same value, independently of c and d. Microspectrophotometry, on the other hand, deals with substances in cells, which cannot be regarded as dilute solution. On the contrary, in the living cell we deal with concentrated solutions and in the fixed cell the water free substances are immersed in and penetrated by a non-aqueous medium. It cannot be taken for granted that the extinction coefficient k of a substance under these circumstances has the same value as the same substance in dilute solution. Consequently, before microspectrophotometry is applied to quantitative analysis of specific substances in living or fixed cells, the extinction coefficient of these substances must be measured. This must be done under conditions resembling as closely as possible those in the cell at the time of analysis (Caspersson, 150).

In order to perform such calibration one has generally to arrange model experiments. If, for example, the substance is present in the living cell at high concentration, it might be adequate to have high concentrations in a special cuvette. Similarly conditions in the fixed cell may be imitated by models consisting of dry droplets of defined substances. After such model droplets have been made, frozen and dried or chemically fixed, measurement involves determination of the total dry weight (total dry mass M) and total extinction E_{tot}.

The mass determination is performed first, either by micro-interferometry or by soft X-ray microradiography. The total extinction of the droplet is then measured by scanning in a micro-spectrophotometer. Usually such droplets are very homogeneous and have a smooth surface which makes the measurement easy and exact. The extinction coefficient can now be determined according to the formula

$$k = \frac{E_{tot}}{M}$$

Measurements done on DNA and RNA according to the methods described
above have shown the extinction coefficient for DNA to decrease
markedly from dilute solution to solid state while the extinction
coefficient for RNA shows no change (Zetterberg 1966)

2.4 Spectrophotometry in the visible region

Since measurements in the UV involve many problems one may
ask why UV-microspectrophotometry was the first method to be
developed. The main reason was that many staining techniques -
including the Feulgen reaction - were not completely understood and
therefore underestimated for quantitative use. Furthermore, it
seemed an advantage that material for UV work had not been in
contact with any chemicals except fixation fluids.

Today the situation has changed and in many laboratories
spectrophotometry in the visible region of the spectrum is preferred to UV spectrophotometry. The best studied and most used method
for DNA determination is the Feulgen reaction, based on hydrolysis
of the biological preparation in 1N hydrochloric acid at 60°C before
staining with Schiff's reagent (Schiff's reagent is fuchsin-
sulphurous acid which reacts with the exposed aldehyde groups to
produce a purple dye). The absorption maximum of the dye is at
546 nm. If the reaction is performed under strictly standardised
conditions it is quantitative (Hale 1966).

The advantages of spectrophotometry in the visible region are
several:

(i)   Expensive quartz optics are not needed. Ordinary optics
of good quality are sufficient

(ii)   The problem of light scatter can be neglected. There is
no need to correct for unspecific absorption

(iii) Whereas the light source is very important for UV
measurements - it must be stable and give enough light -
an ordinary lamp can be used for measurements in the
visible region of the spectrum

(iv)   There is no need for a monochromator. In the case of
the Feulgen reaction an ordinary green filter selects
the light for measurement.

On the other hand measurement of stained preparations is an
indirect method and the operator must be aware of some critical
points which can influence the measurement:

(i)   All steps of the staining procedure for spectrophotometric work must be standardised

(ii)   The linear correlation between the amount of substance
and absorbance after staining must be controlled

(iii) It is of advantage to use standard controls of known values
(iv) The stain must be stable for the duration of measurement

3. DRY WEIGHT DETERMINATION

Dry weight or dry mass determination is of interest for the biologist when he wants to study growth processes in normal tissues or in tumours. It is also of interest for reference as well as calibration for spectrophotometric measurements.

The dry weight of microscopically small structures can be determined in different ways. Two methods used in our laboratory, X-ray contact microradiography and microinterferometry will be discussed.

3. 1 X-ray microradiography

A biological specimen is placed in close contact with a fine grained photographic emulsion and then exposed to soft X-ray radiation in vacuo. The wavelength used is 6 - 20Å. Due to the absorption of the object an image is obtained on the photographic plate. A low optical density on the photograph corresponds to high dry mass per unit area. For quantitative studies each object is photographed together with a standard of known weight and thus the exact weight of the biological object can be determined by comparing the density of the microradiogram of the specimen with that of the standard (Lindström, 1955; Carlson 1970). The obvious advantage of the X-ray method is that the physical background is clearer than e.g. for interference measurements. Furthermore, the measurement is independent of inhomogeneities in the object. The standard (or reference) used in our laboratory is a step wedge of parlodion film pieces of known weight determined on a torsion balance (Carlson 1970b). The main disadvantage of the method is that it requires expensive equipment and takes so much time that it cannot be used for routine studies.

3. 2 Microinterferometry

It had already been shown by Barer (1952) and Davies (1958) that the interference microscope is a quantitative tool which can be used for determination of the dry mass of biological objects. Microinterferometry has gained increasing importance in quantitative cytochemical work ever since that time.

Whilst microspectrophotometry measures specific extinction, the interferometric method gives a non-specific measurement of the total amount of substance. The quantity, directly measured in the interference microscope, is the optical path difference between two rays, only one of which passes through the object. The symbol for the optical path difference is $\emptyset$. This value is proportional to the dry mass per unit area (m) according to the equation

$$\emptyset = \psi . m$$

where ψ is the specific refractive increment or the proportionality factor or the specific optical path difference expressed in cm^3/g. ψ values for solutions of known composition and concentration have been studied and published (Hale 1958, Carlson 1970, Sören 1972). A value of 0.0018 was determined for protein in water and a value of 0.077 for mouse lymphocytes in glycerol. Since a cell is mainly composed of protein and water, the value of 0.0018 seems sufficiently exact for measurement of cells if water is used as embedding medium and the value of 0.077 if glycerol is used as embedding medium. For free fats other ψ values have been determined and this must be taken into account when cells, particularly rich in fats, are measured. "Slow" interferometers with registration and "fast" instruments in which registration is omitted are used for modern interferometry (Caspersson 1979, Caspersson and Kudynowski 1980).

4. MICROFLUORIMETRY

Compared with absorption spectrophotometry, fluorometry is simpler and more sensitive. Measurements are not influenced by uneven distribution of the material, and therefore it is not necessary to use scanning methods as in spectrophotometry. Errors, due to light scatter, are much smaller in fluorometry than in absorption spectrophotometry (Chayen and Denby 1968; Rigler 1966). A microfluorometer is a fluorescence microscope combined with a device for measuring fluorescence. The sensitivity of a fluorescence measurement is dependent on several factors:

(i)   The quantum efficiency of the fluorescent material
(ii) The excitation intensity of the light source
(iii) The sensitivity of the light receiver

The light source should be rich in blue and ultraviolet light. Therefore tungsten lamps are not convenient. Instead mercury or zenon lamps are used. The ideal light source would be a laser, e.g. an argon laser, which is preferentially used for flow measurments. Lasers are very expensive and therefore they have not been used for microfluorometry of individual cells.

For fluorescence measurements the specimen is irradiated with light of short wavelength, selected by suitable filters or a monochromator. These filters are called "excitation filters" and are placed in front of the light source. They eliminate all visible and all red light. The elimination of red light can be a problem

in simple equipment, so a solution of copper sulphate is sometimes
placed in the light path. Such an arrangement also has the
advantage of protecting against overheating. The fluorescent
light can be directed so that it either passes through the object
or illuminates the object from above (epi-illumination). This
latter method is now generally used because it makes the measurements independent of condenser optics. The light which is emitted
from the object after irradiation with the excitation light must
pass a "barrier filter". This is necessary to remove all excitation light so that onlt the fluorescent light enteres the photocell.
If a non-fluorescent object lies on the object stage the microscopic field should appear completely black. If the object emits
light the barrier filter does not absorb all excitation light.

Only a few biological substances have a sufficiently good
natural fluorescence - autofluorescence - to be measurable. Such
substances are vitamin A (green-yellow fluorescence), riboflavin
(yellow), porphyrins (red, have been used for study of chlorophyll),
steroids (blue-white), catecholamines (adrenalin) (green or blue).

Usually it is necessary to stain the specimen with a fluorescent dye such as eosin, fluorescein, acridine orange (AO), quina-
crine mustard, ethidium bromide, Hoechst 33258.

The use of some of these dyes, such as fluorescein and AO
goes back to the early 1940s. AO is a compound of special interest
since it shows a metachromatic effect, which means that it changes
colour depending on the molecular structure of the substance. In
solution it stains double stranded DNA green and single stranded
DNA and RNA red.

Much effort has been spent on developments to utilize the AO
staining method for distinguishing between DNA and RNA (Rigler 1966).
However, the conditions in pure solutions of known composition are
very different from those in a biological object where the protein
component plays an important role in blocking sites for AO reaction.
Other fluorochromes, such as ethidium bromide, proflavine, Chromomycin A and Hoechst 33258 have a weaker fluorescence, but their
excitation wavelength is in the spectral region of argon lasers
and therefore these substances are used preferentially for flow
measurements (see   ). In addition chromomycin A and
Hoechst 33258 are base-specific which makes them very valuable for
studies of base distribution along the DNA spiral.

Compared with absorption measurements which give information
about the total amount of substance in an object, fluorescence
measurements give relative values. This is because absorption
measurements are based on Lambert-Beer's law, so comparison of the
incident light I_o and the transmitted light I allows calculation

of the concentration of the substance. In the case of fluorometry
no simple relation between the intensity of excitation (I_{exc}) and
intensity of the fluorescent light (I_{Fl}) exists. Therefore only
I_{Fl} is measured and hence calculation of the concentration is
not possible. Fluorometry thus is used for comparative studies
between different substances. If absolute values of concentrations
are of interest, it is necessary to compare with a standard of
known concentration or to correlate the fluorescence measurement
with other measurements, e.g. a dry weight determination by interferometry (Rigler 1966).

One of the most valuable properties of fluorescence is its
sensitivity. In absorption photometry the lower limit is given by
the exactness with which small differences in light intensities can
be measured. In fluorescence photometry the limit is determined
mainly by the brilliance of the light source and the sensitivity
of the photomultiplier to small inputs of light. Theoretically,
there is no lower limit for the size of an object that can be
measured. This is the main reason why fluorescence measurements
are so much more sensitive than absorption measurements. Amounts
of 10^{-16} gm DNA can be measured. This corresponds to about 100 genes
(1 gene being about 1000 nucleotide pairs).

Microfluorometry, like other cytochemical methods, has its
limitations and several points must be observed to avoid errors.

(i)   The choice of fluorochrome is of utmost importance.
      Each dye which is introduced must be tested with respect
      to proportionality between fluorescence intensity and
      concentration of substance. Usually model experiments
      on droplets are necessary.

(ii) Fading of the fluorochrome is a severe source of error.
      It is caused by photodecomposition and dependent on the
      object, the mounting medium, the dye itself and the
      intensity of thelight source used for excitation.

(iii) Non-specific fluorescence always occurs and can have
      different origins. A staining method never works 100%
      and it is important to study the non-specific fluores-
      cence when quantitative results are desired.

The thickness of the object and the numerical aperture of the
objective are less important than in absorption spectrophotometry.
In absorption photometry it is important that the thickness of
the object does not greatly exceed the depths of the focus. In
fluorescence cytophotometry the thickness of the object is allowed
to be a multiple of the depth of focus. Objectives with a

numerical aperture of about 1.0 are preferred but if the object is very thick, it may be necessary to use with an objective of lower numerical aperture. In such cases it may be necessary for the excitation to be performed with transmitted light.

5. SUMMARY AND CONCLUSIONS

The development of methods for quantitative cytochemistry has been described. The instrumentation used in different laboratories varies in the degree of automation with machines of the recording type being slower than non-recording automatically scanning and integrating instruments. All the methods discussed in this chapter are based on investigation of individual cells, measured under visual control of the observer.

ACKNOWLEDGEMENTS

This work was supported by the Swedish Cancer Society and by Research Funds of the Karolinska Institutet.

REFERENCES

Barer R. (1952) Interference microscopy and mass determination. Nature 169, 366-367

Carlson L (1970a) A precision stage for two-dimensional scanning movement, In: G.L. Wied and G.F. Bahr (Eds) pp.113-116, Academic Press, NY and London

Carlson L. (1970b) Cytochemical determination of dry mass. Thesis Balder A.B., Stockholm, pp.1-44

Caspersson T. (1936) "Uber den chemischen Aufbau der Strukturen des Zellkerns. Scand. Arch. Physiol. Suppl. 8, 1-151

Caspersson T. (1950) Cell growth and cell function, Norton, New York

Caspersson T. (1979) Quantitative tumour cytochemistry - GHA Clowes Memorial Lecture. Cancer Res. 39, 2341-2355

Caspersson T. and Kudnowski J. (1980) Cytochemical instrumentation for cytopathological work. Int. Rev. Exp. Pathol. 21, 1-54

Chayen J. and Denby E.F. (1968) Biophysical Techniques as applied to cell biology. Methuen and Co. Ltd. The Trinity Press, Worcester and London

Davies H.G. (1958) The determination of mass and concentration by microscope interferometry. Gen. Cytochem. Methods 1, 55-161.

Hale A.J. (1958) The interference microscope in biological research. Williams and Wilkins, Baltimore

Hale A.J. (1966) Feulgen microspectrophotometry and its correlation with other cytochemical methods. In G.L. Wied, Introduction to Quantitative Cytochemistry, pp. 183-192, Academic Press, New York and London

Killander D. (1966) Intercellular variability in normal and neoplastic cell populations in vitro. Thesis Almquist & Wiksell, Uppsala

Köhler A. (1904) Mikrophotographische Untersuchungen mit ultraviolettem Licht X Wiss Mikroskopie 21, 129-165

Lindström B (1955) Roentgen absorption spectrophotometry in quantitative cytochemistry. Acta Radiologica Suppl. 125, 1-206

Rigler R. (1966) Microfluorimetric characterisation of intracellular nucleic acids and nucleoproteins by acridine orange. Acta Physiologica Scand 67 Suppl. 167, 1-122

Sörén L. (1970) Growth and proliferation of human blood lymphocytes stimulated with phytohemagglutinin (PHA) as studied by quantitative cytophotometric and autoradiographic techniques. Linkoping Univ. Med. Dissertations 12, 1-44

Trapp L. (1966) Instrumentation for recording microspectrophotometry. In: Introduction to Quantitative Cytochem. G.L.Wied (Ed) pp. 427-435, Academic Press, New York and London

Zeiss C. (1964) Universal Microspektrophotometer UMSP I, Pamphlet No. 40-811

Zetterberg A. (1966) Nuclear and cytoplasmic growth during interphase. Thesis 1-137, Almquist and Wiksell, Uppsala

TV-CONTRAST ENHANCEMENT METHODS FOR CHROMOSOME ANALYSIS

Lore ZECH

Department of Medical Cell Genetics, Karolinska Institutet,
Medical Nobel Institute, Box 60400, S-104 01 Stockholm,
Sweden.

1. INTRODUCTION

The introduction of banding methods for chromosome analysis
has made it possible to distinguish each individual chromosome in
a human metaphase (Caspersson et al 1971). Depending on equipment
available and experience, different laboratories prefer different
staining techniques for karyotyping, fluorescence (Q)-, Giemsa (G)-
or reversed (R)-banding. Other banding methods, such as DAPI,
staining or nucleolar organizers or centromeric heterochromatin are
only used for more specialised problems.

For conventional chromosome analysis the stained metaphases
must usually be photographed before karyotypes can be prepared.
The chromosomes are then cut out from paper copies and arranged
according to an internationally agreed system (Paris Conference
1971). In Q- and G-banded preparations certain chromosomes are
extremely pale and difficult to distinguish. This is true mainly
for chromosome No. 19 and also for the telomeric regions of most
autosomes, especially chromosome No. 1 (Fig. 1). Therefore it is
often necessary to prepare at least two or three paper copies,
each of which shows the chromosomes at a different contrast. Thus
karyotyping is   time consuming and requires great patience.

2. SYSTEMS FOR KARYOTYPE ANALYSIS

Several systems for semi-automatic and automatic karyotype
analysis have been described. Recently Ernst Leitz, West Germany,
has introduced a programme for semi-automatic chromosome identification of unbanded chromosomes for its image analysis system (TAS-

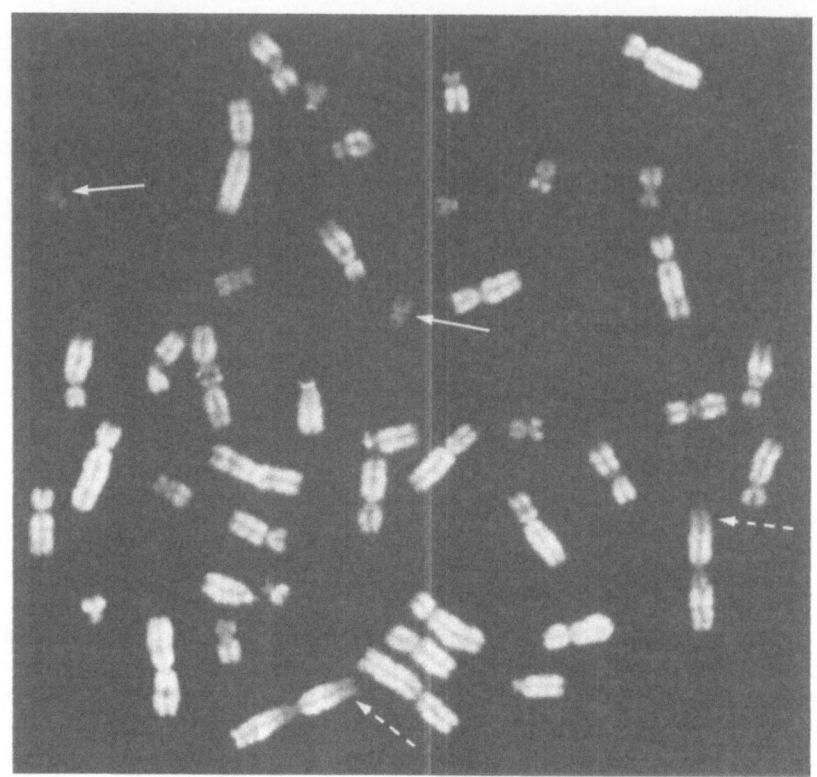

Fig. 1   Q-banded human metaphase demonstrating differences in
fluorescence intensity of different chromosomes.   Solid
arrow: Chromosome No. 19, the most weakly fluorescent
chromosome of the karyotype. Dashed arrow: Short arm of
chromosome No. 1 with pale fluorescence.

plus). Also Carl Zeiss, West Germany, has a programme (for
unbanded chromosomes only) in its IPS-computer for image analysis.
However, these instruments are very complex and in both of them the
chromosome programme is only one of many. Therefore, these systems
are far too expensive for most cytogenetic laboratories. More
recently Joyce Loebl, Gateshead, England, has introduced an
apparatus which is programmed for chromosome analysis only, the
Magiscan 3. It has a programme for unbanded chromosomes which has
now been available for more than a year. A programme for banded
chromosomes has also been developed but this has not been widely
tested. Even though the Magiscan is cheaper than the Leitz TAS-
plus and the Zeiss IPS, it is probably more expensive than most
laboratories can afford. It is also obvious that none of the
automatic and semi-automatic programmes for karyotype analyses is

Fig. 2 Overall view of the TV-equipment used to study negatives
of fluorescence stained chromosome preparations.

yet good enough for more difficult chromosome preparations such as
tumour metaphases or metaphases of generally poor quality. On the
other hand it is desirable to make cytogenetic analyses less tedious
and to find instrumentation which can be at least of some help for
chromosome identification.

I want to describe two simple instruments which have been
developed by T. Caspersson and his group at the Karolinska Institutet in Stockholm. Although these instruments do not perform
automatic karyotpe analysis, they can facilitate and speed up
cytogenetic work and have been used in our laboratory for many
years.

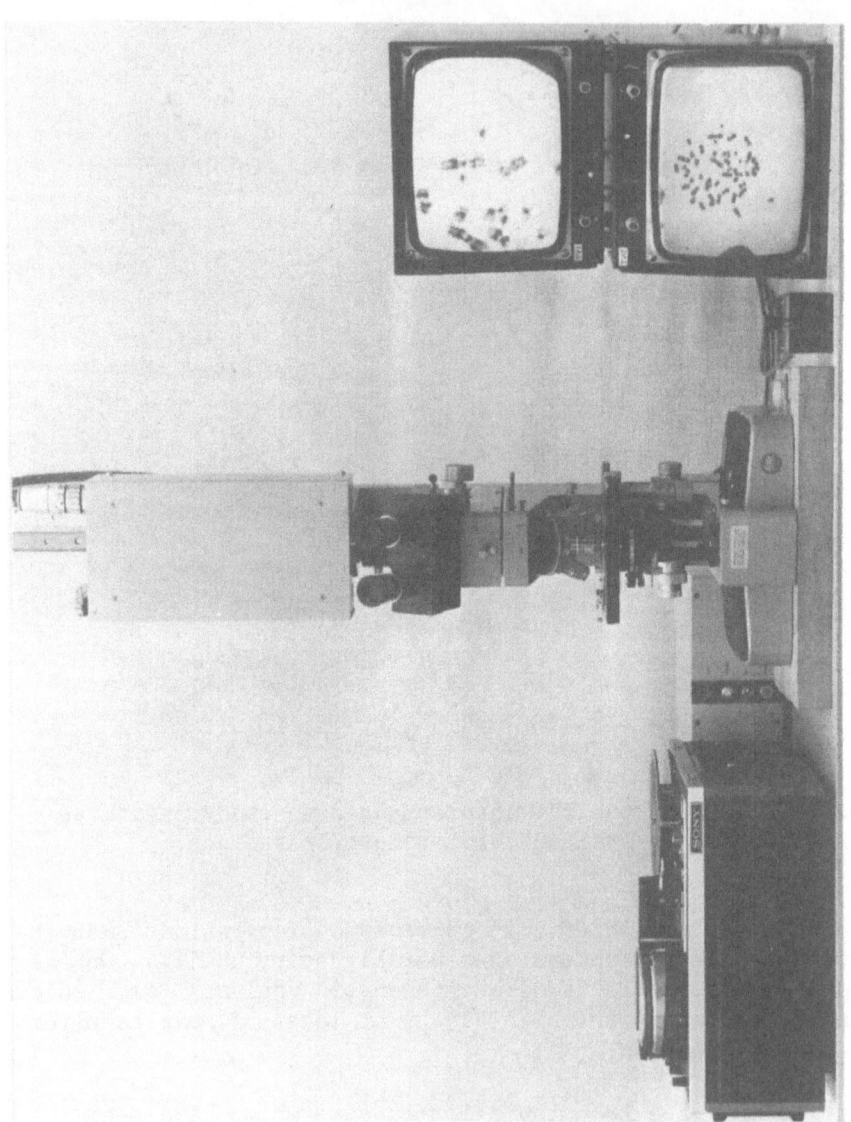

Fig. 3 TV-based equipment used for studying Giemsa-stained chromosome preparations.

Both instruments are based on TV-contrast enhancement. In the first (Fig. 2) photographic negatives of fluorescence stained metaphases are studied (Caspersson et al 1970). The arrangement is as follows:

The TV camera, delivered by "Fernseh-Company, Darmstadt", to T. Caspersson, has a zoom lens. A Plumbicon was chosen as camera tube because it gives a linear relationship between light intensity and video signal. The system works on the European standard, i.e. 625 lines. A special amplifier is introduced in order to amplify the video signal within small intensity ranges. This was necessary because different chromosome regions exhibit very different fluorescence intensities from very bright to very pale (Fig. 1).

The metaphase to be analyzed is presented on a TV-monitor. The contrast on the monitor can be regulated so that differences in fluorescence intensity can be adjusted, i.e. the fluorescence of very dark regions can be enhanced and the fluorescence of very bright regions can be suppressed. In this way each chromosome of the metaphase can be studied individually by selecting the optimal fluorescence intensity in combination with the optimal magnification.

The equipment has been shown to be of great advantage for studying large numbers of normal metaphases or metaphases with only one or a few abnormalities, as is the case in prenatal diagnosis or mutagenicity tests. If the number of aberrations increases, as in tumour cells during the later stages, the identification of these complex rearrangements can be more complicated and it is usually easier to work in the conventional way by cutting out chromosomes from photographs..

In Giemsa-stained chromosome preparations there is no need to photograph the metaphases to be analyzed because the dye does not fade when irradiated by light. Therefore, an instrument was constructed (Caspersson et al 1974) which is similar to the TV-equipment for fluorescence stained material but still more time saving, because the TV-camera is attached directly to the microscope (Fig. 3).

Chromosome analysis with this equipment requires high quality optics. Therefore a zoom system is used as ocular, followed by a TV-camera tube. The contrast enhancement is similar to that described before and permits very good resolution of Giemsa-banded chromosomes on the TV-monitor. On a second TV-monitor, connected with the camera, a survey picture of the whole metaphase is made by means of a video recorder. This picture shows the localization of the chromosomes in the metaphase and remains unchanged during the whole analysis. Each chromosome studied can be marked with a pen on the survey monitor.

The equipment described is not only time saving, it also gives very good pictures on the monitor and is less tiring to work with than conventional karyotyping. Furthermore, it has the advantage that several persons can look at the monitor at the same time and this can be helpful when metaphases with complex rearrangements have to be examined.

This type of TV-set is also very useful for studies of sister chromatid exchanges because the picture on the monitor is richer in contrast than the picture in the microscope.

3. CONCLUSION AND SUMMARY

There is a need for relatively simple pieces of equipment for chromosome analysis. The two instruments described here are based on TV contrast enhancement, are easily handled, and give pictures that are rich in detail. Compared with conventional karyotype analysis the instrumentation described is time saving and less tiring for the operator.

ACKNOWLEDGEMENTS

This work was supported by funds from the Swedish Cancer Society.

REFERENCES

Caspersson T., Lindsten J., Lomakka G., Wallman H. and Zech L. (1970) Rapid identification of human chromosomes by TV-techniques. Exptl Cell Res. 63, 477-479

Caspersson T., Lomakka G. and Zech L. (1971). The 24 fluorescence patterns of the human metaphase chromosomes - distinguishing characters and variability. Hereditas 67, 89-102

Caspersson T., Lomakka G., Zech L., Issler P, Kudynowski J. and Kvarnström K. (1974) Rapid techniques for counting and analysis of chromosome aberrations. Expt. Cell Res. 88, 427-428

Paris Conference (1971) Standardisation in Human Cytogenetics. Birth Defects: Original Article Series VIII, 7 (1972). The National Foundation, New York.

LUMINESCENCE AND FLUORESCENCE METHODS

Dieter ERNST

Institutet für Biophysik, Universität Hannover, 3000
Hannover-Herrenhauser 21, Herrenhauser Strasse 2,
Hannover, G.F.R.

1. INTRODUCTION

The availability of high technical standards of low intensity
light measurement and increasing knowledge concerning fluorescence
and luminescence phenomena have given rise to the development of
numerous new analytical methods. The major improvements concern
sensitivity, specificity and ease of preparation, but the commercial availability of enzyme preparations and the design of particular luminometers have also contributed to a worldwide spread of
these methods.

2. FLUORESCENCE METHODS

2.1 General Remarks

Analytical fluorescence applications tend to compete with
photometric methods. Provided all relevant parameters are
comparable, fluorescence methods show a 10 - 1000 fold higher
sensitivity, less reagent consumption and better selectivity. The
reason is explained in Fig. 1. In double-beam-photometry the
concentration of the substance being analysed is determined from
the difference between the light intensity transmitted by the
sample and the intensity transmitted by a blank control. For low
concentrations, photometry suffers from bad signal to noise ratios
which usually occur when small differences of two nearly equal
values are measured. Conversely the corresponding fluorescence
signal simply becomes smaller for low concentrations and can still
be read with reasonably accuracy. The lower reagent consumption is
due to this higher sensitivity and the specificity of fluorescence

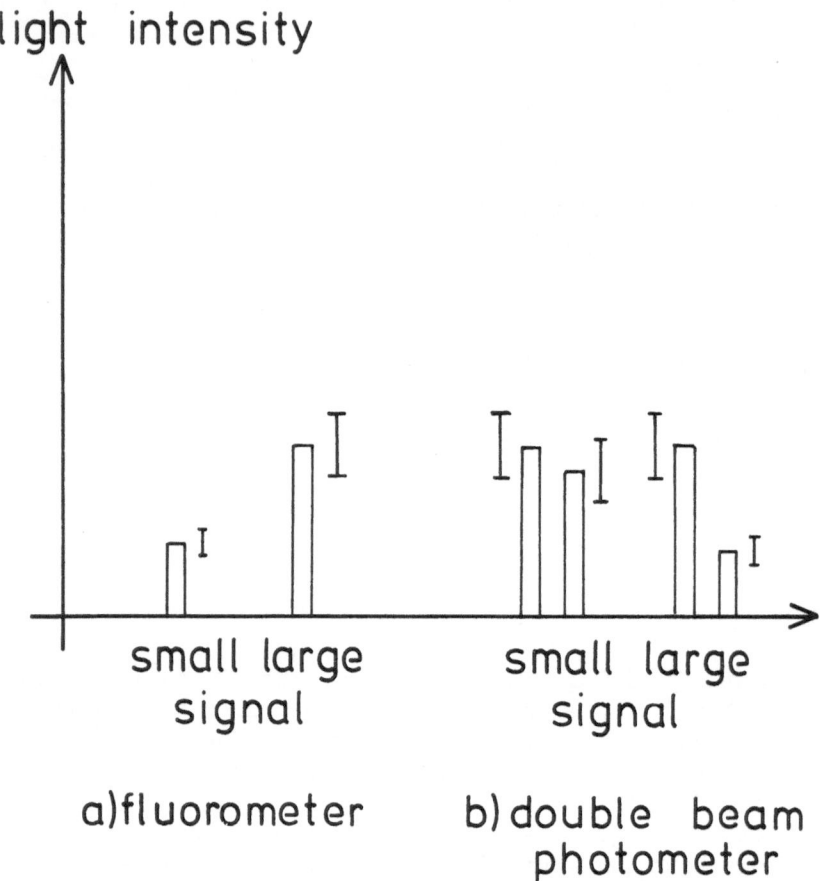

Fig. 1 Comparison of experimental errors in fluorescence and photo-
        metric measurements under small and large signal conditions.
        Columns: light intensities at the receiver. Bars: errors.
        The double columns at the right depict the two beams of a
        double beam photometer. At low analyte concentrations the
        error bars overlap.

methods is higher because in photometry only one parameter can be
specified, the wavelength, whereas in fluorometry both the exciting
and fluorescent wavelengths may be used for selection.

    If quantitative determinations are to be made from microscope
slides, there is a problem of integrating an inhomogeneous signal
distribution over the field of view. In absorption measurements the
concentration is, according to Beer's law, proportional to the logarithm of the signal, whereas in fluorescence determinations the
concentration is proportional to the signal itself except for higher
concentrations. Corrections for this case are given by Ernst (1971).

However fluorescent methods also have potential disadvantages. Fluorescence may be quenched, e.g. by impurities or by the substance being analysed itself fluorescing (self-quenching), or the substance may be decomposed by the fluorescence process. Hence quantum efficiences and the time course of the output signal should be checked. Even a temporary increase of efficiency has been observed (Tiffe 1971).

Temperature has an important influence on fluorescence. Hence all measurements should be made under defined temperature conditions. In deep-frozen samples the sharpness of the spectral lines and the stability of the dye increases. Freeze fluorescence has therefore been developed as a special branch of fluorometry.

One particular effect to mention is fluorescence depolarisation. Measurement of this effect allows conclusions to be drawn on the viscosity of the immediate environment of the fluorescent molecule, and this is important for example in membrane investigations.

Fluorescent dyes have been coupled to antibodies and fluorescence immunoassays (FIA) based on this principle have become an important competitor for radioimmunoassays (RIA). Normally RIAs and FIAs are heterogeneous, that is to say the precipitate (labelled antibody) has to be separated from the supernatant (dissolved label). However, by exploiting the effects of fluorescence depolarisation or quenching, FIAs have been designed which avoid separation of the labelled antibody from the dissolved label. Details and further literature are given by Soini (1979).

2.2 Chlorophyll Fluorescence

Measurement of chlorophyll fluorescence illustrates many typical problems and advantages of fluorescence methods. Therefore it is described here as an example.

The classical method for chlorophyll determination is photometric measurement after solvent extraction (Strickland and Parsons 1972). For chlorophyll estimations in water the material is usually extracted with acetone or alcohol and measured photometrically or fluorometrically. This determination has some drawbacks - in organic solution the absorption peaks depend on the solvent, the material being analysed may quickly decay under illumination or oxygen attack, and the extraction efficiency is not at all constant (Riemann and Ernst 1982).

The chlorophyll concentration in natural waters is too low to be determined photometrically without extraction, but fluorescence measurements are still possible. They also save a lot of preparation work at the expense of some loss of accuracy. Figure 2 shows the energy levels and transitions of chlorophyll a and its spectrum.

120

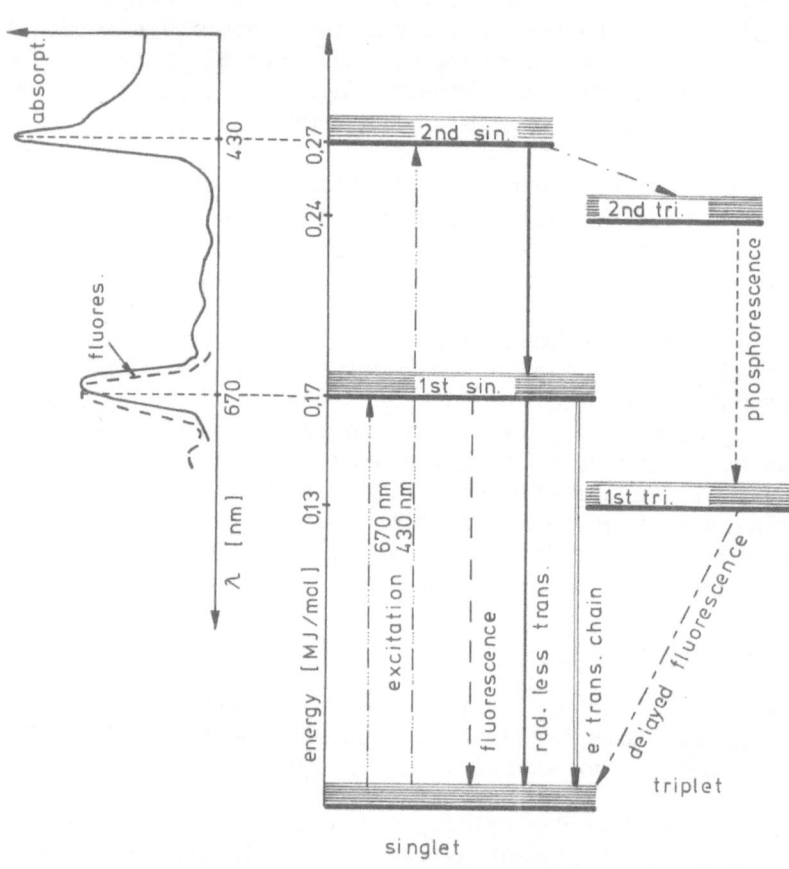

Fig. 2 Term scheme, absorption and fluorescence spectrum of chlorophyll in vivo. Fluorescence occurs from the first excited singlet.

It has to be kept in mind that chlorophyll a is not the only pigment present in plants, but it is the only one which shows fluorescence. As in many other fluorometric applications, the primary absorber may be another chromophore, which, by intermolecular energy transfer, delivers the excitation energy to the fluorescent molecule.

The photosynthetic apparatus of plants consists of two photo-systems (PS I and PS II). The energy collected by PS II is trans-ferred by the photosynthetic electron transport chain to PS I and is then used to produce glucose out of CO_2.

The fluorescence is almost exclusively due to PS II. Since the electron transport chain and certain other effects drain energy away from PS II, they act as quenchers of in vivo chlorophyll fluorescence. Hence the quantum efficiency of this process is not constant but depends on the physiological state of the cell. In vivo fluorescence measurements therefore display rather poor reproducibility unless this variable quenching is blocked. This can be done by several urea derivatives. Among them are certain basic constituents of herbicides, one of the best known being 3 (p-chlorphenyl) 1, 1 dimethylurea (CMU or Monuron). The emission of dark-adapted algae also displays a complex time-dependent behaviour (Kautsky curve, Fig. 3) which requires some standardis-ation to provide a reliable emission signal. Either the peak of the fluorescence curve or the plateau, which is reached after about 2 seconds, should be recorded.

The in vivo fluorescence properties of algae have been used to determine algal biomass (after blocking the electron transport chain), to check photosynthetic turnover (exploiting the quench effect of the transport chain) and to test for herbicides which affect photosynthesis (Lorenzen 1966, Samuelson 1978, Ernst and Schulze 1982).

3. LUMINESCENCE METHODS

3.1 General Remarks

The emission of light after chemical excitation is called chemiluminescence. If it occurs in biological systems,it is known as bioluminescence. This reaction occurs in almost all zoological kingdoms (bacteria, dinoflagelates, crustacea, worms, clams, fish, insects) except higher vertebrates. The firefly and glowworm are the most well-known examples. For reviews see Colowick and Kaplan (1978) and Ward (1981). For analytical purposes two systems are of primary importance:
    i) the firefly (ATP) system:

$$ATP + LH + O_2 \xrightarrow{\text{lase, Mg}^{++}} L + CO_2 + AMP + light$$

(ii) the bacterial (NAD resp. NADP) system:

$$NAD(P)H + RCHO + O_2 \xrightarrow[\text{lase, FMN}]{\text{oxase, L}} NAD(P) + RCOOH + \text{light}$$

LH: reduced luciferin   FMN: flavinmononucleotide
L: oxyluciferin   FMNH: reduced FMN
lase: luciferase   ATP: adenosintriphosphate
oxase: oxyreductase   NAD(P): nicotinamide dinucleotide
                                       phosphate

    The luciferins of the ATP system are compounds which can be oxidised by enzymatic reactions to oxyluciferin in an appropriate environment. The energy liberated by this oxidation is emitted in the form of light. The luciferins of different origin are of identical structure. Luciferases are enzymes mediating this process and may be of different structure depending on their origin. The different colours observed in bioluminescence are caused by these different structures and by changes in pH, ionic strength, temperature and Zn or Cd ions. Reports on specificity are

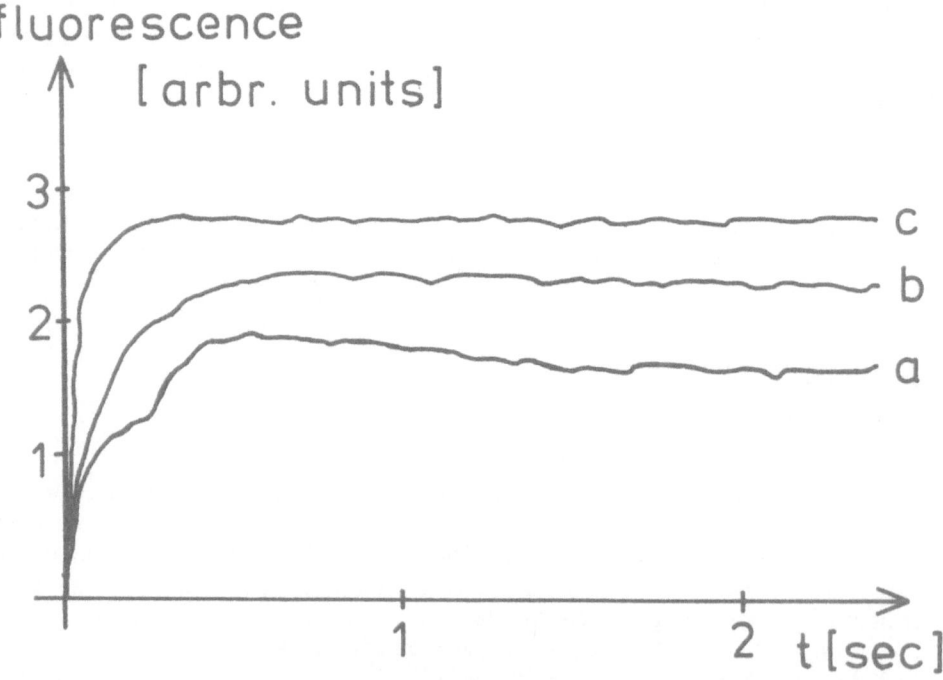

Fig. 3 Time course of in vivo chlorophyll fluorescence
      a) young culture (3 days old)
      b) old culture (7 days old)
      c) electron transport chain of algae blocked by CMU.

contradictory, However, the majority of authors assume that, at least in the usual assays, only ATP serves as a substrate. Activity and storage stability vary with the manufacturing process and temperature but deep-freeze storage (-20ºC) is essential. Raw extracts usually contain other enzymes which compete for the ATP and thus modify the light emission kinetics.

Bioluminescence can be observed in many bacteria. Photobacterium fisheri and Beneckea harveyi are the two most well-known species. The reaction scheme may be interpreted as a side chain of oxidative phosphorylation. The luciferase of these bacteria (scheme 2) co-operates with FMN resp. FMNH (depending on the details of the assay) in a highly specific way. Bacterial bioluminescence assays are thus specific and sensitive tools in determining these substances.

For both the ATP and the NAD(P)H system the total light emission is proportional to each of the reactants provided they are present in limiting concentrations. This is the basis of the analytical application.

These methods can be used for the determination of about 70 compounds, many of which are important for the clinical laboratory (Gorus and Schram 1979, Whitehead et al 1979). For example using the ATP system, ATP, ADP, AMP, cyclic AMP, ATPase creatinin and creatinin kinase may be measured. Using the NAD(P)H system, NAD, NADH, NADP, NADPH, glucose, glucose 6 phosphate, glucose 1 phosphate, FMN, FMNH, and ethyl alcohol may be measured.

The sensitivity of these assays is very high, with detection limits for ATP, NADP and H_2O_2 of 10^{-19} 10^{-16} and 10^{-9} mol/litre respectively. The dynamic range covers all concentrations from the detection limit to where an excess of other reactants cannot be maintained. It is of the order of $1: 10^{-6}$. The speed of luminescent analysis depends on the preparation procedures since the light measurement step requires only seconds or at worst minutes. The chemicals are expensive and their storage time is restricted. However, because of the small amounts required, the cost is low, or at least moderate.

Analytical interference may be caused by enzyme blockers or by competing nonluminescent enzyme reactions. The competitor may be in the material being analysed or in the reagent. The latter can be avoided by choosing highly purified preparations.

If competitors are present, light emission decays quickly and does not follow easily describable kinetics. For example in ATP determinations with crude firefly extract an effective half life of 75 seconds was observed whereas purified reagents yielded a half life of several minutes. In these cases it is recommended to read

just the peak emission or to integrate over the first 6 seconds
(Riemann 1978). With purified reagents first order kinetics,
typified by exponentially decreasing emission, can be observed.
The half-life is long enough to apply an internal standard. That
is to say after reaction with an unknown concentration, a known
amount of pure ATP is injected and the increase in light intensity
may be used directly to quantify the result. The temperature
optima for bioluminescence reactions are usually between 25 and
37°C. The instruments should permit temperature control in this
range.

3.2 Equipment

Three types of instrument have been used to measure
bioluminescence:-
    (i)   fluorometers with the light source switched off
    (ii) liquid scintillation counters
    (iii) specially designed luminometers

The instrument should ideally fulfil the following requirements:-
    a)   high sensitivity (single photon counting)
    b)   a temperature controlled reaction chamber
    c)   quick and complete mixing of reagents in the measuring
        position in the dark
    d)   analogue output to operate a strip chart recorder
    e)   choice of several data evaluation methods (peak reading,
        integration over different times, background subtraction,
        decay rate).

Fluorometers often do not fulfil requirements a, b, c, or e.
They give a measure of the emission spectrum but this is not
essential in routine analytical work.

Liquid scintillation counters may lack requirements b, c, d,
and e. The usual coincidence circuit has to be switched off
because luminescence is a one-quantum reaction which causes non-
coincident pulses in the electron multipliers.

Modern luminometers usually fulfil all the requirements. The
sample is kept in a cuvette in a dark, temperature controlled
chamber and the measurement is started by injecting the reagent.
Instruments differ in data handling, particularly in the freedom
of choice of integration times.

3.3 ATP- measurement

ATP measurement by bioluminescence is one of the most widespread luminometric methods and has gained particular importance in
environmental studies. Since it may be used to illustrate many of

the questions common to all luminescence studies, it is discussed
here in more detail.

The occurrence of ATP is a general feature of living matter.
The determination of living biomass in the presence of detritus is
a problem in soil- and water-research since determinations of
organic carbon or similar common constituents of biomass do not
differentiate between dead and living material. Microscopic cell
counting also does not necessarily differentiate between living
and dead cells, is time consuming and not feasible in the presence
of numerous other particles for example in soil samples.

The occurrence of ATP is a general feature of all living
matter so ATP determinations have been used frequently for biomass
estimations since Holm-Hansen and Booth (1966) introduced this
technique. A detailed literature list is given by Qureshi and
Patel (1976).

Since ATP is enclosed in the cell compartments, the preparation
procedure includes an extraction step. The most frequent method is
to boil in tris-buffer. Karl and Rock (1975) describe further
procedures and Strehler (1968) gives a more general description of
the method.

ATP is a very dynamic cell constituent and the ratio of ATP
to organic carbon has been repeatedly discussed in the literature.
Values between 1: 250 and 1: 400 usually occur in plancton samples.
ATP levels indicate not only biomass per se but also relate to
metabolic activity. The possibility of determining the other
adenosine phosphates ADP and AMP using basically the same technique
- after enzymatic transfer of these compounds to ATP - led to the
concept of "energy charge" (EC), which describes differences in
metabolic activity. Atkinson (1971) defined this term as follows:

$$EC = \frac{[ATP] + 1/2[ADP]}{[ATP] + [ADP] + [AMP]}$$

This relationship gives the ratio of the energy content of
the adenosine phosphates to their total concentration. Prolifer-
ating cultures in log phase reach values of 0.9, normal viable cells
around 0.6 and dying cells have values below 0.4. The energy
charge is an important parameter for describing the vitality of
single-cell organisms or living tissue.

REFERENCES

Atkinson D.E.(1971) Adenine Nucleotides as stoichiometric coupling
    agents in metabolism and as a regulatory modifier. The adenylate
    energy charge. In Greenberg D.M., Metabolic Pathways, Vol 5,
    Academic Press.
Colowick S.P. and Kaplan N.O. (1978) Bioluminescence and chemo-
    luminescence, in: De Luca M.A.: Methods in enzymology. Vol.LVII,
    Academic Press
Ernst D. (1971) Correction for light absorption within the object
    for quantitative microfluorometry: a theoretical approach.
    Applied Optics 10, 1398-1401
Ernst D.E.W. and Schulze E. (1982) Chlorophyll determination in
    the field of fluorometry, Arch. Hydrobiol. Bieh. Limnol. 16, 55
Gorus F., and Schram E. (1979) Applications of bio and chemo-
    luminescence in the clinical laboratory, Clin. Chem. 25, 512-519.
Holm-Hansen O. and Booth C.R. (1966) The Measurement of adenosine
    triphosphate in the ocean and its ecological significance.
    Limnol. Oceanogr. 11, 510-519
Karl D.M., Rock P.A. (1975) Adenosine Triphosphate measurements in
    soil and marine sediments. J. Fish Res. Board Can. 32, 599-607
Lorenzen C.J. (1966) A method for the continuous measurement of in
    vivo chlorophyll concentration. Deep Sea Res. 13, 113-227.
Qureshi A.A. and Patel I (1976) Adenosine triphosphate (ATP) levels
    in microbial cultures and a review of the ATP Biomass estimation
    technique. In: Inland Waters Directorate Canada Centre f. Inland
    Waters Scientific Series, Burlington, Ontario 63.
Riemann B. (1978) Interference in the quantitative determination
    of ATP extracted from freshwater microorganisms In: Proc. Inter.
    Symp. Anal. Appl. Biolumin. Chemolumin.

MICROBEAM DESIGN CONSIDERATIONS

Dieter ERNST

Institutet für Biophysik, Universität Hannover, 3000
Hannover-Herrenhauser 21, Herrenhauser Strasse 2,
Hannover, G.F.R.

1. INTRODUCTION

The purpose of a microbeam apparatus may be either to observe
photobiological reactions in small target areas or to carry out
microsurgical work. Often these two aspects overlap in a particular
application. The first known application was of the photobiological
type: Engelmann monitored the oxygen production of chloroplasts in
Spirogyra by the migration of aerophilic bacteria. When a light
microbeam was directed onto the chloroplast, it developed oxygen
and attracted the bacteria. The microbeam pioneer Chakhotin first
promoted the idea of microsurgery with ultraviolet (UV) microbeams.
He published numerous papers that are discussed in general reviews
of microbeam work by Zirkle (1957), Smith (1964), Moreno (1969) and
Berns (1974).

Microsurgery usually requires high power irradiation, exact
location of the beam, and observation during exposure since the
specimen may be modified by the operation. It does not require
particular wavelengths. Destruction may be caused by light damage,
by thermal or acoustic effects of laser pulses, by the photochemical
effects of UV (e.g. excitation, bond-breaking, membrane damage) or
by the ionization of X-rays or particle beams. Photo- or radiobiological effects require the proper selection of wavelengths and
sometimes control a beam intensity and duration (e.g. for dose-rate
effect experiments). The microbeam often has to be combined with
analytical facilities (microdensitometry, absorption- or fluorescence spectrometry, birefringence measurements) and particular
observation methods, e.g. phase contrast may be required. It is
also important to the design of the microbeam to know whether or

not the specimen has to be observed during irradiation. Usually
one should provide facilities to do this. However, in some cases
it is not feasible as for example in Zirkle's proton microbeam
(Zirkle 1957). Nearly all microbeam tasks require microscopical
work on a living object. This may call for further equipment such
as temperature controlled microscope stages, wet chambers, or an
oxygen supply. The living specimen may also require limited light
intensity. Since staining of living specimens is usually impossible
all sorts of microscopic methods have been used to increase
visibility in microbeam work, e.g.:

(i)   phase contrast (dependent on refractive index)
(ii)   polarization microscopy (dependent on birefringence)
(iii) Normarski interference (dependent on refractive index
       gradients)
(iv)   obtuse illumination of dark field (also dependent on
       refractive index, is much simpler than Normarski optics
       but sacrifices high numerical aperture)
(v)   selection of wavelength (e.g. 255 nm for nucleic acids,
       green light for erythrocytes, red or blue for chloro-
       plasts)

Micromanipulators may be used for surgical manipulations.

An important design consideration is how to define the microbeam, which is usually only a few micrometers in diameter. For all
beams produced by focusing (electromagnetic radiation, electrons)
this is done by projecting the demagnified image of an illuminated
diaphragm into the object plane. For all other radiations the
beam must be shaped by collimating diaphragms. X-rays, protons and
alpha-particles belong to this category. The two latter types of
beam could in principle be focussed because they are charged.
However, appropriate lenses would be impractical and microbeams
described in the literature use simple collimation.

For further discussion a distinction is required between light
(including ultraviolet and the less important infra-red) and
ionizing microbeams. Table I lists the most important features of
each type of microbeam. Because of the ease of construction and
the freedom in choosing selective wavelengths, visible light and
UV microbeams have been used for biological work far more frequently
than ionizing microbeams.

2. MICROBEAMS OF ELECTROMAGNETIC RADIATION

All such microbeams use normal microscope optics, the condenser
or objective of which is used for demagnification of the beam-shaping
aperture. Good imaging quality and good contrast are essential.
Condensers may not fulfil this condition and should then be replaced
by objectives. UV microbeams need a UV objective which may be of

TABLE I

COMPARISON OF DIFFERENT TYPES OF RADIATION FOR MICROBEAM
PURPOSES

Type of radiation	Range compared to specimen	Focusing	Energy or wavelength range	Factors
Infrared	long	Yes	1 μm	diffraction
Visible	long	Yes	400-800 nm	or light scatter
Ultraviolet	dependent on absorption	Yes	250-400 nm	
Soft X-rays	short	No	5 - 10 keV	secondary electrons
Protons (1)		Not used	1 - 2 MeV	scatter
a -particles	short	Not used	3 - 10 MeV	scatter
Electrons (1)	dependent on energy	Yes	30 - 100 keV	scatter at end of track
Heavy ions(1)		Yes	< 20 MeV	better than all others

(1) require optics in vacuo and window to get the beam out

either the refracting or mirror type. The best results seem to be
obtained with the ultrafluar types (Zeiss). The microbeam is
inserted into the normal microscopic light path by a dichroic or
semi-reflecting mirror or through a hole punctured in a very small
mirror. The latter two serve also as beam shaping elements (see
Fig. 1). To protect the observer's eye the microbeam should be
directed away from the observer. This applies particularly for
lasers. However, this is not feasible under all circumstances, e.g.
in Chakhotin's constructions and in a system which incorporates a
polarising microscope used by the author. In such cases and when
lasers are used, the observer's eye should be protected by a filter.

130

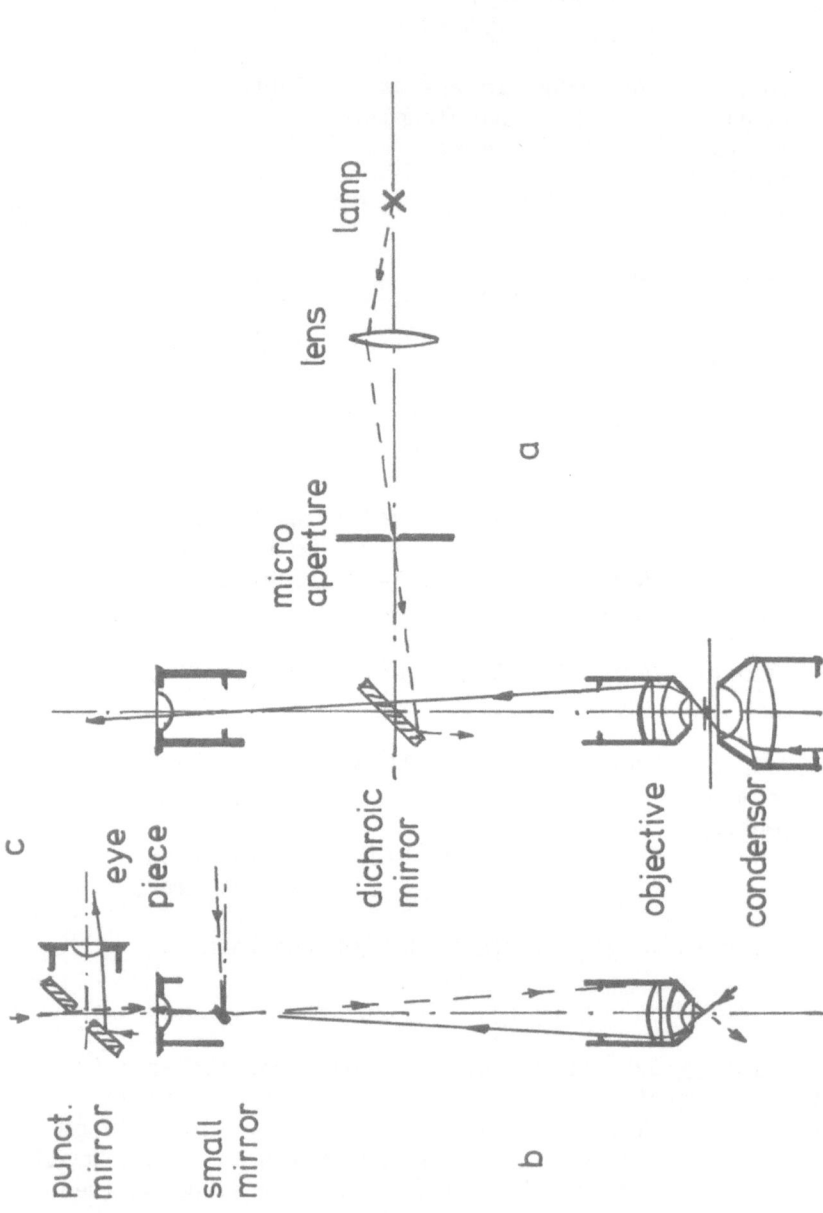

Fig. 1 Light paths for microbeams. a) the microbeam (lamp, lens, microaperture) is inserted into the usual microscopic light path by a semireflecting or dichroic mirror b) a small mirror may serve for shaping the microbeam and inserting it, this mirror must be positioned in the image plane of the objective c) the small mirror may be replaced by a punctured mirror. The upper parts of observation light path and microbeam are then interchanged.

2.1 Basic considerations in geometrical optics

A light microbeam can often be easily added to an existing microscopic facility. To provide the necessary information for such a manipulation certain basic principles in geometrical optics must be applied. Figure 2a shows a light emitter E (surface of a light source or illuminated lens or diaphragm) mounted at distance D in front of a receiving area R (another lens or diaphragm in the light path). The radiant power P entering R is:

$$P = \frac{\pi a^2 . \pi b^2 . d}{D^2}$$

a: radius of emitter
b: radius of receiver
D: distance from receiver to emitter
P: radiant power entering receiver
d: radiant power density of emitter

Using the Helmholtz invariant this can be simplified to

$$P = H^2 . d$$

Figure 2b shows a lamp being imaged by a lens. The lamp and lens may be regarded as emitter E and receiver R and a value for H may be determined. The lens and the lamp's image can again be regarded as a pair of planes. It is now important to note that the terms H calculated for these two emitter-receiver pairs are equal. This follows from the elementary relationships:

$$a_1/D_1 = b_2/D_2 \qquad \text{and} \qquad a_2 = b_1 \qquad \text{(Fig. 2b)}$$

When a series of images follows one other as in a microscope, the term H is constant for all pairs of such corresponding planes, which may be regarded alternately as emitter and receiver. Hence the term Helmholtz invariant. This factor is important for microbeam design.

Figure 2c shows part of a microbeam design. A lamp illuminates the beam-shaping pinhole via a lens. The objective produces an image of the pinhole at the specimen. Typical values for H are around $2 \mu m$ (ultrafluar objective 32/0.40 Glyz, 190 mm tube length, 0.10 mm diameter pinhole yielding a microspot of $3.6 \mu m$). In a corresponding normal microscope without the microbeam pinhole, H is 0.6 mm. For a camera (f = 50 mm, 24 x 36 mm size, field stop f/8) it would be 8.5 mm. Figure 2d shows a monochromator. Light enters the facility via the entrance slit, is made parallel by the first lens, passes the prism and is focussed by the second lens on the exit slit. The slit area and the effective prism area, i.e. its projection on the lens, are the corresponding planes E and R. The

132

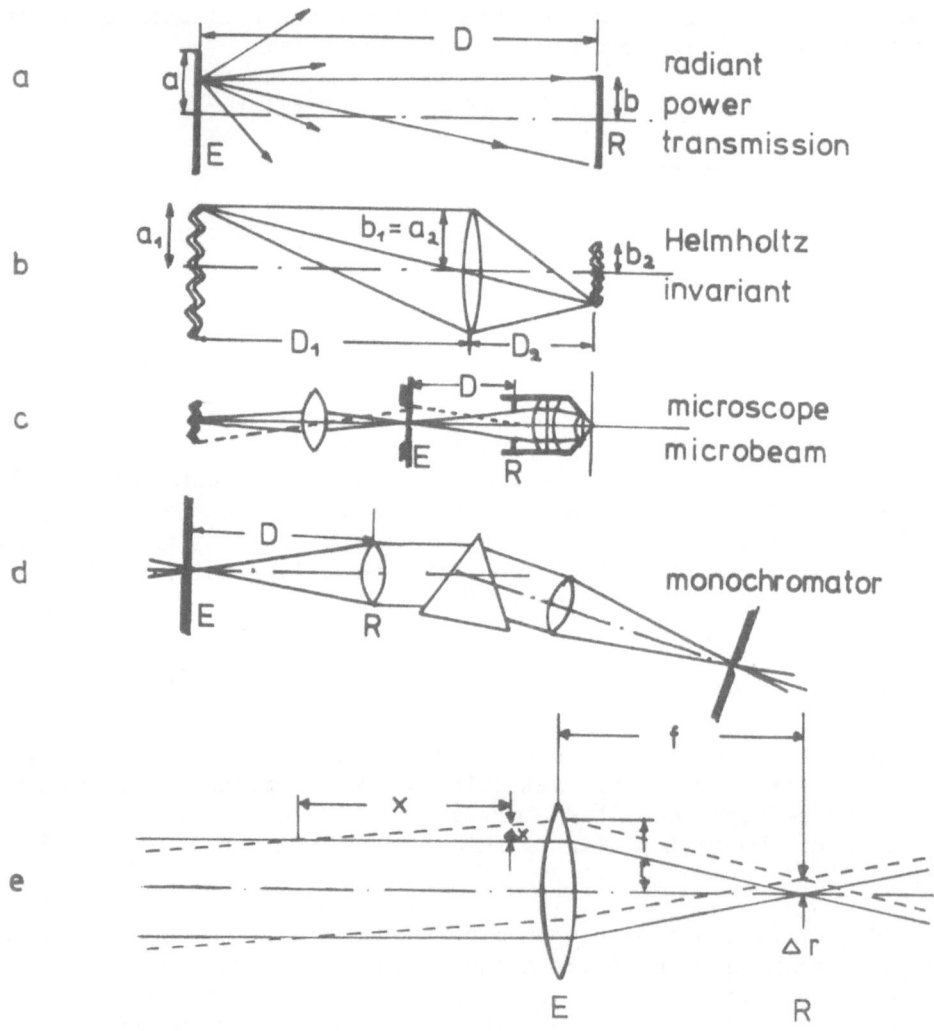

Fig. 2 Helmholtz invariant. a) emitter plane E transmits radiation
to receiving plane R at distance D. b) a lamp filament and
its image. c) principle of microbeam optics: microaperture
and focal diaphragm of objective correspond to E and R.
d) principle of monochromator, the lenses may be replaced
by mirrors, the prism by a grating. e) Helmholtz invariant
for a laser beam.

focal length corresponds to D. For the M4 Q III (Zeiss) H equals
0.104 mm (slit: 0.1 x 14 mm, corresponding to $\Delta\lambda$ = 2nm at λ =
400 nm). It follows that a highly resolving monochromator has a
smaller H than a microscope and would not make full use of its
numerical aperture. However, it will be very suitable for a microbeam.

Figure 3 shows the principle of Kohler's illumination (3a)
which is normally used in microscopes and so called critical illumination (3b). According to Kohler, the condenser has to image the
collimator lens onto the specimen plane. A microbeam field stop at
the position of the collimator would sacrifice a lot of light so
critical illumination is much more effective. It creates an image
of the light source in the specimen. This would not be suitable for
normal microscopy but it is reasonable for a microbeam, which only
makes use of part of the field of view.

2.2 Light sources

It has been shown that geometrical considerations will favour
light sources with low Helmholtz invariant values for microbeam
applications. This applies particularly for lasers, which for the
same reason are very poor light sources for general microscopy
since they cannot exploit at the same time the microscope's field
of view and its high numerical aperture. The derivation of H for
a laser is depicted in Fig. 2e. The laser beam of radius r diverges
very little. (typical values: Δx: x = 1 : 1000 to 1 : 100 000).
It is focussed by the lens (focal length f) to a small image
(diameter Δr). With

$$\Delta x : x = \Delta r : f \qquad \text{we get}$$

$$H = \frac{\pi\, r\, \Delta r}{f} = \pi\, r\, \frac{\Delta x}{x}$$

Typical values are 10μm for a small argon ion laser (Spectra
Physics 30 mW, Δx: x = 1 : 1000, r = 0.3 mm). For Δr's below
0.2 μm the spot size will be diffraction limited. Using an objective of f = 5 mm (magnification approx. 30) this limit would be
reached at $\Delta x : x = 5.10^{-5}$

The most serious disadvantage of lasers is the high price.
Only the HeNe laser can compete with classical lamps, but its red
light is rather unsuitable for photobiological purposes. The blue
HeCd, the green Ar and the frequency doubled UV lasers are one to
two orders of magnitude more expensive. Fixed wavelength lasers are
a limitation in photobiology, but variable wavelength lasers are
again more costly. Incandescent lamps are not very suitable because
of their low radiant power density (high total power does not help

134

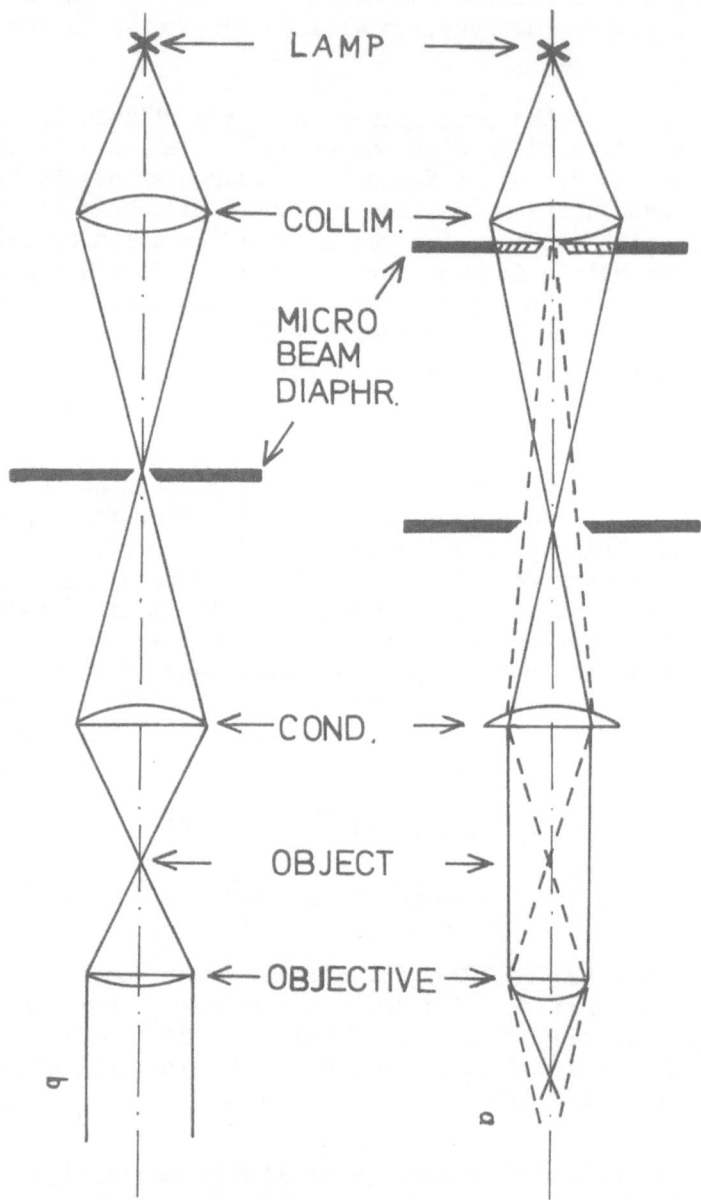

Fig. 3 Principles of (a) Köhler's and (b) critical illumination.

with a low Helmholtz invariant) and their yellowish light. Hg and Xe arc lamps, however, are often used, Hg is particularly suitable especially if one can make use of its spectral lines (254, 265, 497, 302, 313, 355, 408, 439, 541, 579 nm). A Xe arc lamp is better if continuous wavelength selection is required. The power densities are in the order of 0.1 W/cm^2 nm sterad in the visible

region (XBO 150 W, Osram). The power in a 280 nm microbeam of 3.5 μ m is 0.7 erg/min. This can cause cell death within several seconds.

2.3 Focusing, beam localization, shaping and dosimetry

Microbeams of visible light do not create problems for beam localisation and focusing, but since UV rays are invisible, one has to use some facility to visualise them. Usually the red beam of a HeNe laser is inserted into the light path so that it matches exactly the UV microbeam. Some of the methods used to check directly the position and focusing of the UV beam are: UV sensitive TV cameras, UV image converters, oculars with fluorescent glass in the image plane, or fluorescent layers (dye smears) in the object plane. Reflecting objectives have no chromatic aberration and the chromatic aberration of corrected lens objectives is said to be negligible. However, it must be kept in mind that the coverslip introduces a small chromatic error in mirror objectives, and in refracting objectives the remaining focal deviation can be around a few micrometers. Such chromatic aberration can be accommodated since the relationship between focal length and wavelength displays increasing and decreasing portions. Hence it is usually possible to find a visible wavelength with exactly the same focal length as the UV used. Once such a wavelength is found, quick focusing can be made in the visible light.

The shape of the target may require non-circular microbeams. It may even be necessary to alter the shape of the beam frequently. For this purpose, slits with adjustable width and length and slit-like facilities with movable rims have been used. Iris-diaphragms, in addition to their restriction to circular shapes, cannot provide sufficiently small diameters. A convenient method to create all sorts of shapes is to cut the shaping microaperture in aluminium foil.

The dose absorbed by an object is determined in two stages. First one measures the incident dose and this is then multiplied by the absorption of the specimen for that particular wavelength. The latter must be measured by microdensitometry. For details see Ernst (1970) and Seibold (1976). An example of such a measurement in living tissue culture cells is given by Hatfield et al (1970). This measurement may be unnecessary if objects of identical absorption are being compared.

The incident dose may either be measured directly e.g. with a calibrated device (photocell, photomultiplier or photochemically) in the position of the object or it is calculated from the optical data of the apparatus. The latter requires a knowledge of the Helmholtz invariant, the radiant power density of the lamp and the light losses. If the losses are unknown, one may assume 5% loss at each glass-air interface giving a total loss $L = 0.95^n$ where n is the number of interfaces as a rough approximation. Mirrors in the light path, including the reflection gratings in monochromators, are possible sources of error. Manufacturers give data for reflection losses (usually in the range 5 - 10%), but these values change with age. If these data are known the transmitted power follows from the basic considerations in Section 2.1

3. IONIZING MICROBEAMS

Since ionizing rays are not very selective for particular tissues or substances, and far more difficult to handle than electromagnetic radiations, they have been much less widely used for microbeams. The reasons for using them have usually been either their limited range or specific radiobiological questions related to the ionization process.

The beam sources - accelerators or X-ray tubes - are relatively bulky structures and occupy all the space on one side of the specimen. Hence objects can only be observed by incident light microscopy, which is not too favourable for biological specimens such as cells in tissue culture. Methods like dark field, phase contrast or Normarski interference optics are usually advisable. If the radiation source can be built into the condenser as in the alpha-particle microbeam (Smith 1967) the optical conditions are better.

The only ionizing microbeam facility that uses focusing is the electron microbeam of Rickinson and Dendy (1970). All others use collimators. Lead-glass capillaries are generally used as collimators for X-rays. They work like a light guide since the refractive index of the glass for X-rays is somewhat lower than that of air. For particles the collimators are made of microapertures in metal foils 10 - 20 μm thick. Different ways to drill such holes have been described in Table II.

4. DISCUSSION AND CONCLUSIONS

Ionization as a means of damaging biological material attacks all types of substances more or less similarly, whereas light discriminates due to differences in selective absorption. Light microbeams usually penetrate the whole object, whereas the range of ionizing beams may be of the order of the thickness of single cells.

TABLE II

PROCEDURES FOR MANUFACTURING PINHOLES FOR COLLIMATORS

electron beam drilling
laser beam drilling
heavy ion trace etching
puncture of foils by spark or needle
drawing a wire out of a matrix
eching of masked substrates
metal deposition into a larger hole
drawing glass capillaries

Beams of ionizing radiation are broadened by scattering effects which limit the size of microbeams produced in this way to a few micrometres. Soft X-rays of 5 - 10 keV, electrons of 30 - 100 keV, alpha particles of 3 - 10 MeV and protons of 1 - 2 MeV have been used. Diffraction limits the spot size for electromagnetic waves.

Heavy ions have not been used to date because they require extremely expensive equipment. In principle they would be very well suited to microbeam work because they may be focussed, show little scattering due to their high mass and little diffraction due to their short Dirac wavelength.

REFERENCES

Berns M. (1974) Biological Microirradiation. Prentice-Hall
Chakhotin S. (1912) Die mikroskopische strahlenstichmethode, eine
    Zelloperationsmethode. Biol. Zentre.,,32, 623
Chia-Lun Ho (1969) Spherical model of an acoustical wave generated
    by rapid laser heating in a liquid. Rep. from the Journ. of the
    Acoust. Soc. of Amer. 46, No. 3, Part 2 728-736
Ernst D.E.W. (1970) UV Microbeam dosimetry. Advanced Study Institute
    Microbeam Irradiation and Cellular Biology, Stresa, Italy.
Hatfield, J.M.R., Schulze L., and Ernst D. (1970) Measurement of
    the ultraviolet absorption in specific parts of both living and
    fixed mammalian cells, using a specially designed microspectro-
    photometer. Exp. Cell Res. 59, 484-486
Moreno G., Lutz M., and Bessis M. Rev. of Exptl. Pathol. 7, 99-135
Rickinson A.B. and Dendy P.P. (1970) Advanced Study Institute
    Microbeam Irradiation and Cellular Biology, Stresa, Italy.
Salet C., Lutz M. and Barnes F.S. (1970) Parametres Physiques
    characterissant le dommage thermique selectif de mitochondries en
microirradiation par laser. Photochem. Photobiol. 11, 193-205

138

Seibold H.W. (1976) UV-dosimetrie an pinus silvestris pollen und stimulationsuntersuchungen des pollenschluchwachstums nach UV- und ionisirender bestrahlung. Thesis, University of Hannover, Germany

Smith C.L. (1964) Microbeam and partial cell irradiation. Internat. Rev. of Exptl. Pathol. $\underline{7}$, 99-135

Smith C.L. (1967) Alpha-particle microbeams. Advanced Study Instit. Microbeam and Partial Cell Irradiation and Ancillary Techniques, Cannes, France.

Zelles L., Seibold H.W. and Ernst D.E.W. (1979) Localization of the site of action of tube growth stimulation by X-irradiation of single pollen grains of pinus silvestris. Rad. Res. $\underline{79}$, 149-161

Zirkle R.E. (1957) Partial cell irradiation. Advan. Biol. Med. Phys. $\underline{5}$, 104-146

PARTIAL NUCLEAR AND CYTOPLASMIC (MITOCHONDTRIAL) IRRADIATION OF SINGLE LIVING CELLS USING ULTRAVIOLET AND VISIBLE LIGHT

Giuliana MORENO

INSERM U.201, Photophysiologie et Photothérapie, Muséum National d'Histoire Naturelle, Laboratoire de Biophysique, 61 Rue Buffon, 75005 Paris, France.

1. INTRODUCTION

Partial irradiation of a single living cell has been used for two main purposes: (i) to study the function of organelles by their ablation (in this way the technique is an extension of micro-surgery), and (ii) to study the effects of radiation at the level of a single organelle. However, these two aspects are complementary since selective destruction of an organelle contributes to the analysis of the interaction between radiation and living material, and conversely the study of the effects produced by irradiation provides information on organelle functions.

Since the first instrument permitting micro-irradiation with ultraviolet (UV) light proposed by Tchakhotine in 1912, electrons, protons and electromagnetic radiations ranging from X rays to visible wavelengths have been used for partial irradiation (see reviews by Zirkle 1957 and Smith 1964). However, in recent years non-ionizing electromagnetic radiations in the UV and visible range have been mainly employed to achieve micro-irradiation studies. The entire field of partial cell irradiation has been extensively discussed by Berns in a book entitled "Biological micro-irradiation. Classical and laser sources" (1974) and micro-irradiation using UV and visible light has been discussed in more specialised reviews (Moreno et al 1969; Moreno 1972; Berns and Salet 1972; Berns et al 1981) as well as in the present book by Ernst.

The aim of the present report is (i) to summarise the general principles of microbeam devices using UV and visible light (ca. 250-700 nm), the methods used to irradiate living cells and the

techniques for evaluation of micro-irradiation effects; (ii) to
describe some aspects of micro-irradiation related to nuclear and
mitochondrial irradiation of single mammalian cells.

2. MICRO-IRRADIATION TECHNIQUES

Photodestruction or photoalteration of cell structures can be
induced either by photochemical or thermochemical reactions. With
conventional light, the important effects of ordinary doses of
radiation are generally photochemical. Thermochemical reactions
may take place with laser sources because power densities of light
many orders of magnitude greater than those obtained with a conventional source may be achieved in the focal plane of the objective.

Specificity of damage is achieved by two means: (i) specific
absorption by the target structure of radiation with the appropriate wavelength (absorption may be due to endogenous or exogenous
chromophores); (ii) localization of the microbeam to the target
structure. However, structures other than the target especially
above and below the target are necessarily irradiated and damaged,
even with precise selection of wavelength and sharp focus of the
beam.

2.1 Microbeam Devices

Since the first microbeam apparatus (Tchakhotine, 1912) many
instruments permitting partial irradiation of living cells have
been designed. Generally, all these devices focus a beam of light
at a selected wavelength through a microscope objective of high
numerical aperture onto the target specimen. The same objective
is used both to irradiate and to observe the specimen.

For conventional light, the microbeam is produced by focusing
an image of the source onto a small aperture to provide a high
light intensity. A demagnified image of this aperture is then
formed in the plane of the preparation by the objective. For
laser sources, because of the low divergence of the beam, delineation of the beam light by focusing the image of the source onto
an aperture is not necessary. In this case the diameter $\underline{d}$ of the
spot at the site of the irradiation is given by:

$$d = f \, a$$

where $\underline{f}$ is the focal length of the objective and $\underline{a}$ is the divergence of the laser beam.

The minimum diameter of the spot at the target site is
limited by the laws of diffraction. The positioning of the beam
spot upon the target specimen is accomplished by using an auxiliary beam of visible light with the same optical pathway as the
irradiation beam. For lasers, the light of a continuous helium-

neon laser, coaxial with the irradiation beam may be used. Positioning may also be achieved by using fluorescent material.

Phase-contrast observation is necessary in experiments with living cells and this can be achieved in microbeam instruments using an objective containing a phase ring, or for reflecting objectives by placing the phase ring in a more accessible position and forming an image of it in the appropriate plane.

Monitoring the radiation flux may be done at various levels of the focusing system. Photocells or photomultipliers have been used for conventional light microbeams. In laser micro-irradiation, dosimetry is complicated by the short duration of the irradiation and calorimetric devices or photoelectric cells with a rapid response connected to a fast oscilloscope are generally used. The power density at the site of irradiation is calculated taking into account the transmission of the objective. Absolute measurement of the absorbed energy at the site of irradiation is difficult to determine because of the geometry of the focused beam and differences in absorption in different parts of the biological target.

2.2 Preparation of biological samples. Techiques of evaluation of micro-irradiation effects

Cells to be irradiated must be prepared so that they form a mono-layer, since this condition is essential for precise micro-irradiation. All cell preparation must ensure optimal cell survival. When irradiation is followed by cytochemistry, auto-radiography or electron microscopy, it is necessary to employ a technique for location and identification of the irradiated cells which permits their recovery.

All experiments of partial cell irradiation deal with a very small number of cells. Therefore, classical biochemical methods cannot be considered for the study of the effects. Techniques such as cytochemistry, autoradiography, microspectrophotometry, microspectrofluorimetry and polarizing microscopy have generally been used.

3. MICRO-IRRADIATION OF THE CELL NUCLEUS

A large number of microbeam experiments involve micro-irradiation of the nucleus during interphase or mitosis. The book by Berns (1974) and the above-mentioned general reviews have reported in detail the results of these experiments including those on chromosomes, on the nucleolar function, on DNA synthesis and the interaction between the nucleus and the cytoplasm and on location of the nucleolar organizer.

In this chapter the effects of a UV microbeam on one particular aspect of nuclear activity namely molecular damage and its
repair, and the application of these results to the study of
chromosomal arrangement in the nucleus will be considered.

It is well established that UV photons (254 nm) induce photolesions in DNA and in particular pyrimidine dimers. The damaged
bases can be eliminated from the DNA by an enzymatic process (DNA
excision-repair) or unscheduled DNA synthesis, (UDS) leading to
excision of the dimers and repair of the gap by the incorporation
of normal nucleotides. This process, which involves uptake of
tritiated thymidine during phases of the cell cycle other than the
normal S-phase, has been demonstrated by biochemical techniques
and also by cytological techniques using autoradiography. However,
in whole cell irradiation it is necessary to discriminate repair
synthesis from replicative DNA synthesis.

Micro-irradiation can be used to study the effects produced
by UV light in a restricted area of the nucleus, the unirradiated
part of the same nucleus being used as a control (Yatani, 1970;
Moreno, 1971). In cells not in S-phase, ^{3}H-thymidine incorporation was detected in the irradiated spot and by irradiation of
different parts of the cell nucleus it was demonstrated that the
amount of UDS was dependent only on the total number of incident
photons and not on the site of irradiation within the nucleus
(Moreno and Salet, 1974; Cremer et al, 1981). Furthermore, in
cells in S-phase, semi-conservative DNA synthesis was inhibited
at the site of irradiation (Moreno, 1971). The function of trans-
criptional DNA i.e., RNA synthesis, as detected by ^{3}H-uridine
uptake, was also impaired at the irradiated spot (Moreno et al,
1977).

UDS following micro-irradiation of nucleoplasm has been used
to investigate the relationship between chromosome arrangement in
metaphase and in interphase. Irradiation was performed either on
anaphase chromosomes (Sakharov et al, 1976) or on chromosomes
during interphase (Zorn et al, 1979). UDS was then used to
relocate the irradiated chromosomes either in the nuclei of the
daughter cells or in chromosomes of the following metaphase.
These experiments suggest that (i) decondensed interphase chromosomes may occupy rather compact territories and (ii) chromosomes
do not necessarily exhibit a close and permanent association with
their respective homologues.

4. MICRO-IRRADIATION OF MITOCHONDRIA

The first report of selective mitochondrial alteration was
performed using a ruby laser microbeam after supravital staining
of mitochondria with Janus Green B (Amy and Storb, 1965). After
this first report and other studies describing morphological and

cytochemical damage, cell viability and RNA synthesis following
mitochondrial alteration (see Moreno et al, 1969; Berns, 1974),
laser micro-irradiation of mitochondria was extensively developed
in cells without supravital staining. Much of this work has been
conducted in contracting mammalian cardiac cells in culture. The
contractile response was analysed after irradiation of different
cell organelles such as the mitochondria or the nucleus in an
attempt to determine the role of these organelles in the contractile function.

Mitochondria which are rich in pigments (flavins, cytochromes)
have a specific absorption in the visible range. A stimulation in
beating can be produced by irradiating one large or several small
mitochondria in isolated heart cells with a Q-switched laser
(530 nm) (Salet, 1972). An acceleration lasting for nearly 3
minutes was seen in 30% of the micro-irradiated cells before the
stimulated cell returned to its former rate.

In whole cell irradiation, it has also been shown that UV
light is able to stimulate the beating rate of heart cells but
since all components of the cell are able to absorb UV light, it
is difficult to determine the primary target. By positioning the
UV microspot on different parts of a single heart cell it has been
observed that no change of beating rate occurs when part of the
nucleus is irradiated. However, when the cytoplasm is irradiated
in regions that are rich in mitochondria, stimulation of the beating does occur. These experiments (Salet et al, 1976) demonstrated
that the site for triggering an acceleration in the beating rate
is primarily located at the mitochondrial level.

Several mechanisms of stimulation such as membrane damage,
mitochondrial Ca^{2+} release, or ATP photoproduction are suggested.
Further experiments to investigate the mechanism of action have
been carried out in the presence of either an inhibitor of electron
transport (KCN) or ATP (Salet et al, 1979). In both cases no
stimulation occurred. The results of all these experiments
indicate that changes in ATP levels are the most likely mechanism
of stimulation following laser visible irradiation. In the case of
mitochondrial irradiated in the UV range, there is no evidence that
the stimulation mechanism is the same. Nevertheless, UV micro-
irradiation does not trigger any acceleration in cells incubated
with adriamycin, an anti-cancer agent which partially inhibits the
respiration of isolated mitochondria (Lampidis et al, 1979).

5. CONCLUSIONS

Micro-irradiation has contributed to a better understanding
of several problems in cell physiology. A new application of this
technique to the study of organelle structure and function has been
reported recently by Siemens et al (1982). A laser microbeam

was used to stimulate fluorescence in 0.5 μm spots in single mitochondria of myocardial and endothelial cells in culture. Cells were treated with rhodamine 6G or 123 in order to render the mitochondria fluorescent and the fluorescence was recorded from different mitochondria and from multiple sites within the same mitochondrion. The results of these experiments suggest that (i) there are different functional classes of mitochondria in different cell types; (ii) there is a certain heterogeneity and compartmentalization within a single mitochondrion with suborganelle sections estimated to be between 0.25 and 1 μm in diameter. This point is supported by the fact that distinctly different fluorescence patterns can be obtained from different parts of the same mitochondrion.

This type of experiment, combining micro-irradiation and microspectrofluorimetry, can be expected in the future to contribute very precise physico-chemical data on the structure, organization and physiology of the living cell.

Amy R.L. and Storb R. (1965). Selective mitochnodrial damage by a ruby laser microbeam: an electron microscopic study. Science 150, 756-757

Berns M.W. and Salet C. (1972). Laser microbeams for partial cell irradiation. Int. Rev. Cytol. 33, 131-155

Berns M.W. (1974) Biological micro-irradiation. Classical and laser sources. Prentice-Hall Inc., Englewood Cliffs, N.J.

Berns M.W., Aist J., Edwards J., Strahs K., Girton J., McNeill P., Rattner J.B., Kitzes M., Hammer-Wilson M., Liaw L.H., Siemens A., Koonce M., Peterson S., Brenner S., Burt J., Walter R., Bryant P.J., Van Dyck D., Coulombe J., Cahill T. and Berns G.S. (1981) Laser microsurgery in cell and developmental biology. Science 213, 505-513

Cremer C., Cremer T. and Jabbur G. (1981) Laser-UV-microirradiation of Chinese hamster cells: the influence of the distribution of photolesions on unscheduled DNA synthesis. Photochem. Photobiol. 33, 925-928

Lampidis T.J., Moreno G., Salet C. and Vinzens F. (1979). Nuclear and mitochondrial effects of adrimycin in singly isolated pulsating myocardial cells. J. Mol. Cell Cardiol. 11, 415-422

Moreno G., Lutz M. and Bessis M. (1969). Partial cell irradiation by ultraviolet and visible light: conventional and laser sources. Int. Rev. Exp. Pathol. 7, 99-137

Moreno G. (1971). Effects of ultraviolet micro-irradiation on different parts of the cell. II Cytological observations and unscheduled DNA synthesis after partial nuclear irradiation. Exp. Cell Res. 65, 129-139

Moreno G. (1972). Techniques of partial cell irradiation with conventional and laser sources in Techn. Biochem. Biophys. Morph. Vol. 1, pp.47-66 (Glick G. and Rosenbaum M., Eds. John Wilers and Sons, Inc.

Moreno G. and Salet C. (1974). Unscheduled DNA synthesis after ultraviolet micro-irradiation of the cell nucleus. Radiat. Res. 58, 52-29

Moreno G., Nocentini S. and Salet C. (1977). Ultraviolet micro-irradiation on localized areas of the cell nucleus: effects on RNA synthesis. Photochem. Photobiol. 26, 125-127

Sakharov V.N., Voronkova L.N., Pyruzyan L.A. and Lomakina L.Y. (1976). Partial chromosomal set labelling induced by ultraviolet microbeam irradiation of mitotic cells. Nature 260, 784-785

Salet C. (1972). A study of beating frequency of a single myocardial cell. I. Q-switched laser micro-irradiation of mitochondria. Exp. Cell Res. 73, 360-366

Salet C., Moreno G. and Vinzens F. (1976). A study of beating frequency of a single myocardial cell. II Ultraviolet micro-irradiation of the nucleus and of the cytoplasm. Exp. Cell Res. 100, 365-373

Salet C., Moreno G. and Vinzens F. (1979). A study of beating frequency of a single myocardial cell. III Laser micro-irradiation of mitochondria in the presence of KCN or ATP. Exp. Cell Res. 120, 25-29

Siemens A., Walter R., Liaw L.H. and Berns M.W. (1982). Laser stimulated fluorescence of submicrometer regions within single mitochondria of rhodamine-treated myocardial cells in culture. Proc. Natl. Acad. USA 79, 466-470

Smith C.L. (1964). Microbeam and partial cell irradiation. Intern. Rev. Cytol. 16, 133-153

Tchakhotine S. (1912). Die mikroskopische Strahlenstichemethode, eine Zelloperationsmethode. Biol. Zentr. 32, 623

Yatani R. (1970). Induction of unscheduled DNA synthesis by half nucleus irradiation of HeLa cells with an UV microbeam in combination with hydroxyurea treatment. Mie. Med. J. 20, 93-104

Zirkle R.E. (1957). Partial cell irradiation. Adv. Biol. Med. Phys. 5, 103-146

Zorn C., Cremer C., Cremer T. and Zimmer J. (1979). Unscheduled DNA synthesis after partial UV irradiation of the cell nucleus. Exp. Cell Res. 124, 111-119

PHOTOBIOLOGY OF FUROCOUMARINS (PSORALENS) AND RECENT DEVELOPMENTS
IN THE PHOTOCHEMOTHERAPY (PUVA) OF PSORIASIS

Giuliana MORENO

INSERM U.201, Photophysiologie et Photothérapie, Muséum
National d'Histoire Naturelle, Laboratoire de Biophysique,
61 Rue Buffon, 75005 Paris, France.

1. INTRODUCTION

The principle of photomedicine is the use of non-ionizing
electromagnetic radiation in the ultraviolet (UV) and visible
range, in the absence or in the presence of exogenous photosensit-
izers, to treat diseases such as porphyrias, viral infections
(Herpes simplex), malignant lesions and skin diseases. For fuller
discussion of the different aspects of photomedicine, see the
treatise edited by Regan and Parrish, "The Science of Photomedicine"
(1982).

Photodermatology is one of the major aspects of the photomedicine. UV light associated with psoralen-containing plants has
been described more than 3000 years ago for treatment of the
repigmentation of vitiliginous skin, (Scott et al 1976). In 1974,
photochemotherapy with psoralens and UVA (315-400 nm) called PUVA
(psoralen + UVA) was systematically initiated for the treatment of
psoriasis (Parrish et al 1974). Several other skin diseases such
as Mycosis fungoides, eczema, urticaria, Alopecia areata ... have
also been found to respond to psoralen photochemotherapy.

The purpose of the present report is to summarise different
aspects of the biological action of psoralens in combination with
UVA on mammalian cells and to discuss these effects in relation to
PUVA therapy. The reports on photobiological effects of psoralens
and of PUVA are too numerous for them all to be mentioned in the
present review. Therefore, when possible references will be made
to general reviews on the subject.

2. PHOTOREACTIONS WITH DNA

Furocoumarins belong to a group of heterocyclic compounds of which the general formula is shown in Fig. 1. Furocoumarins have been either isolated from plants (parsley, parsnip, bergamot fruits, figs, ...) or synthesised in the laboratory. 8-methoxy-psoralen (8-MOP) and 4, 5', 8-trimethylpsoralen (TMP) have been most frequently used in the clinic for the treatment of skin disease.

In the dark, furocoumarin molecules intercalate between two base pairs of double stranded DNA (Fig. 2). Irradiation of the DNA-intercalated psoralen with wavelengths in the range of 320 - 400 nm produces many photoproducts, among which the most important ones consist of C4 cycloadducts between the 3, 4 or 4', 5' double bond of the furocoumarin and the 5, 6 double bond of pyrimidine bases (monoadducts).

A furocoumarin molecule in which one double bond has reacted with a pyrimidine base, may use its other double bond to react photochemically with a second pyrimidine located one nucleotide apart on the opposite DNA strand (alternating sequences of purine-pyrimidine/pyrimidine-purine) (Fig. 2). The result of this double reaction (bi-adducts) leads to DNA cross-links.

Some furocoumarins cannot form inter-strand cross-links with DNA. Angelicin (Fig. 1), a non-linear furocoumarin, forms only mono-adducts with DNA, for geometric reasons which prevent the formation of interstrand cross-links. Substituted compounds such as

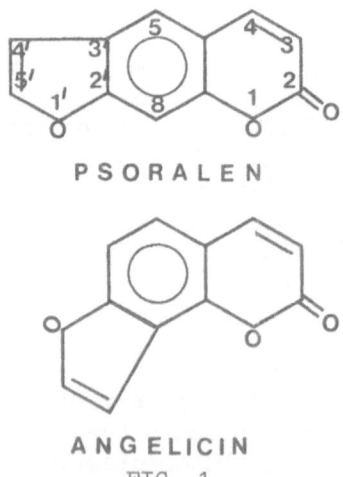

PSORALEN

ANGELICIN
FIG. 1

Fig. 1. Chemical structures of psoralen and angelicin.

149

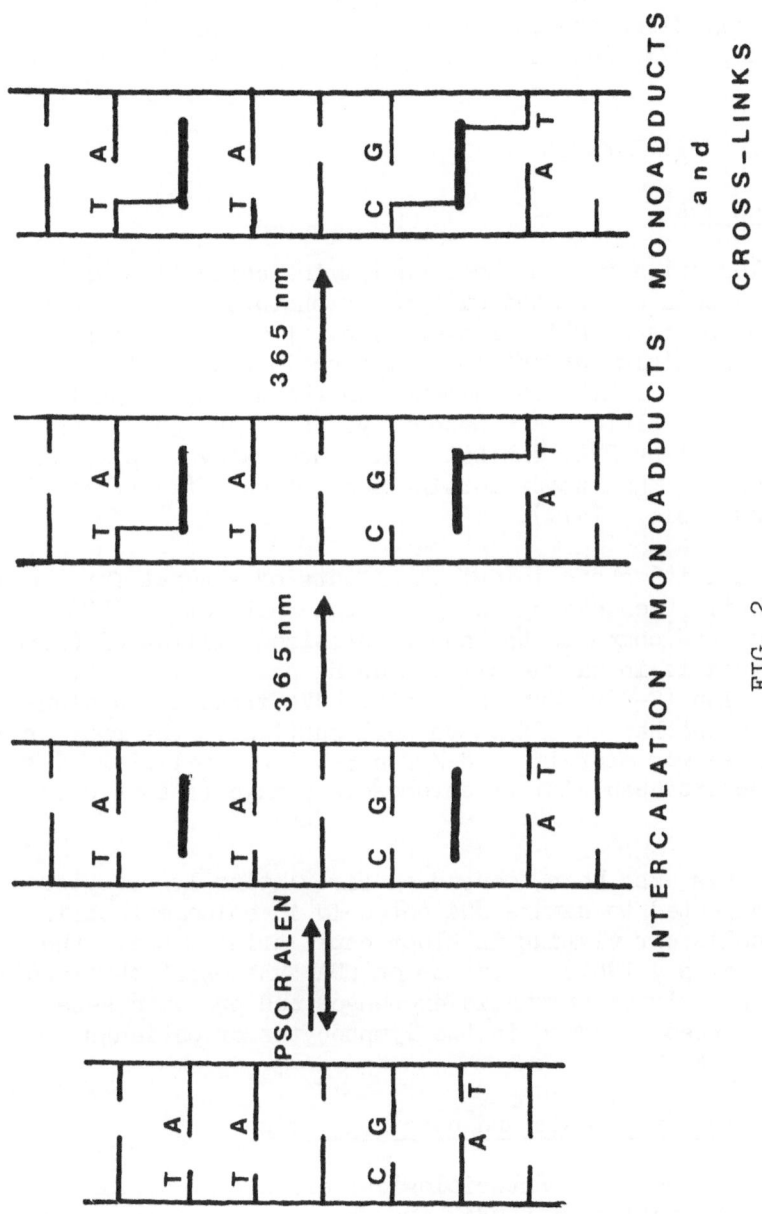

INTERCALATION   MONOADDUCTS   MONOADDUCTS
and
CROSS-LINKS

FIG. 2

Fig. 2: Schematic representation of the photochemical interaction between a bifunctional furocoumarin (psoralen) and DNA (modified from Hélène, 1980).

3-carbethoxypsoralen also form only mono-adducts because of the
presence of the carbethoxy group which prevents the second reaction.

The photophysics and photochemistry of these reactions have
been extensively discussed in several general reviews (Rodighiero
and Dall'Acqua, 1976; Son and Tapley, 1979; Parsons, 1980; Hélène,
1980).

3. PHOTOBIOLOGICAL EFFECTS

3.1 Effects on DNA

The biological and skin-photosensitizing activities of
psoralens have been correlated with their photoreactivity towards
the pyrimidine bases of DNA. Photobinding of furocoumarins to DNA
is largely responsible for the loss of functions of replication
and transcription. Inhibition depends on the degree of photo-
modification of template DNA. Generally, psoralens possessing the
ability to cross-link DNA, inhibit semi-conservative replication
to a higher extent than those forming mono-adducts (Scott et al,
1976; Song and Tapley, 1979).

The comparative photobiological effects of several furocoumarins have recently been evaluated by Ashwood-Smith et al (1981).
These authors have compared the photosensitized killing of Chinese
hamster ovary cells in tissue culture using psoralen, 8-MOP,
5-methoxypsoralen (5-MOP) and angelicin. UVA irradiation alone or
furocoumarin alone had no effect on cell survival. The rank order
of effectiveness was psoralen > 8-MOP > 5-MOP > angelicin. 5-MOP
and 8-MOP also produced similar chromosome damage (sister chromatid
exchanges).

Interest has also been focused on circulating lymphocytes as
PUVA may be expected to damage DNA not only in epidermal cells,
but also in cells circulating in blood capillaries beneath the
skin (Bridges et al, 1981). Effects on DNA synthesis, chromosome
gaps and breaks, sister-chromatid exchanges and presumed gene
mutations have been detected in the lymphocytes of patients
treated with PUVA.

3.2 DNA repair, mutagenesis and carcinogenesis

Basic studies on the photobiological effects of certain
psoralens in combination with UVA have shown a mutagenic activity
in prokaryotic and eukariotic cells (Scott et al, 1976; Song and
Tapley, 1979; Bridges et al, 1981). Mutagenic activity has been
demonstrated in human skin fibroblasts treated with 8-MOP and UVA
(Burger and Simons, 1979). The frequency of mutation increased
linearly with UVA dose for 8-MOP concentrations equivalent to
those found in the skin during PUVA treatment. The expected

number of induced mutants in the human skin per session of photo-chemotherapy was estimated and from a comparison between this frequency and the frequency expected from spontaneous mutations, a higher frequency of mutations during PUVA therapy may be expected.

This mutagenic activity has been correlated with the capacity for elimination of the mono- and bi-adducts and the repair of DNA. The molecular events by which the mono- and bi-adducts are excised and by which the DNA is then repaired have not been completely elucidated. Recently, Hanawalt et al (1981 a and b) have suggested that this process shares at least one step in common with the repair of pyrimidine dimers induced by UV light at 254 nm. This fact has been supported by experiments using human mutant cell lines such as Xeroderma pigmentosum (XP) which have a different capacity for excision-repair of DNA. Fibroblasts from complement-ation group A of XP (completely defective in the excision repair of pyrimidine dimers) have been demonstrated to be totally deficient in both repair replication and cross-link removal after treatment with 8-MOP and UVA. In contrast, in normal human diploid fibro-blasts, repair replication patches for angelicin or 8-MOP photo-adducts were of equivalent size to those observed for pyrimidine dimers. Cross-links were partially removed within 24 hours.

The repair of mono-adducts seems therefore to involve enzymatic processes similar to those involved in repair of pyrim-idine dimers. In this process repair replication consecutive to the excision of the damaged base (mono-adduct) by an endonuclease or an enzyme complex (glycosylase and endonuclease) replaces the excised region with normal nucleotides using the undamaged complementary DNA strand as template. In the case of the repair of inter-strand cross-links, the damage is present also in the opposite DNA strand at sites only one nucleotide apart. Both excision-repair and recombinational strand exchange (like in bacteria) have been proposed for this type of repair, but the enzymatic mechanisms of this process have not yet been elucidated. However, it is admitted that during the various steps of the repair process, errors may be introduced in the base sequence of the repaired patches of DNA resulting in mutations (Hanawalt et al, 1979).

At the present time, it is difficult to evaluate the role of mono-adducts and bi-adducts in mutagenesis. According to Hanawalt et al (1981 a and b) the interstrand cross-links are less easily repairable lesions, resulting in more effective blockage of DNA functions and leading to a higher lethality, so mono-adducts, like pyrimidine dimers, are probably the more mutagenic lesions. On the contrary, Dubertret et al (1979) suggest that mono-addition lesions are more easily repaired than cross-links and are consequently less mutagenic.

This mutagenic activity of psoralens plus UVA has raised the
suspicion that PUVA therapy may be carcinogenic in human skin.
Induction of skin tumours has been demonstrated in mice treated
with PUVA using 8-MOP (Bridges et al, 1981). Until recently, it
was considered that only bifunctional furocoumarins were carcinogenic, but in a recent report Pathak et al (1982) demonstrated
that monofunctional furocoumarins (derivatives of angelicin) were
as carcinogenic as bifunctional molecules.

Another possibility is that PUVA, although potentially mutagenic and carcinogenic by itself, is acting as a promoter through
an effect on immunocompetent cells rather than on the epidermal
cells. Many studies on lymphocytes, on delayed hypersensitivity
and rejection of skin and tumours are in favour of this hypothesis
(Bridges et al, 1981; Grekin and Epstein, 1981).

In a study on the incidence of tumours in human psoriatic
skin during PUVA therapy, Stern et al (1979) reported a higher
incidence of skin cancers, especially in patients who had been
exposed to ionizing radiation or who had a previous history of
cutaneous carcinoma. However, because of the relatively short
period of intense use of PUVA therapy, this question of long-term
risks is still controversial (Grekin and Epstein, 1981).

4. PSORALENS IN PHOTOCHEMOTHERAPY OF PSORIASIS

In spite of the many results on the photobiological effects
of furocoumarins associated with UV light, identification of the
intracellular molecular targets responsible for the therapeutic
effect in PUVA therapy is an open question.

Psoriasis is a common disease of unknown etiology affecting
1 to 3% of the world population. It is a disorder which is
characterized by disturbed epidermal maturation (keratinization)
associated with hyperproliferation of cells in the basal compartment leading to hyperplasia of the epidermis.

It is admitted that the therapeutic effect of psoralen plus
UVA in psoriatic skin is due to the photochemical reactions of
these molecules with epidermal DNA leading to inhibition of DNA
synthesis and cell division. However, there is no direct proof
that the effects on DNA are the only molecular events involved in
the therapeutic effect of psoralens plus UVA in psoriatic skin.
The presence of bi-adducts between DNA and furocoumarin molecules
(8-MOP and TMP) has not been demonstrated in guinea pig skin
treated with clinical doses used in PUVA therapy (Cech et al, 1979)
or in human psoriatic skin during PUVA therapy (Lerche et al, 1970).
On the other hand, DNA repair has not been detected in psoriatic
skin during PUVA therapy using irradiation sources without UV B
(wavelengths below 315 nm) (Bishop, 1979; Bioulac et al, 1980 ;

Hönigsmann et al, 1981).

It is thus possible to suggest other cellular targets such as proteins and lipoproteins for the therapeutic effect. Studies by microspectrofluorimetry on the localisation of different psoralens in living cells in culture indicated that the psoralens penetrate all cell compartments (Moreno et al, 1982). Autoradiographic analysis of psoriatic skin after treatment with tritiated 8-MOP and irradiation showed that the molecule was localized in the nucleus and in the cytoplasm of all the cells in the different layers of the epidermis and also in proteins like keratin and in lipoproteins of the fatty globules (Bertaux et al 1981). Therefore psoralens, because of their ubiquitous localization, may photosensitize reactions in cell compartments other than the nucleic acids by production of singlet oxygen generated by furocoumarin excitation. For example, it has been recently suggested that singlet oxygen produced by energy transfer from excited psoralens can induce oxidative reactions in human epidermis (Dubertret et al, 1982) and in cell membranes (Salet et al 1982).

In conclusion, it is possible that more than one mechanism is involved in the remission of psoriasis. Even if some adverse effects such as long-term hazards concerning mainly cutaneous cancers and alterations of the immune system have been reported, one cannot negate the usefulness of the PUVA, but the potential risks should be carefully assessed before beginning such therapy.

Ashwood-Smith M.J., Natarajan A.T. and Poulton G.A. (1981), Comparative photobiology of psoralens in "Psoralens in Cosmetics and Dermatology", Proc. Int. Psoralens SIR, p.117-131 (J. Cahn, P. Forlot, C. Grupper, A. Meybeck and F. Urbach, Eds). Pergamon Press France.
Bertaux B., Moreno G. and Dubertret L. (1981) Autoradiographic localization of 8-methoxypsoralen in psoriasis skin in vitro. Acta Dermatovener, 61, 481-485
Bishop S. (1979), DNA repair synthesis in human skin exposed to ultraviolet radiation used in PUVA (psoralen and UVA) therapy for psoriasis. Brit. J. Dermatol. 101, 399-405
Bioulac P., Denechaud M., Dubuisson L., Doutre M.S., Ducasson D. and Beylot C. (1980). Unscheduled DNA synthesis in psoriatic skin after ultraviolet irradiation and the effects of a combined treatment with 8-methoxypsoralen and longwave ultraviolet radiation: a clinical study. Brit. J. Dermatol. 102, 285-295
Bridges B.A., Greaves M., Polani P.E., and Wald N. (1981). Do treatments available for psoriasis patients carry a genetic or carcinogenic risk? Mutat. Res. 86, 279-304

154

Burger P.M. and Simons J.W.I.M. (1979). Mutagenicity of
8-methoxypsoralen and long-wave ultraviolet irradiation in
diploid human skin fibroblasts. An improved risk estimate in
photochemotherapy. Mutat. Res. 63, 371-380

Cech T., Pathak M.A. and Biswas R.K. (1970). An electron microscopic
study of the photochemical cross-linking of DNA in guinea pig
epidermis by psoralen derivatives. Biochim. Biophys. Acta 562,
342-360

Dubertret L., Averbeck D., Zajdela F., Bisagni E., Moustacchi E.,
Touraine R. and Latarjet R. (1979). Photochemotherapy (PUVA)
of psoriasis using 3-carbethoxypsoralen, a non-carcinogenic
compound in mice. Brit. J. Dermatol. 101, 379-389.

Dubertret L., Santus R., Bazin M. and Sa e Melo T. (1982). Photochemistry in human epidermis. A quantitative approach. Photochem
Photobiol. 35, 103-107

Grekin D.A. and Epstein J.H. (1981) Psoralens, UVA (PUVA) and
Photocarcinogenesis. Photochem. Photobiol. 33, 957-960

Hanawalt P.C., Cooper P.K., Ganesan A.K. and Smith C.A. (1979).
DNA repair in bacteria and mammalian cells. Ann. Rev. Biochem.
48, 783-836

Hanawalt P.C., Liu S.C. and Parsons C.S. (1981a). DNA repair
responses in human skin cells. J. Invest. Dermatol. 77, 86-90

Hanawalt P.C., Kaye J., Smith C.A. and Zolan M. (1981b). Cellular
responses to psoralen adducts in DNA. "Psoralens in Cosmetics
and Dermatology", Proc. Int. Psoralens SIR p. 133-142. (J. Cahn,
P. Forlot, C. Grupper, A. Meybeck and F. Urbach, Eds) Pergamon
Press France

Hélène, C. (1980). Furocoumarines et Photochimiotherapie. Pathol.
Biol. 28, 281-285

Hönigsmann H., Jaenicke K.F., Brenner W., Rauschmeier W. and
Parrish J.A. (1981). Unscheduled DNA synthesis in normal human
skin after single and combined doses of UV-A, UV-B and UV-A with
methoxsalen (PUVA). Brit. J. Dermatol. 105, 491-501

Lerche A., Sondergaard J., Wadskov S., Leick V. and Bohr V. (1979)
DNA interstrand cross links visualized by electron microscopy in
PUVA-treated psoriasis. Acta Dermatovener. 59, 15-20

Moreno G., Salet C., Kohen C. and Kohen E. (1982). Penetration
and localization of furocoumarins in single living cells studied
by microspectrofluorimetry. Biochim. Biophys. Acta 721, 109-111

Parrish J.A., Fitzpatrick T.B., Tanenbaum L. and Pathak M.A. (1974)
Photochemotherapy of psoriasis with oral methoxsalen and long-
wave ultraviolet light. N. Engl. J. Med. 291, 1207-1211

Parsons B.J. (1980). Psoralen photochemistry. Photochem. Photobiol. 32, 813-821

Pathak M.A., West J.D. and Mullen M.P. (1982). Photosensitizing
and carcinogenic properties of mono- and bi-functional psoralens.
Proc. 10th Ann. Meeting Ann. Soc. Photobiol. Vancouver 1982.
abstract MAM-D9, p. 67

Regan J.D. and Parrish J.A., The Science of Photomedicine, New York, Plenum Press, 1982, in press

Rodighiero G. and Dall'Acqua F. (1976). Biochemical and medical aspects of psoralens. Photochem. Photobiol. 24, 647-653

Salet C., Moreno G. and Vinzens F. (1982). Photodynamic effects induced by furocoumarins on a membrane system. Comparison with hematoporphyrin. Photochem. Photobiol. 36, 291-296

Scott B.R., Pathak M.A. and Mohn G.R. (1976). Molecular and genetic basis of furocoumarin reactions. Mutat. Res. 39, 29-74

Song P.S. and Tapley K.J. (1979) Photochemistry and photobiology of psoralens. Photochem. Photobiol. 29, 1177-1197

Stern R.S., Thibodeau L.A., Kleinerman R.A., Parrish J.A. and Fitzpatrick T.B. (1979). Risk of cutaneous carcinoma in patients treated with oral methoxsalen photochemotherapy for psoriasis. N. Engl. J. Med. 300, 809-813

EFFECTS OF IONIZING RADIATION ON THE STRUCTURE AND PHYSICAL
PROPERTIES OF THE SKIN

Harcharan Singh RANU

Department of Biomedical Engineering, Louisiana Tech.
University, Ruston, Louisiana 71272, U.S.A.

1. INTRODUCTION

Skin is a viscoelastic material, whose properties modify with
age, some diseases and certain therapeutic treatments.

In the radiobiological treatment of benign and malignant
disease of the skin, considerable variations in dose, number of
fractions and total treatment time occur in clinical practice.
Thus, although benign skin diseases are usually given small doses
once weekly or fortnightly and malignant tumours are treated daily
or on alternate days, it is not uncommon for the time-intervals
between successive treatments to be one, two, four, seven or more
days. The cure rate is over 90% for malignant skin diseases and
therefore the residual effects of radiotherapy are of considerable
clinical concern. After cure, therefore, the most important
consideration is the effect of radiation on the normal tissue and
the success with which radiation injury is repaired.

This paper reports a series of experiments which attempt to
assess the effect of fractionation on the structure and physical
properties of skin in the hope that it will contribute towards an
understanding of the role of fractionation in the radiotherapy of
skin diseases. These experiments were designed with a view to
monitoring long-term skin response to low energy X-rays commonly
used in radiotherapy of the skin.

2. MATERIAL AND METHODS

2.1 X-Irradiation

Nine groups of 2 month old female Sprague-Dawley rats with twelve animals in each group, each weighing about 265 grams, were irradiated with 50 kV X-rays at approximately 3800 R per minute. The calculated dose at the hair follicle base was 0.997 rad per roentgen of exposure at the skin surface.

The rats were anaethetised with ether and two areas 5 cm in diameter were irradiated on the left flank. On the first day every rat in each group received a single exposure of 1500 R on the anterior flank and 1000 R, the first of two exposures, on the posterior flank. Previous work by Ranu (1975) had indicated there was no detectable difference in the radiosensitivity of the anterior and posterior flank of the rat. Second exposures of 1000, 2000 and 3000 R were given to the posterior flank after intervals of 1, 4 and 7 days.

Since the investigation was designed to study long term effects of radiation on the skin, observations were made several months after exposure. Previous work (Ranu et al 1975) had indic- ated that, approximately three months after completion of the irradiation, tissue response has reached a stable level and little, if any, further recovery occurs. Therefore, half of the rats in each group were sacrificed at 50 days and half at 120 days after irradiation. Sagittal strips 3.2 cm long and 1 cm wide were cut from the skin of each animal so that the centre of the section was concentric with the 5 cm diameter circle of irradiated skin. A similar section was cut from the non-irradiated portion of the same flank and all specimens were placed in normal saline before subseq- uent mechanical testing. Specimens were tested within an hour of excision.

2.2 Testing

A schematic arrangement of the apparatus used for measuring the tensile mechanical properties is shown in Fig. 1 (Mayes hydrau- lic testing machine). The specimens were mounted in specially designed jaws (Ranu 1981a), in saline to ensure constant hydration, at $\pm 0.5^{\circ}$C. The skin thickness was measured with a Mercer dial gauge and the inaccuracy due to specimen compression was estimated to $\pm 5\%$. The increase of tensile load was set to 0.25 mm/s corres- ponding to the rate at which skin recovers when stretched to blanching in vivo (Ranu 1975). The skin was stretched to rupture and the load-extension characteristics were recorded on an X-Y plotter. For each specimen the following physical parameters were measured (1) skin thickness, (2) load at rupture, (3) extension at

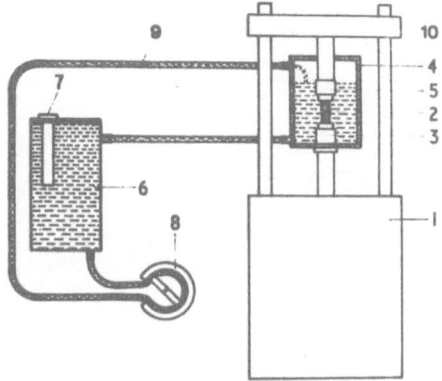

Fig. 1   Schematic arrangement of the apparatus used for measuring
tensile mechanical properties. (1) Mayes hydraulic servo-
controlled testing machine (2) Skin specimen (3) Movable
clamp (4) Perspex immersion tank (5) Fixed clamp (6) Fluid
reservoir (7) Heater (8) Peristaltic pump (9) Silicon
tubing (10) Upper crosshead

rupture (4) mechanical stress at rupture (this was calculated from
the load at rupture divided by the cross-sectional area of the
specimen), and (5) stiffness, S. The first part of the load-
extension curve (Fig. 2) which is primarily associated with alignment of the collagen fibres (Ranu 1979a) was referred to as the
skin elasticity by Ranu et al, 1975. The final part of the load-
extension curve is primarily associated with stretching of the
aligned collagen fibres (Ranu 1979a and b). This property will be
referred to as the stiffness of the skin and is defined by Ranu
(1981 a and b) as

$$S = \frac{\Delta F}{\Delta E} \; X \; \frac{L}{A} \quad \text{where} \quad \frac{\Delta F}{\Delta E}$$

is the slope of the last quarter of the load-extension curve, L
is the original gauge length and A is the cross-sectional area of
the specimen.

2.3 Structural analysis of skin

The scanning electron-microscope (SEM) combines high resolution with depth of focus to give a three-dimensional image of the
surface structure of a relatively large and therefore a significant
piece of tissue (Ranu 1979a). This instrument does not require
ultra-thin sections of the specimen, as the images evolve only from
the emission of electrons from its surface. When applied to tissues

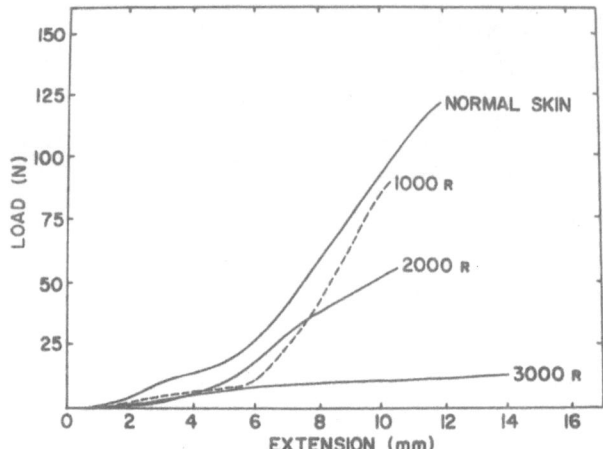

Fig. 2 Typical load against extension curves for skin

such as skin, its main contribution has been an understanding of fibrous structure and the elucidation of the response of tissue to mechanical stress and disease.

2.4 Specimen preparation and viewing

Samples were fixed in buffered glutaraldehyde, freeze sectioned and dessicated through a series of ethanol water mixtures of increasing concentration and finally for 24 hr in a dessicator. The samples were then attached to SEM brass stubs with Durofix, allowed to dry and painted with conducting silver dag (Dag 915, Acheson Colloids Co., Plymouth, England) in order to allow the charge of the electron beam to flow to ground. The specimens were then carbon coated (60-80nm) in vacuo and viewed and photographed in the Cambridge Scientific Instruments, Steroscan, Mk I, Scanning Electron Microscope.

3. RESULTS

3.1 Physical Parameters

For each animal three specimens of skin were tested to rupture and the parameters mentioned earlier were obtained. The results for the parameters measured were normalized against the results of non-irradiated (normal) skin. Thus, for example, the average value of extension at rupture (50 days after the first exposure) for skin irradiated by two fractions of 1000 and 3000 R with a one day interval was 10.6 mm (See Table 1). This is divided by the average value of extension at rupture for non-irradiated skin in that particular group, 16.3 mm, to give the normalised extension at rupture 0.65. The graphical representation of these results is shown in Fig. 3, where the normalised extension at rupture is plotted as a percentage (65%) against fractionated exposure. Only the second exposure of 1000, 2000 and 3000 R is recorded on the absissa since the initial exposure of 1000 R was constant in all these fractions. Thus the zero represents a single exposure of 1000 R. This zero point was obtained by the same normalisation process, using data from the single exposure experiments of Ranu (1975). Since the zero exposure point was obtained from a different group of animals, a broken line joins this point to the other points.

To investigate the role of recovery in fractionated treatments of the skin, a single exposure of 1500 R was also given to each animal and the biological effect recorded. This is shown as a straight line on the graph. The size of the second fraction, following an initial exposure of 1000 R, necessary to produce the same effect as the single exposure could be measured from the graphs. This is referred to as the "matching exposure". Where this line is intercepted by the graphs, the biological effect is the same for the fractionated exposure as for the single exposure and therefore the matching exposure for a particular time interval between fractions can be read off from the intercepts on the appropriate graph. Figure 3 shows matching exposures of 1725 R for a one day interval, 1625 R for four days, and 1525 R for seven days.

The matching exposure was also determined by considering the effect of a single exposure of 1500 R for the three different time intervals (1, 4 or 7 days) between fractions separately. Each number was normalised as described above and the average value for each group, as portrayed in Fig. 3 corresponds to the three arrows against the ordinate axis labelled to represent the time intervals of 1, 4 and 7 days. The matching exposure was determined by noting where the one day curve intercepts the projection of the "1 day" arrow and similarly for 4 and 7 days. With this technique o of matching,exposures of 1600, 1625 and 1750 R were obtained for intervals of 1, 4 and 7 days between fractions. The magnitudes of the matching exposures obtained by the two methods do not differ greatly.

Figs. 3 - 6 (see following pages) Variation of normalised parameters with exposure 50 days after an initial exposure of 1000R. Time between the two fractions for the three groups: O-1 day, x-4 days, ●-7 days. The horizontal line represents the normalised extension at rupture at 50 days following a single exposure of 1500 R averaged for all the groups, while the arrows represent the average for the three different groups.

162

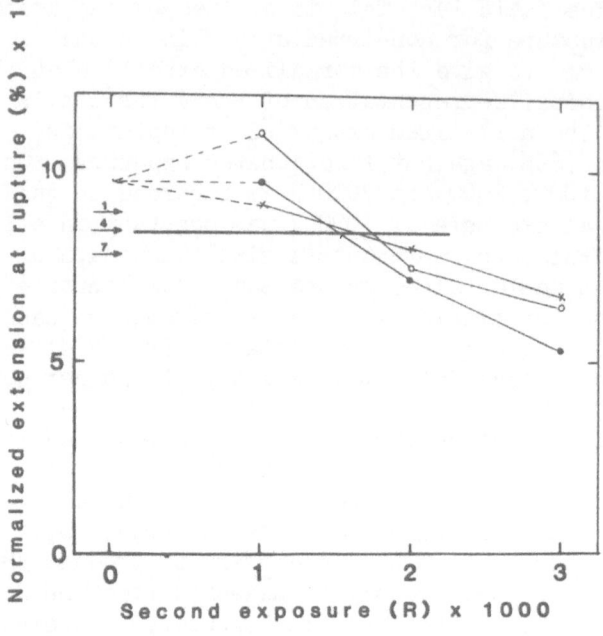

Figure 3

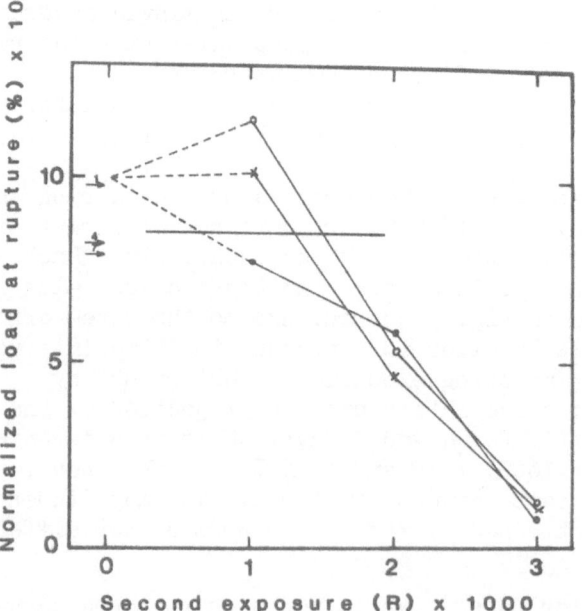

Figure 4

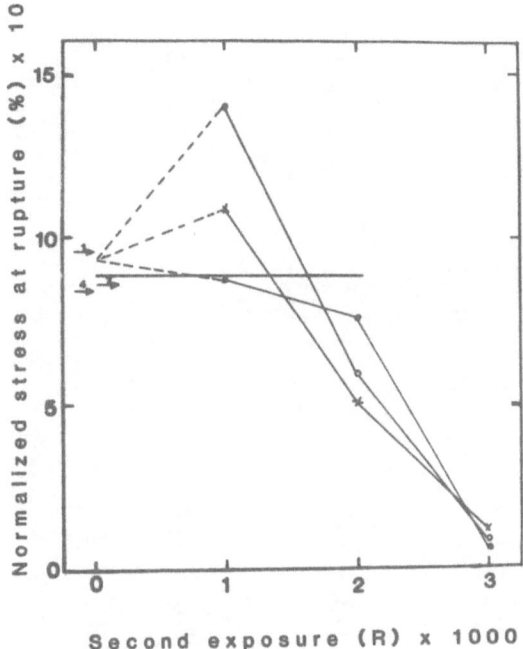

Figure 5

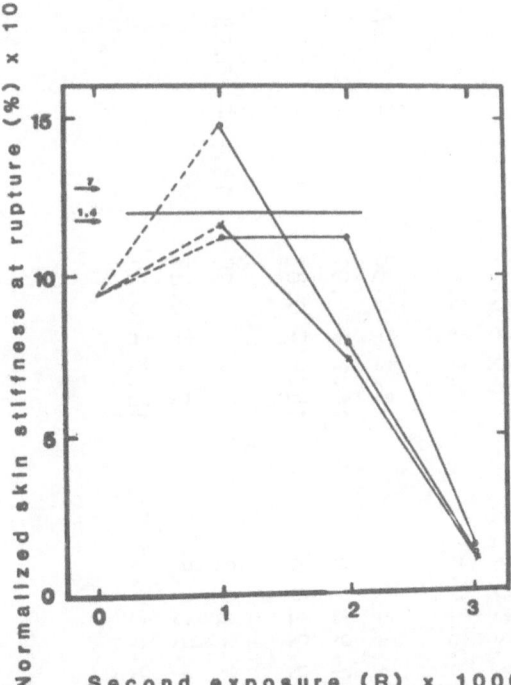

Figure 6

164

TABLE I

EXTENSION AT RUPTURE, MEASURED IN MM, 50 DAYS AFTER THE INITIAL RADIATION

NONIRRADIATED	SINGLE EXPOSURE 1500 R	FRACTIONATED EXPOSURES 1000 R + 1000 R	FRACTIONATED EXPOSURES 1000 R + 2000 R	FRACTIONATED EXPOSURES 1000 R + 3000 R	DAYS BETWEEN FRACTIONS
15.0 ± 0.52	15.3 ± 0.57	16.9 ± 0.34	13.0 ± 1.3		1 DAY
17.2 ± 0.9	14.0 ± 0.31				
16.3 ± 0.67	13.5 ± 0.55			10.6 ± 0.73	
19.1 ± 0.31	17.2 ± 0.38	17.4 ± 0.47			4 DAYS
17.9 ± 0.61	16.1 ± 1.01		14.3 ± 1.13		
19.9 ± 0.64	14.5 ± 0.83			13.4 ± 1.34	
19.2 ± 0.35	14.7 ± 0.58	18.5 ± 0.98			7 DAYS
21.5 ± 0.59	17.2 ± 0.47		15.6 ± 1.7		
19.6 ± 0.68	14.9 ± 0.34			10.6 ± 0.84	

TABLE II

MATCHING EXPOSURES AT 50 DAYS AFTER THE INITIAL RADIATION

Parameters Analysed	Intercept with single Exposure of 1500 R	Matching Exposures (R) for Different Time Intervals Between Fractions		
		1 Day	4 Days	7 Days
Extension at rupture	Ave of all groups	1725	1625	1525
	Ave of different group	1600	1625	1750
Load at rupture	Ave of all groups	1500	1275	1450
	Ave of different group	1300	1350	1575
Skin thickness	Ave of all groups	1700	1225	2200
	Ave of different group	1700	1450	2100
Stress at rupture	Ave for all groups	1625	1350	750
	Ave of different group	1550	1425	1050
Skin stiffness	Ave for all groups	1400	1275	
	Ave for different group	1425	1325	1825
Matching Exposure Averaged for all the Parameters		1552	1393	1612

The extension at rupture, load at rupture, stress at rupture and skin stiffness are plotted against second exposure in Figs. 3, 4, 5, and 6 respectively. The change in skin thickness is shown in Fig. 7. The six matching exposures obtained from each of the figures are shown in Table II. In a few cases where the graph did not intercept the appropriate horizontal line, an artificial intercept was made by extrapolating the experimental measurements of 2000 R and 3000 R onto the line. Similar experiments were carried out 120 days after irradiation and the average matching exposures for all parameters for each time interval between fractions are plotted against the time interval in Fig. 8.

3.2 SEM Results

All sections were taken through the mid-dermis parallel to the epidermis. Figure 9 shows that the collagen fibres of the normal dermis consist of multi-directional networks of wavy interlacing fibres. These fibres are grouped in large and small bundles which are linked by networks of fine, threadlike fibrils, crossing the fibre bundles.

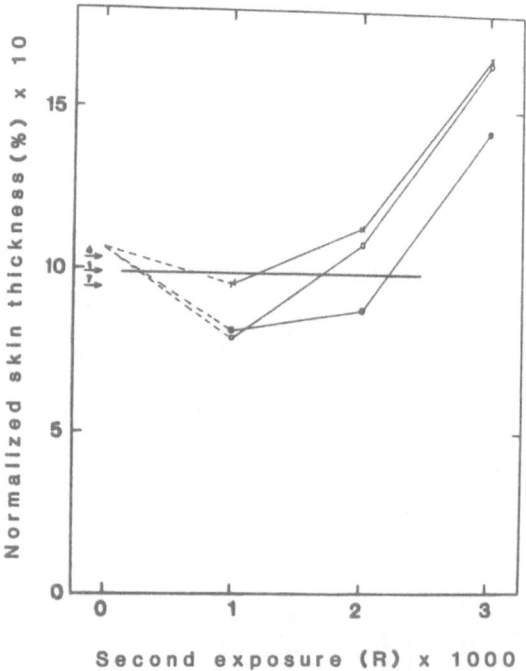

Fig. 7 Normalised skin thickness against exposure using the same symbols as in Figs. 3 - 6.

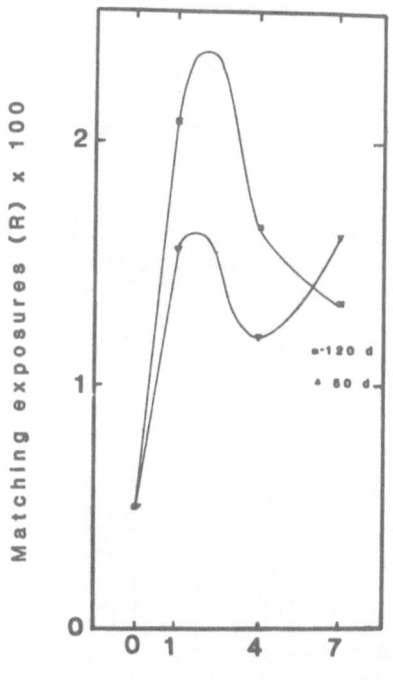

Fig. 8   Variation of matching exposure with the time interval
         between fractions.

These networks of fibrils can be seen to be interwoven and appear
to be anchored between the components of the large fibre bundles.
A coiled configuration is exhibited by some of these connecting
fibrils.   The diameters of collagen and elastic fibres in skin
have been found to vary from 4 - 7 μm and 0.25 - 0.5 μm respectively.
Elastic fibres having a diameter greater than 2 μm have an
amorphous core with a corrugated and undulating surface.

In teased out specimens, the fibres are very long, in
comparison with their diameter, and no free ends can be seen (Ranu
1979 a).   They are thicker at the bottom surface of the dermis
than at the top and at higher magnifications the fibres appear to
be smooth.

Figure 10 is a micrograph of normal rat dermis taken in the
derivative mode using a derivative processor (Scanning Electron
Microscope Derivative Processor, Type SDP, Science Data Technology

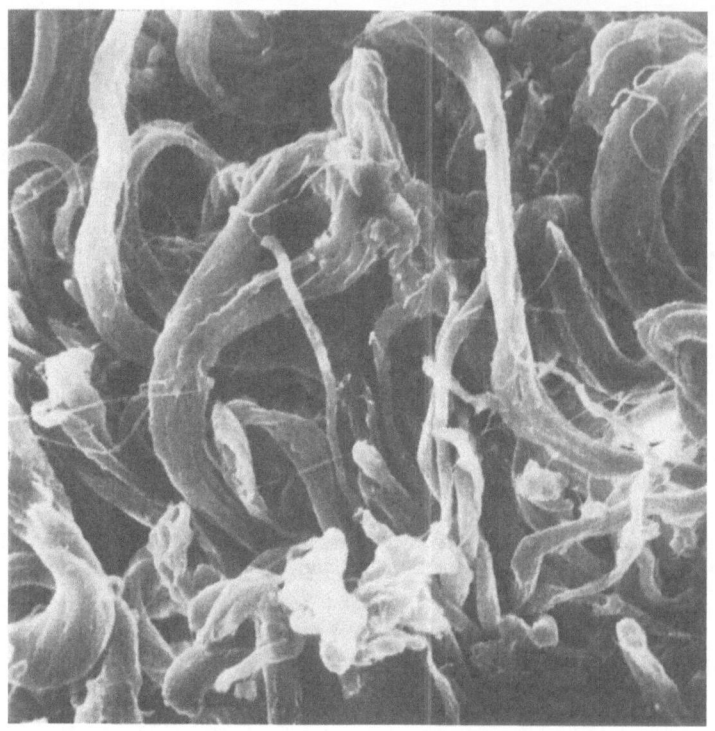

Fig. 9   Section of normal dermis, taken from 6-month old rat skin.
Magnification X 1000.

Ltd.) This technique enhances contrast in the bright areas, but
the three dimensional effect observed in the normal mode micrograph
is lost.

One hundred and twenty days after exposure to radiation, low
power SEM pictures show that the dense mesh work of collagen has
cracks running in it and these cracks are interlinked with one
another. When this structure is compared with normal dermis, it
can be seen that the multidirectional networks of fibres have given
way to a matted configuration. This configuration suggests that
there is a loss in the interstitial fluid.

Figure 11 shows a higher power micrograph 120 days after an
exposure of 1500 R. This suggests that structural changes within
the collagen fibres of the dermis have taken place, i.e. the collagen fibres appear to be thick and irregular.

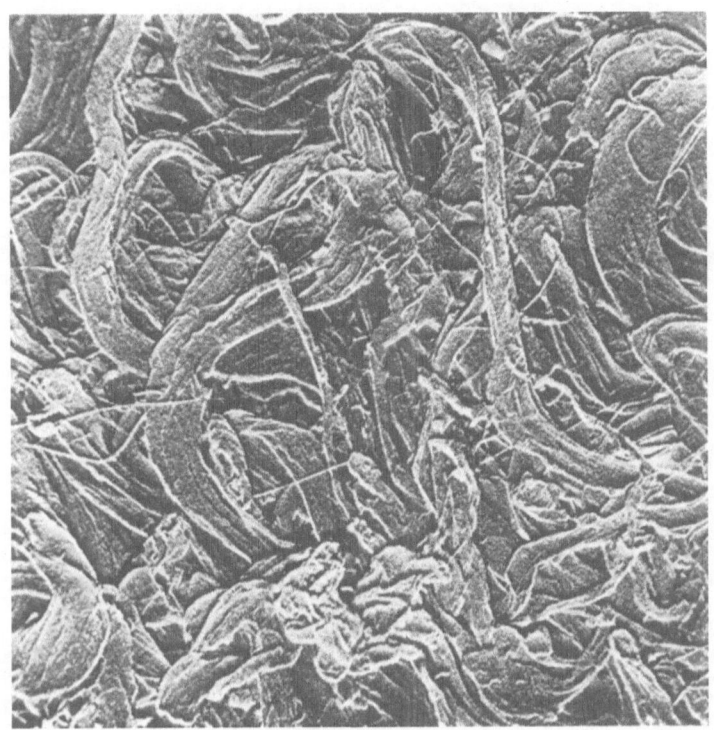

Fig. 10 Derivative mode of normal dermis. Magnification X 1000.

Figure 12 is the micrograph of rat dermis, 120 days after receiving exposures of 1000 + 3000 R, the time interval between the first and second exposure being 4 days. The irradiated collagen has now changed so much that it has lost its original well-defined fibrous structural properties.

4. DISCUSSION AND CONCLUSIONS

There is evidence of a shoulder in the exposure response curve for many of the parameters studied. After an exposure of 1000 R followed by a second exposure of 3000 R, for the load at rupture, the stress at rupture and to a lesser extent the extension at rupture, the value of the parameter is almost the same whether the time interval between exposures is 1, 4 or 7 days. The value for every parameter except for the extension at rupture was about 10% of the value for non-irradiated skin.

There is no significant diminution in skin thickness and after a second exposure of 3000 R there is a marked increase in thickness.

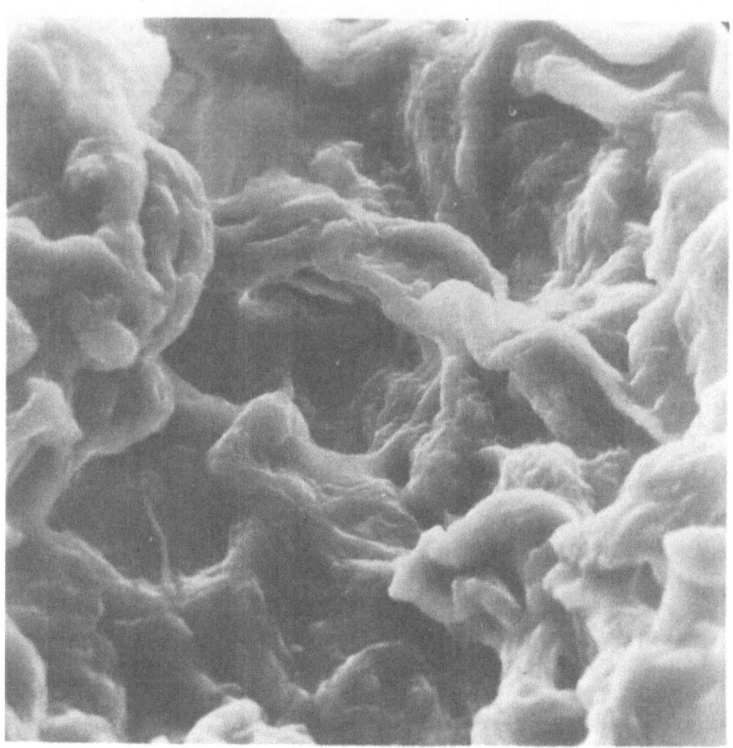

Fig. 11   Section of dermis 120 days after an exposure of 1500 R.
Magnification X 1120

This, together with scanning electron micrographs, suggests that
the collagen content does not decrease following exposure to irradiation. Because of the convergence of the results after a second
exposure of 3000 R, it is deduced that the skin is in essentially
the same condition whether the time interval between exposure was
1, 4 or 7 days. The condition of the skin after this treatment
was essentially a more or less solid mat of collagen somewhat
thicker than the non-irradiated skin (Fig. 8). The structure of
this matted collagen or the nature of the collagen, or possibly
both, is such that the skin is much weaker and more liable to
break. This may be due to cracks running through the structure.
Further support for this view that the skin after higher exposures
of irradiation consists of a fibrous mat of poor quality friable
collagen is provided by inspection of the load-extension curves.
As indicated by Ranu et al (1975) and Ranu (1979a), at low exposures
there are two components of these graphs, one of which has been
associated with the alignment of the collagen fibres (elasticity)
and the other with the actual stretching of the fibres themselves

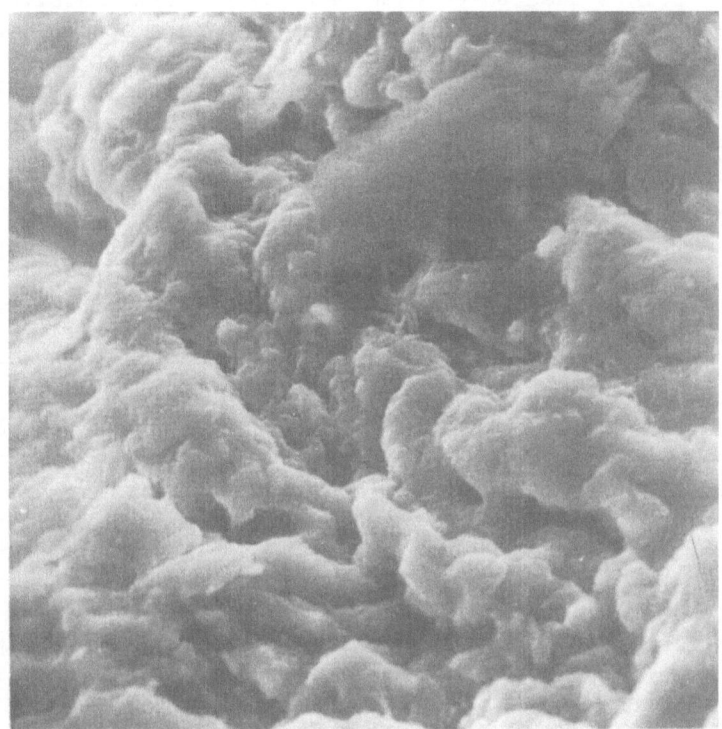

Fig. 12 Section of dermis 120 days after an exposure of (1000 +
         3000 R). Time interval between the first and second
         exposure being 4 days. Magnification X 1170.

(stiffness). This type of extension curve is apparent for the
lowest exposure (1000 R plus 1000 R second exposure). For the
highest exposure (3000 R second exposure) there is no evidence of
two components. Indeed, the extension curve runs along the axis,
almost but not quite parallel to it. Such an extension curve
would be compatible with a matted structure of low strength. This
suggests that once this "matted" condition is induced, the effect
of irradiation on the physical properties of skin is the same what-
ever the fractionation pattern. The influence of the time interval
between fractions manifests itself largely up to a second exposure
of 2000 R where the skin retains an open mesh structure.

     The properties of skin which were measured were chosen on the
basis that they represented the spectrum of characteristics
indicating its mechanical properties. Matching exposures have been
averaged over all these parameters so as not to reflect excessive
dependence on one particular parameter. For instance, at zero days

the matching exposure to be added to 1000 R to give the same mechanical effect as 1500 R is obviously 500 R as shown in Fig. 7.

Three points need to be compared (a) the time interval between
fractions at which the maximum matching exposure occurs can only be
between 0 and 4 days (Denekamp et al 1969, Burlin et al 1973).
Therefore, drawing a curve through experimental points in Fig. 7
suggests the maximum of the matching exposure occurs at about 2
days. (b) The value of the maximum of the matching exposure obtained
in these experiments was 1552 R at 50 days and 2083 R at 120 days.
Similar matching doses have also been obtained by Denekamp et al
(1969). Also, the second fraction following an initial exposure of
1000 R can exceed the single exposure required to produce the same
biological effect. (c) The results also exhibit cyclic variation of
skin sensitivity with time one to two days after the first irradiation. This could be due to the initial exposure which might have
sterilised a high proportion of cells in the more sensitive phases.
Those remaining would be in the more resistant phases and hence a
partially synchronised population would be produced which might
persist through several cell cycles. Another possible explanation
of the cyclic sensitivity is that this interval permitted more
recovery and repair.

These results suggest that the maximum long term skin sparing
effect is likely to be obtained with an interval of two days rather
than one day between fractions. This observation is also supported
by an in vivo clinical survey (Ranu et al 1976; Ranu and Ranu 1977;
Ranu 1983) of female patients receiving radiotherapy of the breast,
thus confirming that radiotherapy given on alternate days causes
least damage to the skin, because this number of fractions a week
permits a preferential recovery of the tissue.

ACKNOWLEDGMENTS

The author wishes to thank Dr Duane F. Bruley, Dr C. Wimberly
and the Louisiana Tech Foundation for their excellent support. The
assistance of Dr C. Russ Greer, Dr H.K. Ranu and Mr Miles Kanne,
Cook Inc., Bloomington, Indiana, U.S.A. is also appreciated. I am
grateful to Nathan Busch, Paul Copeland, Edwin Irwin and Barry
Sands for their help concerning this chapter.

REFERENCES

Burlin T.E., Challoner A.V.J., Magnus I.A. and Szur L. (1973) The
  effects of divided doses of X-rays on the regrowth of hair in
  the mouse. Int. J. Radiat. Biol. 23, 121
Denekamp J., Ball M.M. and Fowler J.F. (1969) Recovery and
  repopulation in mouse skin as a function of time after X-irradiation. Radiat. Res. 37, 361
Ranu H.S. (1975) Effects of ionizing radiation on the mechanical
  properties of skin. Doctoral Thesis, CNAA, London, England.
Ranu H.S., Burlin T.E. and Hutton W.C. (1975) Effects of X-irradiation on the mechanical properties of skin. Phys. Med. Biol.
  20, 96
Ranu H.S., Hutton W.C., Burlin T.E. and Ranu H.K. (1976) Radiation
  and its effect on the elastic properties of the human skin. In
  Digest 11th International Conference on Medical and Biological
  Engineering, Ottawa, Canada, p. 52.
Ranu H.S. and Ranu H.K. (1977) Effects of ionizing radiation on the
  physical properties of skin. In Digest 1st Mediterranean
  Conference on Medical and Biological Engineering, Sorrento, Italy,
  p.19.5
Ranu H.S. (1979a) The mechanical and structual response of skin to
  irradiation. J. Biomech. 12, 601
Ranu H.S. (1979b) Effects of X-rays on the fibrous structure of
  skin as revealed by SEM. 6th International Congress of Radiation
  Research, Tokyo, Japan, p. 100
Ranu H.S. (1981a) Effects of fractionated doses of X-irradiation on
  the mechanical properties of skin - a long term study. In Proceed-
  ings of 2nd International Symposium on Bioengineering and the Skin.
  Eds. R. Marks and P.A. Payne, M.T.P. Press, Boston p. 15.
Ranu H.S. (1981b) Radiotherapy response of human skin evaluated in
  terms of structural and physical characteristics. 3rd International
  Symposium on Bioengineering and the Skin. Philadelphia U.S.A. (In
  press)
Ranu H.S. (1983) Effects of different radiotherapeutic fractionation
  regimes on the skin of patients. 7th International Congress of
  Radiation Research (In Press)

ELECTROMAGNETIC FIELD DOSIMETRY IN BIOMEDICAL INVESTIGATIONS

Hubert TRZASKA

Institute of Telecommunications and Accoustics, Technical
University of Wroclaw, Wroclaw, Poland.

1. INTRODUCTION

There has been rapid growth recently in the number and power of
high frequency (HF) generators, e.g. in the following applications:
telecommunication; radio- and TV-transmitters; industry; HF heating;
medicine; short- and microwave therapy; and household equipment
including microwave ovens. In all these cases electromagnetic
radiation (EM) is radiated into space either intentionally e.g.
telecommunications, radar or unintentionally. Sometimes there are
less evident sources of EM radiation, such as high power ultrasound
equipment or NMR equipment.

Presently electromagnetic field (EMF) is not the most danger-
ous factor polluting the natural environment. Water pollution is
a much more serious problem. Unlike other factors, however, the
natural EM environment is changed on a global scale and EMF induced
in the heart of the African jungle by transmitters located in other
continents, exceeds by more than 1000 times the natural EMF level.

Investigations into the biological activity of electricity
have followed those on electricity itself and the experiments of
Galvani, Volta and D'Arsonval are well known. Because of the global
range of pollution by EMF of the natural EM environment, investigations into bioeffects have always been considered important and
were intensified in the late sixties for political reasons.

The simplest effect accompanying exposure of a body to EMF is
a change in its temperature or "thermal effect". Much attention
is devoted, however, to the evaluation of biological changes after

EMF irradiation when there are no thermal effects. General biological processes influenced by EMF include changes in the rate of
growth, effects on the hearing senses, quantitative changes in the
ionic content of tissues, changes in the bioelectric function of
the nervous system and the heart. The development of these changes
is significantly affected by the complexity of EMF rather than by
its intensity (Mikolsjczyk 1981). Reversible changes in behaviour,
physiological function and microstructure are reported at EMF levels
far below those at which thermal effects occur.

2. SAFETY STANDARDS

To protect the general population against EM radiation hazard,
many countries have introduced occupational exposure and/or general
public exposure standards. The most significant difference in
attitude between East and West, with regard to biological mechanisms
of EMF effects, concerns the question of thermal versus non-thermal
effects. As a result of this, Eastern and Western standards differ
considerably. In the United States of America and some Western
European countries, the maximum permissible value for continuous
exposure, is 10 mW/cm^2.   It is believed that 100 mW/cm^2 is the
lowest level at which significant biological damage could occur, so
10 mW/cm^2 is one tenth of this level.

The traditional Soviet and East European view has been that
EMF can functionally, and even morphologically in some cases, alter
the organism at power densities well below the maximum permissible
thermal exposure level adopted in the West.

In Table I the American ANSI proposed revisions of the above
thermal threshold level of 10 mW/cm^2, proposals of the Massachusetts
Non-Ionizing radiation regulations, and Polish occupational exposure
and general public exposure standards are given. ANSI occupational
exposure is averaged over any period of 0.1 hour, Massachusetts
General Public Exposure is averaged over any 0.5 hour period, and
Polish standards define the maximal values for continuous irradiation. Polish standards at frequencies below 300 MHz use the E-field
limits. In Table I these values are presented in mW/cm^2 or μW/cm^2
for simple comparison using the formula

$$E = \sqrt{120\,\pi\,P}$$

It is worth emphasising that the trend is to lower permissible
EMF levels. It has been suggested that it may be necessary to
reduce EM radiation at some "resonant" frequencies (Presman 1970).

TABLE I

RADIO FREQUENCY OCCUPATIONAL AND GENERAL PUBLIC EXPOSURE
STANDARDS IN USA AND IN POLAND

	U.S.A.		Poland	
Frequency MHz	ANSI mW/cm^2	Gen.Pub.Exp.* mW/cm^2	Occ. Exp. mW/cm^2	Gen. Pub. Exp. $\mu W/cm^2$
0.1				
0.3				
1	100	20	0.1	6.6
3				
10	$900/f^2$	$180/f^2$		
30				
100	1	0.2	0.013	1.1
300 / 1500	$f/300$	$f/1500$		
100 GHz	5	1	0.01	2.5
300 GHz				

(* Special report - Microwave Journal Vol.25, pp. 170-171, 1980)

Similar standards are used to limit irradiation of electro-
explosive devices, ignition of fuel, and to ensure undisturbed
operation of electronic devices in EMF (e.g. cardiac pacemakers).

3. WHAT TO MEASURE?

In the far field it is sufficient to measure the electric
field intensity (E) or power density (P). In the near field it is
necessary to measure separately the E and the magnetic (H) field
intensities as their ratio can vary over a wide range, depending
upon the structure of the radiation source.

The most important parameters of the EMF which should be
measured are:

(i)   EMF intensity. To a first approximation, the EM energy absorbed by a body is proportional to EMF intensity - this parameter is usually measured and is of primary importance from the point of view of thermal effects.

(ii) Frequency. The EMF frequency affects interaction with living bodies in two ways:
  a.   electrical properties of tissues change markedly with frequency, affecting the depth of EMF penetration and reflection coefficients.
  b.   due to changes of the"electrical" dimensions of a body, tissues and cells may exhibit resonant absorption at certain frequencies.

(iii) Polarisation. The relative positions of a body and the E vector are important for EMF energy absorption. Absorption is different for linear and elliptic polarizations.

(iv) Modulation. As a result of non-linear properties of tissues, the bioeffects of two EMFs of equivalent power density could be different and dependent upon maximal peak values.

(v)   Impedance. The impedance $Z = E/H$ in the near field varies depending on radiation source structure, $0 < |Z| < \infty$ whereas for the far field $Z_o = 120\pi$

In laboratory investigations of bioeffects, the Specific Absorption Rate (SAR) measurement has been accepted internationally. This represents power absorption in unit mass. SAR is defined by

$$SAR \; = \; \frac{c\,\Delta T}{t} \; = \; \frac{P_a}{m} \; [W/kg]$$

where c = caloric constant $(kJ/kg.^{\circ}C)$; t = heating time (s) T = temperature change $(^{\circ}C)$; P_a = absorbed power (W); m = mass (kg).

From the above formula, SAR could be measured in two ways; i.e. by measuring P_a or T. In the latter case, however, heating time could be   estimated using:

$$t \; = \; \frac{c\,\Delta T\,\rho}{2\pi\,\epsilon_o \epsilon_r f\,\tan\delta\,E^2} \; [s]$$

where ρ = density (Kg/m^3); $\epsilon_o \epsilon_r$ = dielectric constant (F/m); f = frequency (Hz); $\tan\delta$ = loss factor.

If, in this formula, we use mean values of c, ρ, ϵ_r and $\tan\delta$ for tissue, the minimal detectable temperature change $T = 0.1^{\circ}C$ and $E = 10$ V/m, and we have for radio frequencies : $t \approx 10^5 - 10^{10} s$.

This example shows the limitations of thermal measurements in EMF dosimetry. In some cases, however, it is the only method available for strict P_a measurements.

It is necessary to remember that the above formulae are only valid for a homogeneous medium and lossless EM energy transfer to heat. They also assume thermal insulation of the body.

4. EMF MEASUREMENTS

4.1 Laboratory methods (exposure systems)

In laboratory methods, the parameters of EMF, frequency, modulation, polarisation, are known "a priori" and can be changed by an experimenter.

(i) Standard EMF. The principle of standard EMF generation is shown in Fig. 1. It could be generated by using a Standard Transmitting Antenna (STA) or a Standard Receiving Antenna (SRA). In the first case EMF intensity is calculated on the basis of the STA excitation measurement, e.g. current I, parameters of the antenna and geometry of propogation between STA and the Object Under Test (OUT). In the other method, the EMF is generated in an arbitrary way and then is standardised using an SRA.

Fig. 1 The principle of standard EMF generation

The standard can be placed in an open space, a natural cavity, or if the influence of external EMF has to be reduced, in a shielded enclosure. Simultaneous use of STA and SRA, as shown in Fig. 1, may improve accuracy of the standard EMF.

This method is the simplest from a conceptual point of view and requires little technical equipment. Therefore, it is widely used in biomedical as well as technical investigations. A comparatively large space with a quasi uniform EMF distribution makes it possible in biomedical experiments to keep animals in quite large enclosures and to allow them to move freely. The most important disadvantage of the method is low accuracy of the absorbed power estimation and this is important when thermal effects are investigated.

(ii) TEM transmission line (Crawford cell) This method was developed by Crawford (1981) at the National Bureau of Standards. The apparatus consists of a matched strip transmission line working in the TEM mode (Fig. 2). The EMF intensity inside the cell is calculated on the basis of the cell excitation measurement and its dimensions. The measurement of fed, reflected, and transmitted power makes it possible to calculate power absorbed by OUT and then by SAR.

(iii) Circular waveguide. This method was developed especially for biomedical investigations by Guy and Chou. The technical concept is similar to that presented above, but a circular waveguide is used to get a more uniform, whole body irradiation of the animal. In both of these methods, however, animals are kept in a comparatively small space. The main disadvantage of both methods, as well as any other, is that it is not possible to localise "hot spots" or local hyperthermia.

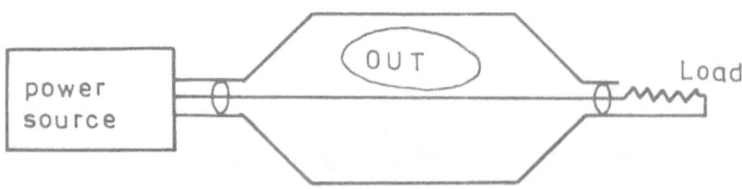

Fig. 2 The Crawford cell arrangement.

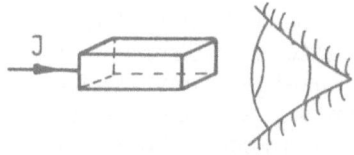

Fig. 3 Open waveguide.

(iv) Open waveguide. The method was developed by Rosenthal
(1980) for eye cataract investigations (Fig. 3). In this method
the power radiated by an open ended waveguide is estimated and the
approximate power density in its vicinity is found. It is difficult,
however, to estimate which part of the radiated power was absorbed
by the eye.

(v) Theoretical models. Many methods can only estimate EM
energy absorption with limited accuracy and assume uniform energy
distribution inside the body. Temperature distribution measurements
in vivo are almost impossible, and difficult even in vitro where
they are in any event of only limited applicability.

To estimate EM energy absorption and distribution inside the
body, on the basis of external EMF or SAR measurements, theoretical
methods have been developed. Complicated models of a human body,
containing two hundred cells with different electric parameters
and illuminated by TEM or a spherical wave were developed by Gandhi
et al (1979) (Fig. 4)

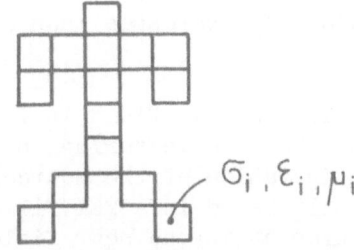

Fig. 4 Simple model of a man

They used advanced analytical and computing methods to calculate power absorption in different parts of the model as a function of frequency and polarization of the incident EMF.

Experimental work in this field has been done by Guy et al (1979). He used simple, homogeneous phantoms (cylinders, spheres, ellipsoids, and human body shapes), calculated power absorption and temperature distribution inside the phantom, and then measured them with a thermographic camera. These methods made it possible to discover places of increased EM energy absorption in the human body as well as providing a better understanding of technical and bio-medical problems of dosimetry.

4.2 Open field methods

Open field EMF measurement methods are based upon the use of a EMF sensor which responds to EMF for example by a change in temperature, voltage or current in a known manner. Semi-conducting sensors, e.g. Hall-cell, Josephson junction, magnetodiode are very sensitive, small and work over a wide frequency range. However, non-linear characteristics, the earth's magnetic field and difficulties with designing wide-band probes limit their practical applications.

Presently, sensors in the form of small "electric" and "magnetic" dipoles are used most widely. An example of the electric field meter is shown in Fig. 5. The meter contains a short dipole, placed a distance d from a radiation source, at which voltage is measured. As shown in the equivalent network of the system (Fig.5), the antenna output voltage U_L is a function of the mutual impedance of antennæ and source Z_{as}. The voltage could be expressed in the form:

$$U_L = e \frac{Z_L}{Z_a + Z_L + \Delta Z_a}$$

where e is the electromotive force induced in the antenna by the EMF, Z_a and Z_L are the input impedance and load impedance of the antenna respectively, and ΔZ_a expresses changes of Z_a due to Z_{as}.

In the far field ΔZ_a = 0 whereas in the near field it is a complicated function of the distance d between measuring antenna and source of radiation as well as configuration of the source. It may be concluded that if the meter is calibrated in a TEM field (the normal procedure) its accuracy of measurement in the near field will be a function of d. In the case of multipath propagation, the polarisation in the near field is not known "a priori". If the EMF is measured with a single antenna (as in Fig. 5), the maximum value

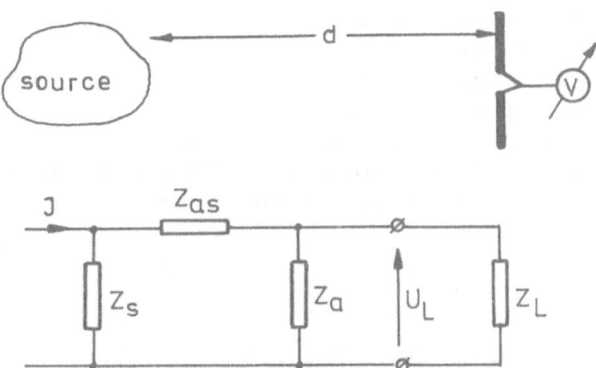

Fig. 5   The EMF meter and its equivalent network

or amplitude is independent of the polarisation. If we use, however,
a probe with three perpendicular dipoles,the results of the same
amplitude measurements would be in the ratio $1:\sqrt{2}:\sqrt{3}$ for linear,
circular and spherical polarisations respectively. As mentioned
above, in the near field E and H should be measured separately.
Moreover, E should be measured with an "electric" antenna and H
should be measured with a "magnetic" antenna.

   It is extremely difficult to measure power density in the near
field and results are uncertain. In the far field P is often calculated on the basis of E measurement. This is a very simple and
exact method when used in the far field. Commercially available
power density meters designed for near field measurements are based
on the same concept; E or H field meters are frequently calibrated
in power density units and this is acceptable in many cases,
especially at frequencies above 300 MHz. Below 300 MHz, in the
vicinity of small-sized radiation sources the use of this calibration without strict analysis, may give highly erroneous results.

   The near field power density measurement needs simultaneous
measurement of E and H fields and amplitude and phase measurements
are necessary. Detailed analysis of the accuracy of measurement of
near field EMF is given by Trzaska (1981).

5. CONCLUSIONS

   The development of near field EMF dosimetry for biomedical
and technical purposes has not been satisfactory and further work
is required on both the concepts and on technical aspects of

measurement. In addition to long-standing problems with near field measurements, new problems have been discovered recently.

Theoretical considerations (Presman 1970, Sedlak 1981) show the possibility of a major role for EMF in information transfer within the living body and between living bodies. Up to now EMF radiation has been used primarily at low frequencies in remote sensing for EEG and EKG purposes. It is proposed, however, that similar radiation could appear at quantum levels over a wide frequency range. To examine this hypothesis it will be necessary to construct more sensitive instruments or find ways to stimulate living organisms to increase their radiation output.

REFERENCES

Crawford M.L. (1981) Intl. EMC Symp. Proc. paper No. 70, Zurich
Gandhi O.P., Hagmann M.J., and D'Andrea J.A. (1979) Radioscience
   vol. 14, No.6S, Nov-Dec. pp.15-23
Guy A.W., Chou C.K. (1977) HEW Publication FDA/77-8011. Vol. II
   pp.389-410
Guy A.W., Wallace J., McDougall J.A. (1979) Radioscience Vol. 14,
   No.6S, Nov-Dec pp. 63-74
Mikolajczyk H., (1981) Prace Instytutu Medycyny Pracy w Lodzi
   No. 4/8/1981, p. 5-33
Presman A.S. (1970) Electromagnetic Fields and Life. Plenum Press,
   New York
Rosenthal S.W. (1980) Private communication
Sedlak W. (1981) Prace Instytutu Medycyny Pracy w Lodzi No.4/8/1981
   pp.55-65
Trzaska H. (1981) Prace Instytutu Medycyny Pracy w Lodzi No.4/8/1981
   pp.168-267

III. TECHNICAL ADVANCES APPLICABLE MAINLY IN VIVO

PHYSICAL PRINCIPLES OF RADIONUCLIDE EMISSION IMAGING USING SINGLE PHOTON TECHNIQUES

P.P. DENDY

Department of Biomedical Physics and Bioengineering,
University of Aberdeen, Foresterhill, Aberdeen AB9 2ZD,
Scotland, UK.

1. INTRODUCTION - THE DETECTION PROBLEM

Radionuclide imaging seeks to derive information on the distribution of radioactivity within the body. For a good over-view see Mallard (1972).

In many examinations it is not possible to achieve a high degree of selective incorporation in the region of interest and the problem reduces to one of detecting either hot or cold spots in a uniform background. To gain some insight into this problem and to investigate how the detectability of a lesion may vary with the many parameters involved, consider the simple model shown in Fig. 1. The following assumptions are made:-

(i) there is background region of total depth L, containing a radionuclide concentration (activity/unit volume) C_b and a target consisting of a spherical region of diameter d at depth t with a different concentration C_t,

(ii) the radionuclide emits gamma rays, which are attenuated with coefficient μ in the medium, assumed to be unit density tissue,

(iii) there is pure exponential attenuation of gamma rays in the medium

(iv) a detector that accepts all gamma rays incident normally with equal efficiency is used,

DETECTOR

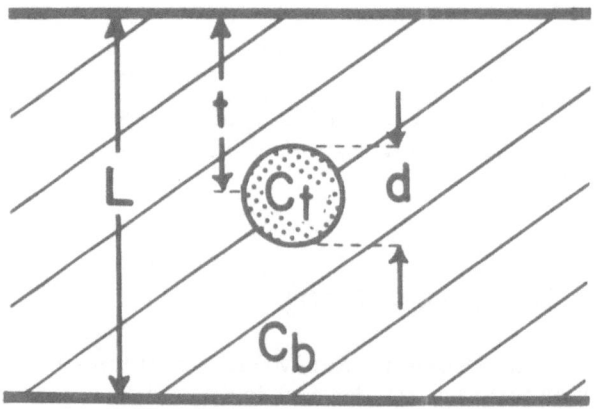

Fig. 1 A simple model showing a small region of high activity in a uniform background of lower activity.

(v)   the signal to noise ratio is an adequate measure of detectability of the target

Consider a narrow cylinder of cross-sectional area A in the background region. The number of gamma rays dN reaching the surface per unit time will be

$$dN \, \alpha \int_{x=0}^{L} C_b . A . e^{-\mu x} . dx$$

where C_b A dx is the activity in an element dx at a distance x from the surface.

Since the count rate in the detector is proportional to this,

$$R_b = \frac{kAC_b(1 - e^{-\mu L})}{\mu}$$

The change in count rate due to the signal is

$$S = kA(C_t - C_b) \int_{t-\frac{d}{2}}^{t+\frac{d}{2}} e^{-\mu x} dx = \frac{kA(C_t - C_b)}{\mu} \left[e^{-\mu(t-\frac{d}{2})} - e^{-\mu(t+\frac{d}{2})} \right]$$

Now the noise is proportional to $R_b^{\frac{1}{2}}$ due to Poisson Counting Statistics so

$$\frac{S}{N} = \frac{k^{\frac{1}{2}} A^{\frac{1}{2}} (C_t - C_b)}{C_b^{\frac{1}{2}}} \left[\frac{e^{-\mu(t-\frac{d}{2})} - e^{-\mu(t+\frac{d}{2})}}{\mu^{\frac{1}{2}}(1 - e^{-\mu L})^{\frac{1}{2}}} \right] \tag{1}$$

This equation shows that for a fixed geometry

$$\frac{S}{N} \alpha \frac{(C_t - C_b)}{C_b^{\frac{1}{2}}} \qquad \text{and thus to visualise}$$

tumours for example we need a good differential concentration between tumour and surroundings, $(C_t - C_b)$ and a low background concentration C_b.

If equation 1 is rearranged as a concentration ratio,

$$\frac{S}{N} \alpha\, C_b^{\frac{1}{2}} \left(\frac{C_t}{C_b} - 1 \right)$$

and this shows that for a fixed concentration ratio, $S/N \propto C_b^{\frac{1}{2}}$, in other words to double the signal to noise ratio requires four times the amount of activity.

Equation 1 may also be used to investigate the effect on signal to noise ratio of different values of μ and hence gamma ray energy. This shows that the best S/N is generally obtained with small values of μ (higher energy gamma rays) but for superficial targets in a deep background, best results are obtained for gamma rays of intermediate energy.

The model may also be used to show that for a fixed concentration, as one might expect, S/N increases as the diameter of the target, i.e. the value of A, increases and decreases as the depth increases. Sometimes a particularly useful insight may be obtained by simple modelling. For example as shown in Fig. 2, when concentration ratio is plotted as a function of target diameter, for a fixed S/N, the curve rises very steeply indeed at small target diameters.

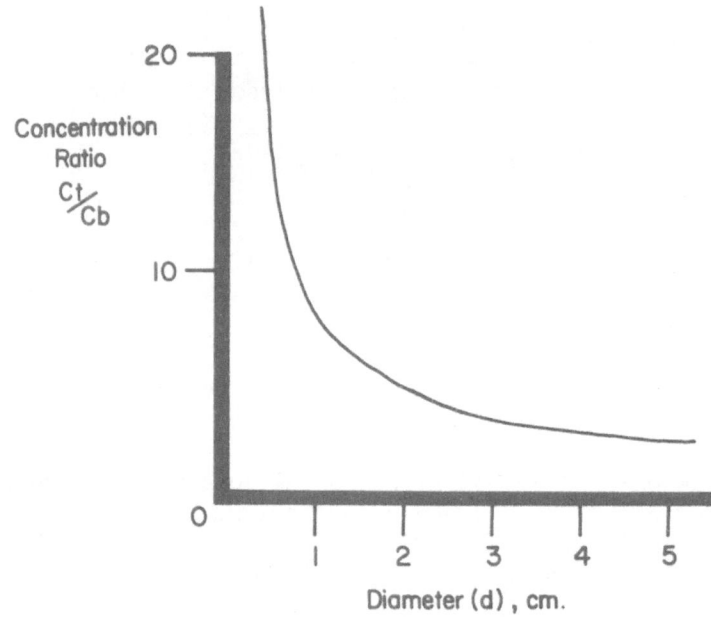

Fig. 2   Graph showing how the concentration ratio required to
produce a fixed signal/noise ratio varies with target
diameter.

Precise quantification is not possible, both because of the
limitations of the model, for example practical detectors do not
have point source invariance in air, and because of other relevant
factors in the total imaging system such as collimator design.
Nevertheless such an idealised model can provide a lower limit on
which to base detectability of targets and serves to emphasise that
this is very much a signal to noise ratio problem with reduction of
the noise almost as important as increasing the signal.

2.   HIGH PERFORMANCE SCINTILLATION IMAGING SYSTEMS

All commercial imaging devices available at the present time
use thallium activated sodium iodide $\underline{/}NaI(Tl)\underline{/}$ crystals as the
basic detecting system, so the properties that make these crystals
particularly useful for clinical imaging will now be summarised:-

(i)   The high density $(3.7 \times 10^3$ kg m$^{-3})$ increases the
       probability that an incident gamma ray will undergo an
       interaction in the crystal thereby increasing sensitiv-
       ity.

(ii)   The high atomic number $(Z = 53$ for iodine) favours the
       photoelectric interaction, thus a pulse is generated
       which represents the full energy of the gamma ray. This
       pulse will be accepted by the pulse height analyser but
       for monoenergetic gamma rays, at least in theory, the
       analyser can be set so that all pulses of lower energy
       are rejected. This is of particular importance in
       discriminating against gamma rays that have been Compton
       scattered in the patient and can no longer provide use-
       ful positional information.

(iii) Thallium gives a high conversion efficiency of the order
       of 10% so a 140 keV secondary electron gives a scintill-
       ation containing 14 keV of light energy or about 4200
       photons. This high photon yield is of particular impor-
       tance in the gamma camera.

(iv)   The decay time of the light flashes has a very short
       half life - only about 0.2 μ s. This is important when
       studies requiring very high counting rates are envisaged.

2.1 The rectilinear scanner

The basic equipment (Harbert and Neto 1978) consists of
collimator, scintillation crystal and shield, photomultiplier (PM)
tube, HT supply, amplifier, pulse height analyser and display. (Fig.
3a and 3b)

2.1.1. Detector and collimator - a NaI(Tl) scintillation crystal
8-13 cm in diameter and 5 cm thick fitted with a lead shield and
collimator is an efficient detector for gamma rays up to 500 keV.
Since positional data is derived simply from the position of the
detector as it scans in a raster motion over the region of interest,
spatial resolution and sensitivity are determined predominantly by
the design of the collimator and the dimensions of the scintillation crystal.

A focussed collimator is used for almost all scanning applications (Fig. 4) and on purely geometrical considerations, using the
notation on this figure, the radius of the field of view at the focus
R_F = 2 rf/t and if the collimator has N holes, its efficiency (gamma
rays transmitted/gamma rays emitted from a point source)

$$E_F = N^2 r^2 / 4 \ (f + t)^2$$

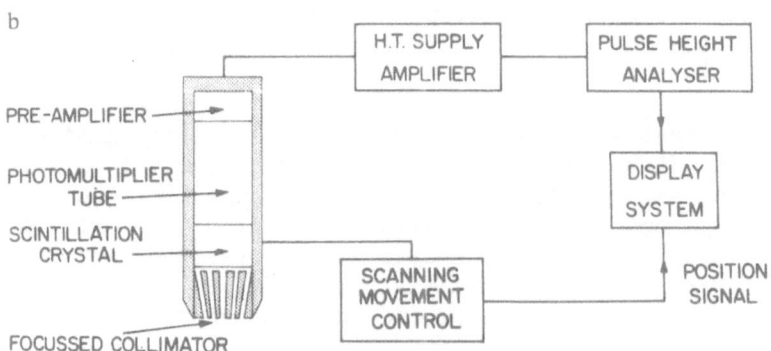

Fig. 3a) A view of the El Scint whole body scanner with control and
        display units
    3b) A simple block diagram of the principle components of a
        scanning device.

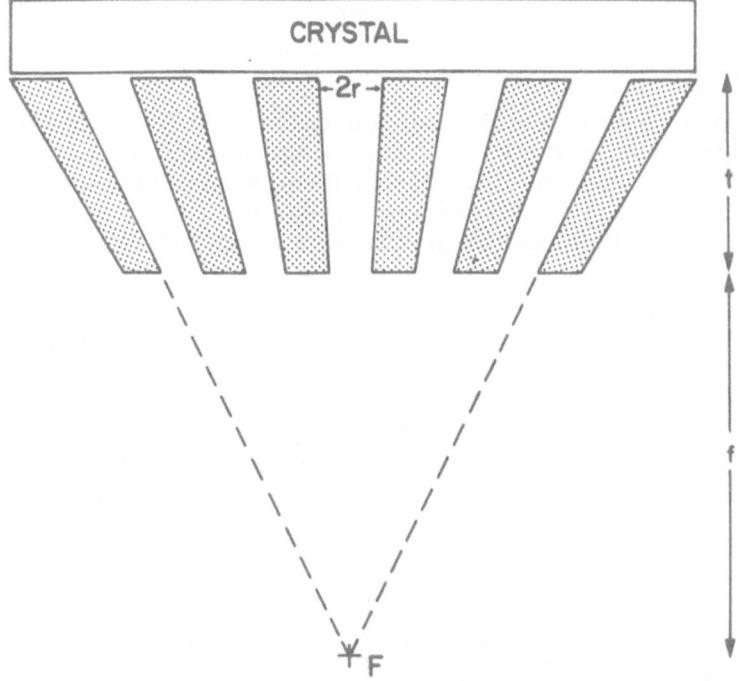

Fig. 4 A section through a focussed collimator showing the
dimensions that determine the radius of the field of view
at the focus and efficiency

    The conflicting requirements for good spatial resolution and
high sensitivity are confirmed by these equations. If $\underline{N}$ is fixed,
efficiency is increased by increasing $\underline{r}$ but resolution deteriorates. A typical compromise might be a resolution in the focal
plane of 12 mm but a sensitivity of only 2.2%.

2.1.2 A pulse height analyser may be used to reject gamma rays
which have been scattered into the detector from outside the field
of view because they will lose energy in the scattering process.
However, as discussed later, this process is not very effective.

2.1.3 Two main display systems evolved following the simple dot printing apparatus of the original scanners. The photo-scan was produced by a light source moving over and exposing an X-ray film in synchrony with the detector moving over the patient. The light intensity was controlled by the output of the pulse height analyser, either directly as a series of pulses or via a ratemeter. Alternatively, the ratemeter output could be used to position a multi-coloured inking ribbon in an electromechanical colour printer or to activate a colour TV display to produce an instant colour map.

2.2 The gamma camera

The gamma camera consists of two units (Fig.5), the collimated detector mounted on a stand to allow it to be manoeuvred around the patient, and the console containing pulse processing electronics and displays. Only brief details can be given about the various components, for further information see Erickson and Brill (1978).

Fig. 5   The detector and console of an Ohio Nuclear Sigma 410 gamma camera.

2.2.1 Components of the detector system are shown in Fig. 6. When
an incident gamma ray passes through the collimator and interacts
with the scintillation crystal (an event), a large number of visible light photons is generated. The number of photons reaching
each PM tube is proportional to the solid angle subtended by that
PM tube at "the event", so by examining all the PM tube signals,
the position of the original event may be deduced. Also by summing
all the signals, the energy of the incoming gamma ray may be found
and, as with the scanner, pulse height analysis may be used to
discriminate against scattered rays. If the gamma ray energy falls
within the PHA window, the position signals are applied to an
oscilloscope display to produce a brief spot of light on the screen
in a position related to that of the scintillation event. A permanent image is produced by integrating these spots onto photographic
film.

2.2.2 Intrinsic resolution. Since the number of light photons
reaching each tube is low, the signal is susceptible to large statistical variations which the camera electronics interprets as
resulting from changes in the spatial position of the scintillations.
This lack of reproducibility in spatial positioning is known as the
intrinsic resolution of the camera and is about 5 mm.

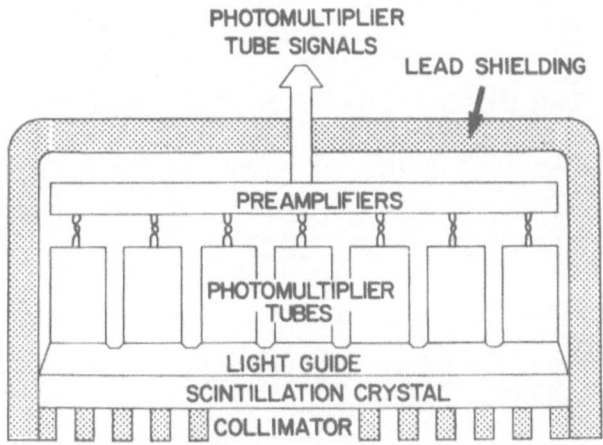

Fig. 6   A section through the detecting system of a gamma camera
         showing the main components.

196

2.2.3 System resolution. Gamma rays are emitted isotropically so a collimator is required to define the direction of those rays which make up the image. This usually consists of an array of parallel holes separated by lead septa running perpendicular to the crystal face. As for the converging collimator used with the scanner, resolution and sensitivity are mutually exclusive requirements. A range of collimators is generally used and the total system resolution, which now incorporates both the camera resolution and the collimator resolution may be anything from 7-15 mm depending on the collimator chosen.

Figure 7 illustrates the differences in performance of a high resolution collimator and a collimator which has a 2.3X higher sensitivity, but correspondingly poorer resolution. This figure also shows how performance deteriorates when scattering medium is interposed between the test object and the camera face.

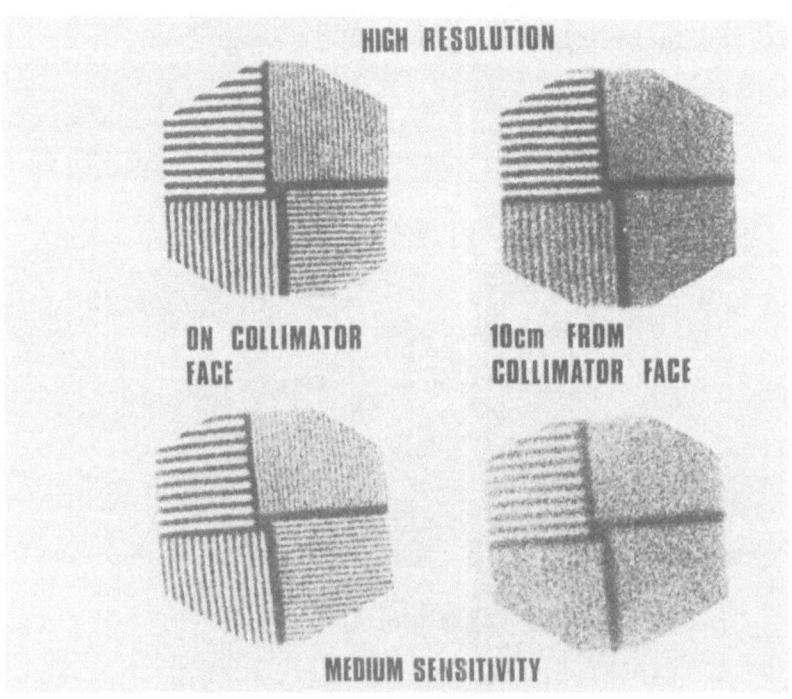

Fig. 7   Subjective impression of resolution obtained with a bar phantom having bar widths and separations of 4.0, 4.8, 6.4 and 9.1 mm.

Diverging, converging and pinhole collimators have been designed for use with gamma cameras for specialised imaging procedures. For all such collimators, magnification, resolution and sensitivity all vary with the depth of the organ being imaged.

2.2.4 Crystal thickness. A compromise is required here. If the crystal is too thin, sensitivity is unacceptably low. If the crystal is too thick, multiple interactions may occur and positional information is lost. Modern camera crystals are usually between 9-13 mm thick and stop about 90% of 140 keV gamma rays from Tc99m but only about 25% of 360 keV gamma rays from I131.

2.2.5 Uniformity. Even the very best cameras show some evidence of variation in image count density when a uniform distribution of activity is imaged. Practically, non-uniformities may be expressed in terms of integral uniformity $(C_{max} - C_{min})/(C_{max} + C_{min})$ where C is the count per picture element, or sometimes C_{rms}/C_{mean} where

$$C_{rms} = \sqrt{\sum_{i=1}^{N} \frac{\left[C_i - C_{mean}\right]^2}{N}}$$

Differential non-uniformity measures the contrast between a picture element and any adjacent element. It is given by $\frac{C_{max} \cdot D}{C_{mean}}$ where C_{max} is the maximum difference in count between adjacent elements a distance D apart.

These methods of measurement can give somewhat different figures depending upon the type of non-uniformity. For example quite big changes across the camera face may result in only a low value for differential non-uniformity, if they occur gradually.

The standard method to correct for non-uniformity is to collect a digitised image of a flood source. For each pixel a correction factor f(i) is calculated so that multiplication of the original pixel count density C(i) by this factor returns the count density to the mean value m, i.e. f(i) = m/C(i).

Micro-processor based systems now perform a similar correction procedure on line but this does result in some counts being discarded, perhaps as many as 10%, so care is required in precise quantitative studies.

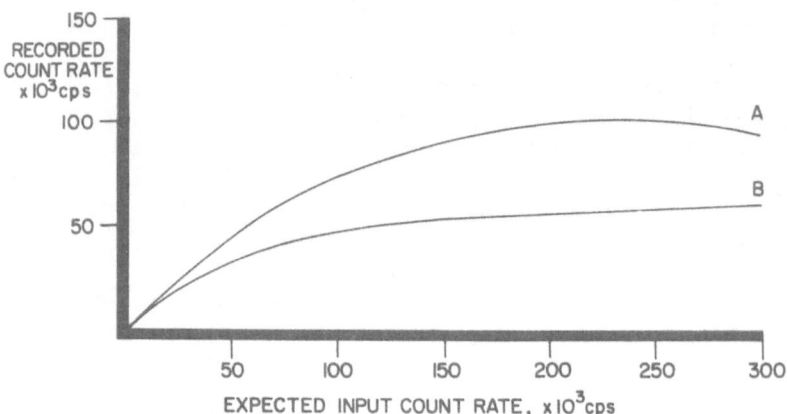

Fig. 8 Comparison of the observed and expected count rates for a
gamma camera using either (A) a flood source with no scatter
or (B) a BSI phantom (Hospital Physicists' Association Topic
Group Report No. 27) with scatter. In each case the energy
window was 20%.

Four causes of non-uniformity have been suggested

    (a) Spatial non-linearity - this is apparent on re-examination
        of Fig. 7
    (b) Spatial variation in detector sensitivity
    (c) Spatial variation in energy sensitivity
    (d) Spatial variation in resolution

The first of these is thought to be the most important (Todd-
Pokropek et al 1976, Wicks and Blow 1979) and since this results
from an inability to assign the correct spatial co-ordinates to
detected gamma rays, recent attempts to solve the non-uniformity
problem have concentrated on correction matrices which are used to
shift the recorded position of each gamma ray to compensate for this
error. Such a process involves no loss of counts.

2.2.6 Performance at high count rate. The limitations on performance at high count rates result from both the processing time
required by various system components, such as pulse height analyser
and ratio circuits, and pulse pile-up where two successive scintillations in the crystal occur so close together that one has not
decayed before the other appears so they are registered as one
event. (Strand and Larsson 1978). Typical results are shown in
Fig. 8. The marked difference in response with and without scatter

TABLE I

PERFORMANCE OF A MODERN GAMMA CAMERA AT HIGH COUNT
RATES - IN ALL INSTANCES A 20% ENERGY WINDOW WAS USED

Source of data	With or without scatter	With or without collimator	Value of expected count rate where 10% losses occur	Maximum observed count rate
Manufacturers' specification	w/out	w/out	60	100
Flood	w/out	w/out	45	106
BSI phantom	w/out	with	34	79
BSI phantom	with	with	17	62

is a consequence of pulse pile up.   This effectively adds two
scattered gamma rays to produce a total energy signal that falls
within the PHA window. Therefore it is only possible to make a
valid estimate of count rate losses using a realistic phantom,
under practical working conditions, with the correct scattering
conditions. Failure to do so will produce spurious results as
shown in Table I.

2.2.7 Dynamic Studies. The gamma camera is capable of producing
high quality static images (Fig. 9) and plays an increasingly
important role in dynamic investigations. For this purpose a
suitable data processing system is necessary.

    Many cameras incorporate dedicated data processing units.
These often provide the cheapest system and may add no more than
$35,000 to the price of an $85,000 camera. However, they tend to
lack flexibility and many larger centres buy the camera and data
processing system separately. In either case it is worthwhile
listing some of the features that will make a system attractive
although it must be emphasised that no one system will provide all
these facilities:-

(a) an operating system that allows background manipulation of
     data collected from a previous study during collection of a
     new study

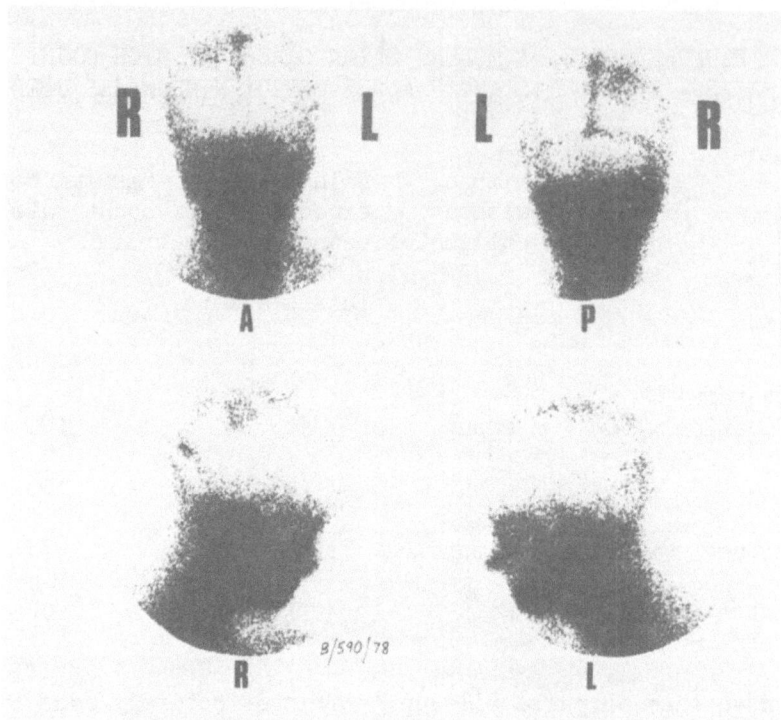

Fig. 9   Gamma camera images of the brain (A= anterior, P = posterior
          R = right lateral, L = left lateral) showing a right
          parietal tumour

(b)   adequate memory (say 32K)

(c)   a patient monitor system that is accessible at all times for
       patient positioning

(d)   well-tried software comprising a simple, straightforward
       sequence of commands for each investigation. The facility to
       set up pre-defined studies for standard investigations is
       useful.

(e)   a good colour television display system for defining regions
       of interest with flexible display manipulation

(f)   a wide range of data analysis programmes

(g)   good hard copy output for images and graphical data

(h) rapid access to studies on disc and cheap long-term storage of scan data

(i) the facility for the user to insert new programmes - e.g. correction for patient movement

3. LIMITATIONS OF CONVENTIONAL RADIONUCLIDE IMAGING

Typical performance figures for a modern gamma camera are shown in Table II. Although such a camera provides high quality images there are still a number of technical limitations with this design of instrument.

(i) Resolution or reproducibility is largely governed by the intensity of light from the scintillation incident on the photo-cathodes. At low intensities the small number of photons reaching a PM tube will have an intrinsic statistical variability. If N is the mean number of light photons, then the actual number received will have a standard deviation of $N^{\frac{1}{2}}$. Hence for good intrinsic resolution, each PM tube should receive a high light flux.

Spatial linearity is optimised by sharing the light amongst many PM tubes so the requirement for good linearity is at variance with that for good resolution and a compromise must be made.

(ii) One way to improve both factors would be to make more visible light photons available and the relatively poor conversion efficiency for the NaI(Tl) crystal - the energy required to create a visible light photon is about 35 eV (or 4×10^3 photons per 140 keV gamma ray) - is a major limitation.

(iii) Discrimination against scatter. Ideally, the use of pulse height analysers (PHA) can eliminate completely scattered gamma rays (which will have a lower energy). However, because of the statistical problems just discussed, the signal from a 140 keV gamma ray is very variable - i.e. it behaves like a gamma ray of variable energy as shown in Fig. 10. Thus in order to collect most of the unscattered rays, a range of energies, say from 125 keV to 155 keV must be collected. However, it emerges from the Compton Theory of photon scattering that a 140 keV gamma ray may be scattered through quite a large angle without suffering much change in energy. Therefore by using a wide window, PHA discrimination against scatter is much less effective than one would expect (Kouris et al 1982).

(iv) As discussed earlier choice of crystal thickness is a compromise. If the crystal is too thin it will be insufficiently sensitive, particularly at higher energies. If the crystal is too thick, there is an increasing probability that multiple inter-

TABLE II

PERFORMANCE PARAMETERS FOR A MODERN GAMMA CAMERA

Field of view	Hexagonal 37 cm inscribed circle
Energy resolution	12 – 14%
Resolution	
Intrinsic	4.5 mm
High resolution collimator 10 cm mix D	9.7 mm
Intrinsic sensitivity	5×10^{-2} cps Bq^{-1} steradian^{-1}
Uniformity (after on line correction)	5%
Count rate for 10% losses	17 kcps

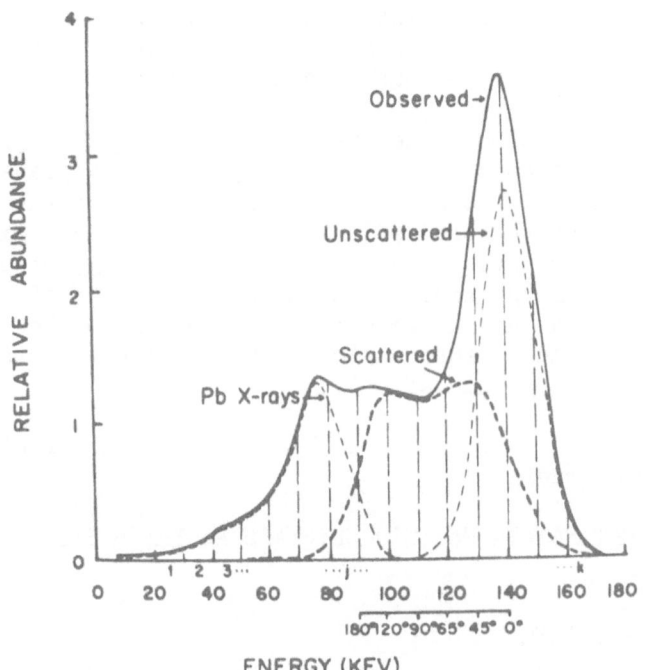

Fig. 10 Gamma ray spectrum of events detected by a gamma camera.
If a 20% window is used on the pulse height analyser, even
gamma rays scattered through quite large angles (shown on
the lower scale) are detected. (Reproduced with permission from Atkins et al 1972)

actions will occur, for example if a gamma ray is first Compton
scattered and then absorbed by the Photoelectric (PE) effect it will
lose the same total energy as a single PE event and this will be
acceptable to the pulse height analyser. However, the position
assigned to the event will be the centre of gravity of the two interactions and intrinsic resolution will be degraded.

(v) The conflicting requirements of resolution and efficiency
for collimator design have been discussed. The poor geometric
efficiency of collimators utilised for imaging, ranging from 2% for
a focussed collimator to as little as 0.01% for a parallel hole
collimator, represents a major loss of gamma rays. Since the number
of gamma rays available for collection is strictly limited by the
amount of activity that may be administered to the patient, this is
a major problem.

(vi) The gamma camera can perform rapid, sequential imaging but in such work high photon fluxes are required to produce reasonable photon densities. Above about 20 kcps there is a marked loss in recorded counts, leading at high count rates to both inaccuracy in data quantification and image distortion. Since the maximum input count rate in clinical imaging may be as high as 70 kcps, this is clearly a potentially serious practical problem.

(vii) Conventional radionuclide imaging represents a projection of the radionuclide concentration of an object onto a plane and image interpretation is an attempt to reconstruct visually the spatial distribution from the projection. However, particularly when dealing with a distribution that varies continuously throughout a volume, the final image represents the superposition of activity distributions in many different planes normal to the direction of the projection. Information on deep-lying structures may be severely degraded by this effect.

Ways in which one can attempt to solve some of these problems will now be discussed briefly.

4. MODIFICATIONS THAT RETAIN NaI(Tl) CRYSTALS AS THE BASIC DETECTING DEVICE

4.1 The intensifier camera

If an image intensifier and electron multiplier are used, a gain in excess of 10^4 can be achieved on the 4×10^3 light photons generated by a 140 keV Tc99m gamma ray.

It can be shown (Roux et al 1972) that for the demagnified images produced by intensifiers, four PM tubes are sufficient to calculate accurately the spatial co-ordinates of the intense output signals. Thus a single, collimated NaI(Tl) crystal may be connected directly to the photocathode of a two-stage image intensifier and the 2 cm diameter output screen may be viewed by 4 PM tubes whose signals give conventional X, Y (position) and Z (energy) pulses. (Driard et al 1976). In a commercial device a 32 cm field of view is achieved with an intrinsic resolution of 4.5 mm FWHM at 140 keV and a maximum count rate of 10^5 cps is claimed.

4.2 Multicrystal detectors

This is a relatively simple concept in which the single large crystal of the gamma camera is replaced by a mosaic of small crystals. The size of the individual crystals (typically ~ 8mm x 8 mm) determines the resolution of the system and each crystal is individually addressed by a pair of PM tubes which determine the X and Y position co-ordinates.

Advantages include the ability to handle extremely high count rates (up to 4×10^5 cps) and improved performance for high energy gamma rays (because the crystals are thicker) but the field of view is small (~ 15 cm x 22 cm) and light loss in the light guides which couple each crystal to the PM tubes degrades energy resolution.

4.3 Hybrid scanners

These devices obtain their name because they combine some of the principles of scanners with those of cameras. A long, narrow bar of NaI(Tl) or a linear array of smaller crystals is used as the detector. The position of an event along the long axis of the crystal is normally measured by an array of PM tubes but in one version just two PM tubes are used, one at each end of the crystal looking along its length. Provided simple linear attenuation of the light flash along the crystal is assumed, both the gamma ray energy and the position of the scintillation may be deduced. The orthogonal co-ordinate is found from the position of the detector as it moves in a direction perpendicular to the crystal axis.

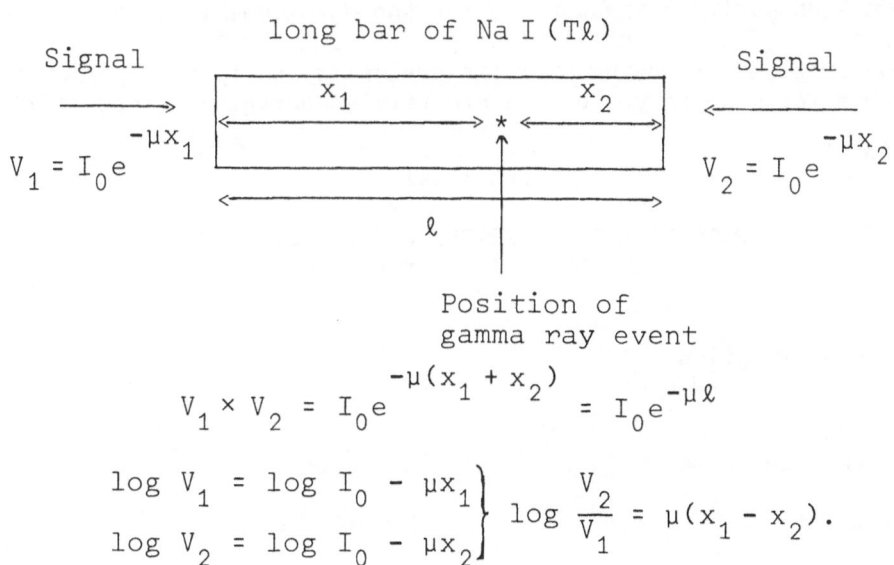

$$V_1 \times V_2 = I_0 e^{-\mu(x_1 + x_2)} = I_0 e^{-\mu \ell}$$

$$\left.\begin{array}{l} \log V_1 = \log I_0 - \mu x_1 \\ \log V_2 = \log I_0 - \mu x_2 \end{array}\right\} \quad \log \frac{V_2}{V_1} = \mu(x_1 - x_2).$$

The hybrid scanner has a number of design features which are attractive in less well developed countries (Crawley et al 1980). For example - very rugged construction; easy maintenance; easy to operate; relatively cheap; can be made independent of power supplies. In addition, it can be used with the higher gamma ray energy from In 113m (392 keV) as well as with the 140 keV ray from Tc 99m. The former generator has a half life of 115 days compared with 67 hr for Tc99m.

5. ALTERNATIVE IMAGING SYSTEMS

5.1 The semiconductor camera

In a high purity germanium detector, only 2.9 ev is required to produce one electron/hole pair (a signal) so one 140 keV gamma ray will produce about 50,000 electrons (contrast the 4000 light photons). The statistical fluctuation of this signal is very low so energy resolution of the order of 1% may be achieved and with it much better discrimination against scattered radiation.

Major problems with this approach have been caused by lack of ultra high purity germanium and for the development of cameras, a lack of very large crystals (see e.g. Kaufman 1978) but both of these problems are being overcome gradually and typical performance figures for a prototype camera are very encouraging (Table III).

TABLE III

PERFORMANCE PARAMETERS FOR A PROTOTYPE
SEMI-CONDUCTOR GAMMA CAMERA

Spatial resolution	2 mm [a]
	5 mm [b]
Maximum count rate	250 kcps
Sensitivity	14×10^{-4} cps Bq^{-1} [c]
Energy resolution	3% [d]
Detector area	64×32 mm^2

(a) intrinsic resolution
(b) at 10 cm depth in tissue using a high resolution collimator
(c) with a high resolution collimator
(d) at 140 keV

5.2 Compton effect camera

An interesting potential development of semi-conductor devices is a cammera in which the Compton effect is used to predict the origin of an emission (Everett et al 1976). In Fig. 11 the path of an incident photon which suffers two interactions is shown, one at A and one at B. If the co-ordinates of the interactions $(x_1 \ y_1 \ z_1)$ and $(x_2 \ y_2 \ z_2)$ can be found, the trajectory of the photon after first interaction is defined.

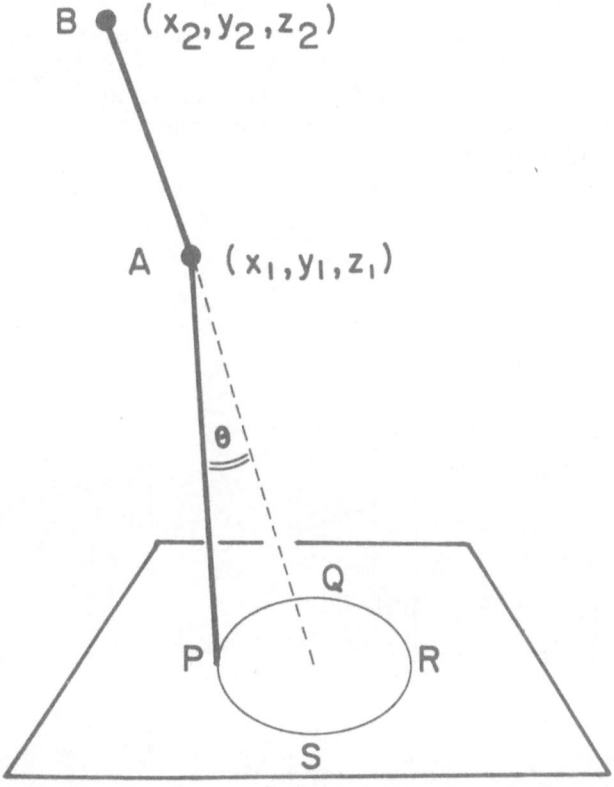

Fig. 11   Measurements that permit accurate calculation of the scattering angle θ provide the basis of the Compton effect camera.

Provided the fractional loss of energy at A can be found very precisely (i.e. with a solid state detector) the angle   through which the photon was deviated may be calculated. Provided A is the first interaction, the surface of the cone on which the photon originated can be computed and if this intersects a particular plane of interest at PQRS, the locus of the source will be a conic section. Numerous emissions from a point source of activity will produce several ellipses which should intersect at an emission point. Extension of the principle to many sources in different planes is then a computational problem.

5.3 Coded Aperture Imaging

Since the weakest link in conventional imaging systems is the collimator, a technique that dispenses with it is clearly very attractive and it has been suggested that the collimator could be replaced by a zone plate (Barrett 1972). This will consist of concentric annuli, each of equal area, and made up, alternately of gamma ray opaque and transparent material. Each source within the object will then project a shadow onto the detector (Fig. 12) and this shadow will contain information about the amount of activity (intensity) its position (centre of circle) and distance from the zone plate (size of rings).

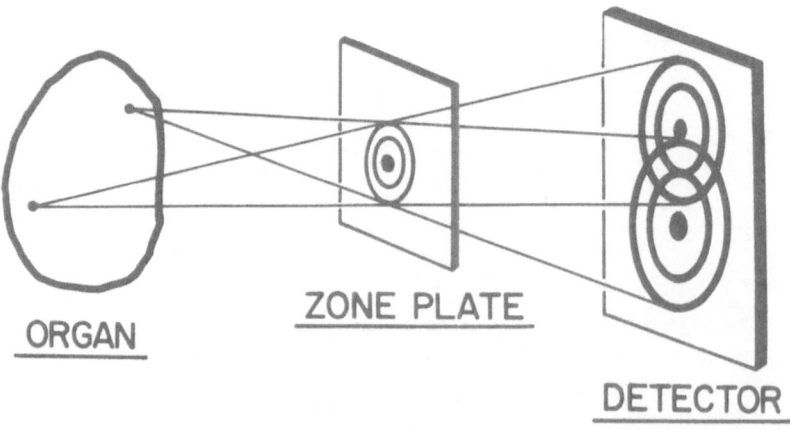

ZONE PLATE

ORGAN

DETECTOR

Fig. 12 Diagram showing how a zone plate can generate a ring
pattern corresponding to each point in the object.

The count rate will now be very high (perhaps 5×10^3 cps Bq^{-1}).
X-ray film might provide a cheap, simple and highly portable imaging
system.

De-coding the image involves unscrambling the data contained in
the pattern of concentric rings. In effect highly coherent light
may be used to create the Fourier Transform of the hologram and as
the viewing screen is moved backwards and forwards, light from
rings of different diameters is focussed corresponding to different
planes in the object. There are a number of technical problems
associated with decoding which will not be discussed, but one of
the most important is the signal to noise ratio. If this ratio for
a zone plate, R_z, is compared with that for a camera R_c

$$\left(R_z / R_c \right) \propto \left(C / N \right)^{\frac{1}{2}}$$

where C = counts/image resolution element and N = number of elements
in the image.

As the number of elements in the image increases, C varies as
$N^{\frac{1}{2}}$ so this ratio varies as $N^{-\frac{1}{2}}$ and for an object of fixed size,
there is a minimum element size below which the ratio becomes less
than one. Thus the Zone Plate system is best for small objects - a
4 mm element may be used for the thyroid - but resolution is rather
worse for larger objects.

6. HIGH SENSITIVITY, LOW RESOLUTION WHOLE BODY IMAGING

The conflicting demans of resolution and sensitivity have
already been discussed. In the modern gamma camera collimator
design has a big effect on resolution but in order to achieve a
typical resolution of the order of 10 mm, the design of the collimator is such that only about one in ten thousand of the gamma rays
reaches the crystal and activities in the region of 4-10 mCi (150 -
350 MBq) of Tc99m are required to give good images in a reasonable
imaging time.

In a number of investigations, for example into the metabolism
of trace elements such as copper and zinc in the body, it is quite
impracticable to administer such quantities. Therefore a scanning
whole body counter is designed to optimise sensitivity. The Aberdeen system consists of four detectors, each comprising a 100m thick
x 150 mm diameter sodium iodide crystal situated two above and two
below the patient. Each crystal is fitted with a very coarse multi-
hole collimator with a focal length of 300 mm chosen so as to focus
in the middle of the body when the patient is laying on the scanning
bed. The detector system is surrounded by 100 mm lead and 20 mm

iron to provide an effective shield against extraneous radiation.
The bed is controlled digitally and moves longitudinally through
the shadow tunnel, stepping laterally at the end of each passage by
a predetermined amount.

The spatial resolution of the system has been determined for
line sources in air at the focal distance and figures for the full
width half maximum are 10 cm for Cr51 (0.32 MeV) and 12 cm for Fe59
(1.10 and 1.29 MeV). In view of the resolution and storage capacity
of the system, a pixel size of 25 x 25 mm has been found suitable
for most applications and has been used as standard. Under these
conditions the detection limits for extended sources of humanoid
shape in supine position (to get a significantly distinguishable
image at a scanning speed of 1 pixel/second) are 13 μCi of Cr51
and 1.8 μCi of Fe59. For a standard format of 16 x 52 pixels of
25 mm x 25 mm size, useful distribution patterns may be obtained
from total body burdens as small as 0.5 μ Ci of Zn65 in scanning
times of 30 min. (Gvozdanovic et al 1981)

The scanning whole body counter has been used to study metabolism of zinc which is required in minute quantities as part of
many essential co-enzyme systems. If there is a deficiency, this
can result in keratinisation of skin, erythema and slow wound healing. This is a condition frequently associated with cirrhosis of
the liver.

The question to be answered was whether the deficiency resulted
from malabsorption of dietary zinc or some other cause. By way of
comparison, patients suffering from coeliac disease were also
studied. This is one of the classical malabsorption syndromes
characterised by malnutrition, abnormal stools and varying degrees
of oedema, skeletal disorders, peripheral neuropathy and anaemia.

0.5 μ Ci Zn - 65 was administered to different groups of
patients and whole body scans were performed at frequent intervals
for many weeks. There was little difference in the observed overall retention of Zn-65 but when region of interest curves were
plotted for the liver and for the skeleton (the sacro iliac region)
they showed slow liver clearance and slow bone uptake in normals,
(Fig. 13a) and in patients with coeliac disease but much more rapid
liver clearance and bone uptake in cirrhotic disease (Fig. 13b)

In conclusion, simple whole body counting would have shown
little or no difference between the total retention of Zn-65 for
normal and cirrhotic patients. However, even with very poor resolution, the scanning machine demonstrates the different patterns of
retention. In terms of the original problem this suggests that
when zinc is in the liver it is available for incorporation into
enzymes but when it is incorporated into the skeleton (e.g. in

cirrhosis) it is irreversibly bound and not available for further
metabolic incorporation.

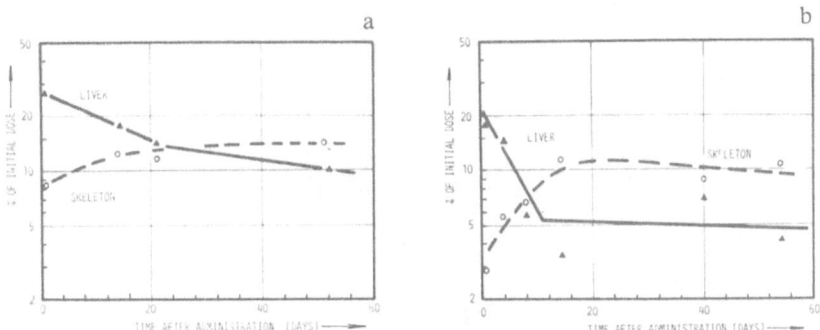

Fig. 13 Dynamic curves showing the content of Zn-65 tracer in
liver and skeleton obtained from integration over areas
of interest (a) normal   (b) patient with cirrhosis
(Reproduced with permission from Gvozdanovic et al 1982)

7. SINGLE PHOTON COMPUTED TOMOGRAPHIC IMAGING

Single photon emission computed tomography (SPECT) has been
developed to solve the problem of activity in overlying structures.
The principle is very similar to transmission tomography except
that now a transverse section of the body may be scanned using a
pair of opposed detectors, a single traverse being obtained at each
of a series of angles around the section of interest. A gamma
camera rotating around the patient can also be used to obtain such
a series of angularly spaced profiles.

As shown in Fig. 14, a projection essentially consists of a
line integral of a two dimensional function g(xy) where g(xy) here
represents the amount of activity in one pixel in the plane of
interest. All pixels lying along the line L will project to the
point P but when the distribution is viewed from a different angle,
the element g(xy) will contribute to a different point on the
profile. Hence by back-projecting all the profiles it is possible
to estimate the contribution from the element g(xy). The method
of reconstruction used in Aberdeen has been described fully elsewhere (Keyes 1976).

212

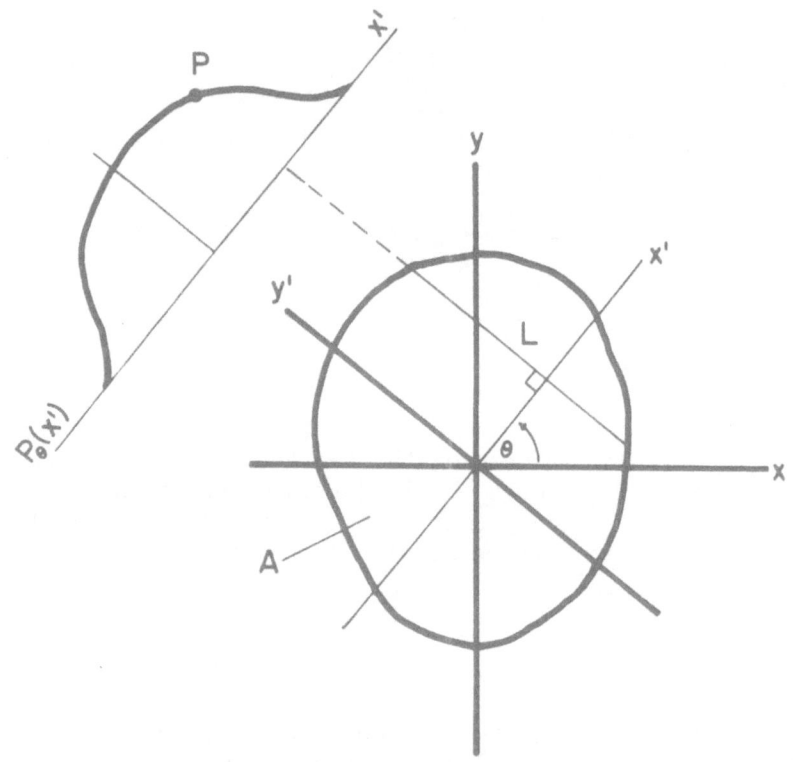

Fig. 14   Sketch showing how the profile generated at angle θ,
$P_\theta(x')$ represents the projection of contributions from
a number of elements along lines such as L.

We have developed three systems:-

(i)   The Aberdeen Section Scanner (MK1) comprises two
cylindrical sodium iodide crystal detectors each 50 cm in area
which produce opposite conventional views simultaneously.   To
obtain a transverse section view, the detectors rotate through 180°
around the chosen plane, performing a single scan at each angular
interval of 6 degrees.

(ii)   The tomographic camera comprises a Nuclear Enterprises
large field of view detector with a medium sensitivity collimator.
The camera is mounted in a heavy rotating gantry and is interfaced

to a DEC gamma-11 data processing system (Chesser and Gemmell 1982).
Each of 60 gamma camera images spaced at 6 degree intervals is
stored in a 64 x 64 matrix of element size 7 mm and any one of 64
sections can be displayed.

(iii) The Aberdeen Section Scanner (MK2) has been designed to
expose the maximum area of sodium iodide crystal to the tomographic
plane of interest (see Table IV), thereby giving greatly increased
sensitivity which may be used either to give better statistics in
a single section or to obtain multiple sections in a time comparable with that of the camera. We have retained the facility for
conventional imaging. The system consists of four banks of six
scintillation detectors arranged at the sides of a square in a
translate - rotate configuration. Each detector consists of a
60 x 100 x 120 mm NaI(Tl) crystal with its own photomultiplier.
The four scanning heads are mounted on a heavy annular frame and
the separation distance of the detector faces may be varied between
31 cm and 41 cm for brain or body scans.

TABLE IV

AN ESTIMATE OF THE TOTAL CRYSTAL AREA USEFULLY
EXPOSED TO GAMMA RAYS IN DIFFERENT SPECT SYSTEMS

System	Crystal area
Camera (single section of effective thickness 2 cm)	80 cm^2
Aberdeen Section Scanner Mk 1	120 cm^2
Cleon CT Imager	3100 cm^2
Aberdeen Section Scanner Mk 2	1440 cm^2

Under optimum conditions, the resolution of a modern SPECT
system is about 10 mm (FWHM). However, some of the physical
criteria normally used to assess imaging performance are no longer
entirely appropriate because they take no account of the process of
reconstruction.

If a tomographic system is to reproduce faithfully a three-
dimensional distribution of radioactivity, it must be capable of
providing reliable images of both an extended source of uniform
activity and of a distribution of localised sources of equal
activity. Some of our recent work (Gemmell et al 1982) has been
directed towards establishing how closely tomographic systems,
especially cameras, come to achieving these aims and how any
deficiencies might be overcome.

To obtain an accurate sectional image of a uniform extended source of radioactivity, great care is required at every stage in both data collection and image reconstruction. There are many possible causes for artefacts and in particular the requirements of camera uniformity or satisfactory on-line or off-line uniformity correction are much more stringent than for conventional gamma camera imaging (Rogers et al 1982).

Figure 15(a) illustrates another major problem. When sources of equal activity are placed at different radial distances from the centre of rotation in a body-sized phantom of tissue equivalent material (Rando phantom), the apparent activity in the reconstructed image shows strong radial variation. Furthermore, this error cannot be eliminated by arbitrary choice of attenuation coefficient. Recent work, both by our group and by others, has attempted to eliminate this radial variation by iterative methods. Our approach has been developed by Dr Noel Evans and follows that of Walters et al (1981). A first approximation image is obtained by conventional filtered back projection of the original profiles using no attenuation correction. A set of attenuated profiles is then calculated from this image and compared with the actual profiles to form "error" profiles. These are used to form an "error" image which is subtracted from the first approximation image to give a second approximation. This is the starting point for the next iteration. Using two iterations, thereby extending the data processing time from 30 sec to 4.5 min, the radial variation in apparent activity has been largely eliminated (Fig. 15b)

Many other questions that are currently of importance in SPECT are considered in a review by Keyes (1982)

There have been many reported clinical investigations in which emission tomography has been employed and the majority show some evidence of improved detectability or additional information on lesion shape and position (see e.g. Berche 1981). When detailed comparisons are made with conventional imaging in a prospective trial however, the improvement is small and results for a large number of patients must be compared to demonstrate a statistically significant improvement with the tomogram (Dendy and Gemmell 1983).

8. CONCLUSIONS

During the past 15 - 20 years, the wide availability of a range of artificially produced radionuclides together with significant improvements in the design of imaging equipment based on the sodium iodide detector have led to a dramatic increase in the nuclear medicine service. It may readily be shown that the question of detection depends on achieving an adequate signal to noise ratio so the development of new organ-specific radiopharmaceuticals is also essential.

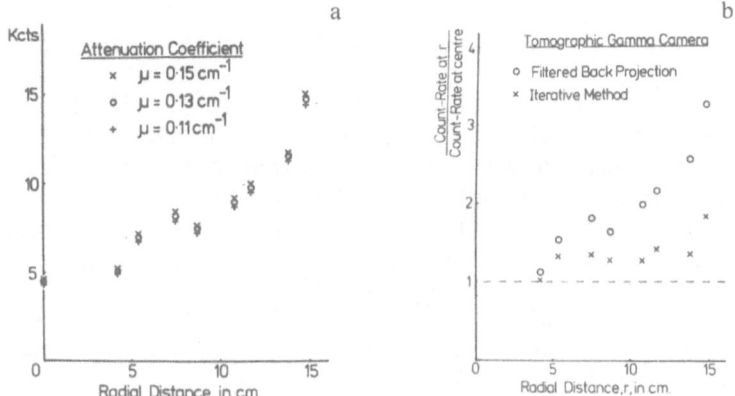

Fig. 15 Radial variation in apparent activity in the tomographic
reconstructed image of point sources of equal activity
(a) the effect cannot be removed by any selection of
attenuation coefficient
(b) the effect can be largely removed by an iterative
approach

Static imaging using either a scanner or a gamma camera is still
important for some investigations notably bone scans, but recent
trends have been towards dynamic studies, making greater use of the
physiological nature of the investigation. Since the gamma camera
is capable of both static and dynamic imaging, it has become the
instrument of choice in most centres.

Notwithstanding, there are still a number of technical limitations to the normal gamma camera, notably poor discrimination
against scattered radiation, the low sensitivity of the collimated
system and poor performance at high count rates. Several more
recent devices which set out to overcome one or other of these
problems have been described. In the longer term, the best prospect
of significant improvement of the overall performance of a standard
camera system would appear to be the replacement of the NaI(Tl)
crystal by a large crystal solid state detector.

REFERENCES

Atkins F.B., Beck R.N., Hoffer P.B. and Palmer D. (1977) Dependence
of optimum baseline setting on scatter fraction and detector
response function. Medical Radionuclide Imaging Vol. 1, IAEA
Vienna pp. 101-117.

Barrett H.H. (1972) Fresnel zone plate imaging in nuclear medicine
J. Nucl. Med. 13, 382

Berche C., Aubry F., Langlais C. Vitaux J. et al (1981) Diagnostic
value of transverse axial tomoscintigraphy for the detection of
hepatic metastases. Results of 53 examinations and comparison
with other diagnostic techniques. Eur. J. Nucl. Med. 6, 435-452.

Chesser R. and Gemmell H.G. (1982) The interfacing of a gamma camera
to a DEC Gamma-11 data processing system for single photon emission
tomography. Phys. Med. Biol. 27, 437-441

Crawley J.C.W., Ajdukiewicz A.B., Bassett N. et al (1980) A radioisotope scanner for use in developing countries in Medical Radionuclide Imaging Vol. 1, IAEA Vienna, pp.73-82

Dendy P.P. and Gemmell H.G. (1983) An evaluation of the contribution
of single photon emission computed tomography (SPECT) to radionuclide imaging of the liver. Annales de Radiologie 26, 72-81

Driard B., Verat M. and Rozieres G. (1976) A large field image
intensifier tube for scintillation cameras. IEEE Trans. Nucl. Sci.
NS-23, 502

Erickson J. and Brill A.B. (1978) Scintillation Cameras in Text
Book of Nuclear Medicine - Basic Science. Eds. A.Fernando,
G. Rocha, J.C. Harbet. Lea and Febiger, Philadelphia, pp.264-284.

Everett D.B., Fleming J.S., Todd R.W. and Nightingale J.M. (1976)
A camera using Compton interactions in Medical Images, formation,
perception and measurement. Ed G.A. Hay. Institute of Physics and
John Wyley, Bristol pg.89

Gemmell H.G.,Dendy P.P., Pitt W.R., et al (1982) Physical criteria
for the performance of tomographic imaging systems. Nuclear
Medicine and Biology - Proceedings Third World Congress, Paris.
Pergamon Press pp. 476 - 479.

Gvozdanovic D., Ettinger K.V., Smith D.B. et al (1981) Investigations of long-term in vivo tracer distribution patterns using an
ultra-high sensitivity scanning system in Medical Radionuclide
Imaging 1980. IAEA Vienna Pg.83

Gvozdanovic D., Gvozdanovic S., Crofton W.W. et al (1982) Ultra-
high sensitivity imaging in study of long term distribution of
Zn-65 tracer. World Congress on Medical Physics and Biomedical
Engineering, Hamburg. Eds. W.Bleifeld, D.Harder, H-K Leetz, and
M.Schaldoch. Published by MPBE Hamburg, Abstract 21.26

Harbert J.C. and Neto A.D. (1978) Rectilinear Scanners in Text Book
of Nuclear Medicine - Basic Science. Eds. A. Fernando, G. Rocha
and J.C. Harbert. Lea and Febiger, Philadelphia pp.248-263

Kaufman L., Lorenz V., Hosier K. et al (1978) Two-detector 512-
element high purity germanium camera prototype IEEE Trans.Nuc.
Sci. NS-25, 189.

Keyes J.W. (1982) Perspectives on Tomography. J. Nucl. Med. 23, 633-640

Keyes W.I. (1976) A practical approach to transverse section gamma camera imaging. Br. J. Radiol. 49, 62-70

Kouris K. Spyrou, N.M. and Jackson D.F. (1982) Progress in medical and environmental physics. Vol. 1. Imaging with ionising radiations. Surrey University Press and Blackie & Son.

Mallard J.R. (1972) The radionuclide imaging process and factors influencing the choice of an instrument for brainscanning, In Progress in Nuclear Medicine. Eds. E.E. Pochen and V.R. McCready, Karger Basel and University Park Press, Baltimore pp.1-114

Rogers W.L., Clinthorne N.H., Harkness B.A. et al (1982) Field flood requirements for emission computed tomography with an Anger camera. J. Nucl. Med. 23, 162-168

Roux G., Gaucher J.C., Lansiart A. and Lequais J. (1972) Detecteur photoelectronique analogique de la position de scintillations faiblement lumineuses in Photo-Electronic Image Devices. Academic Press New York p.1017

Strand S.E. and Larsson L. (1978) Image artefacts at high photon fluence rates in single crystal NaI(Tl) scintillation cameras. J. Nucl. Med. 19, 407

Todd-Pokropek A.E., Erbsmann F. and Soussaline F. (1976) The nonuniformity of imaging devices and its impact on quantitative studies in Medical Radionuclide Imaging. Vol. IAEA Vienna, pp. 67-84

Walters T.E., Simon W., Chesler D.A. et al (1981) Attenuation correction in gamma emission computed tomography. J.Comp. Assist. Tomog. 5, 89-94

Wicks R. and Blau M. (1979) Effect of spatial distortion on Anger camera uniformity correction. J. Nucl. Med. 20, 252

PHYSICAL LIMITATIONS TO THE QUALITY OF X- AND GAMMA RAY IMAGES

Peter F. SHARP

Department of Biomedical Physics and Bioengineering,
University of Aberdeen, Foresterhill, Aberdeen AB9 2ZD,
Scotland, UK.

1. INTRODUCTION

Compared with conventional photographic images, both X-ray
and nuclear medicine images have poor visual quality. If we
consider the number of photons involved, then in conventional photography images have photon densities of the order of 10^{11} to 10^{12} cm^{-2}
whilst in radiography the figure is about 10^{7} cm^{-2} and in nuclear
medicine between 10^{2} and 10^{3} cm^{-2}. At these low photon densities
noise, resulting from the discrete nature of photon emission, will
be evident. Spatial resolution in radiology can be quite good, of
the order of a millimetre, but in nuclear medicine it is usually
between one and two centimetres.

The quality of an image, however, should not be judged simply
on its similarity with a photograph but rather on how effectively
it can be used for its intended purpose. For example, although
nuclear medicine images are noisy and have poor spatial resolution,
their value lies in their ability to show how a pharmaceutical is
handled by the body. Thus bone imaging, Fig. 1, in which the
uptake of the radiopharmaceutical indicates metabolic activity in
the skeleton, provides an extremely sensitive way of detecting
early metastatic deposits perhaps showing them several months
before bone density has altered sufficiently to be seen on a radiograph.

Reducing noise and improving spatial resolution will of
course improve the effectiveness of the imaging technique. Unfortunately it is frequently necessary to choose either to reduce noise
or to improve resolution and there have been several investigations

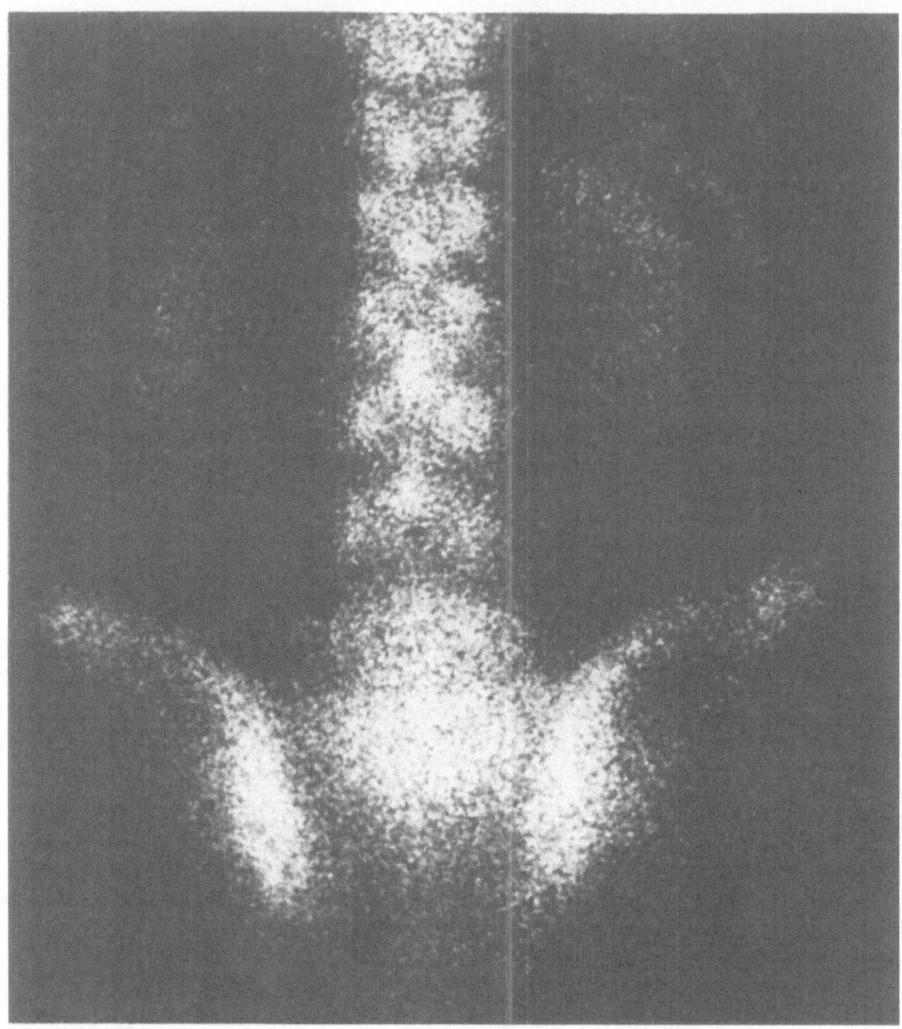

Fig. 1. The posterior view of a normal bone image.

into which offers a greater gain. (Keller and Coltman 1968; Tsui et al 1978).

The following sections will show how spatial resolution and noise can be described, and consider the physical factors influencing them and how they interact to determine image quality.

2. SPATIAL RESOLUTION

Perhaps the oldest measure of spatial resolution is that based on the ability to resolve two parallel lines. Although the minimum necessary separation will depend upon a number of factors, the most commonly made assumption, known as Taylor's criterion (Longhurst 1967), is that it is equal to the width of a line image at half its maximum intensity, the full width at half maximum height (FWHM).

The concept of using line images to define resolution was initially adopted to determine the resolution of spectral lines, but the idea has remained useful since it fits into the framework of linear systems theory.

At the most general level the spatial response of an imaging system can be expressed by its point spread function (PSF) or impulse response, this being the image of a point source or delta function input. However, there may be practical difficulties in creating such an input, for example in nuclear medicine there is the problem of concentrating sufficient radioactivity into the necessary small volume. Alternatively, measurements can be made with a line source which is narrow enough to act as a δ function across its width but effectively infinite in length. The line spread function, LSF, and PSF are related by the expression

$$LSF(x) = \int_{-\infty}^{\infty} PSF(x,y)dy$$

Thus a profile taken across the line image will be identical to the one through the PSF.

a third possibility is to measure the profile across a sharp edge, the edge spread function (ESF). In this case,

$$LSF(x) = \frac{d}{dx} ESF(x)$$

In practice the need to differentiate the ESF to give the LSF may outweigh any advantage of this approach.

The FWHM of the PSF or LSF is frequently used in nuclear medicine as the measure of system resolution. Whilst it indicates the smallest detail appearing in the image, this is not the same as the smallest detail faithfully reproduced by the system. The optical transfer function (OTF) indicates how effectively spatial information, defined in linear systems theory by the amplitude and phase associated with a particular spatial frequency, is transmitted by the imaging device.

The OTF is equal to the Fourier transform of the PSF normalised by the area under the PSF. (Dainty and Shaw 1974).

$$\text{OTF}(f_x, f_y) = \frac{\iint \text{PSF}(x,y)\exp[2\pi i(f_x \cdot x + f_y \cdot y)]dxdy}{\iint \text{PSF}(x,y)dxdy}$$

The relationship between an object $O(x, y)$ and the resulting image $I(x, y)$ can be described very simply in frequency space since the normal convolution relationship becomes

$$I(f_x, f_y) = O(f_x, f_y) \times \text{OTF}(f_x, f_y)$$

Also if the imaging system consists of several components effectively acting in series, then the OTF of the complete system can be calculated from the product of the OTFs of these components.

$$\text{OFT}_T(f_x, f_y) = \prod_i \text{OTF}_i(f_x, f_y)$$

Thus the use of the system transfer function considerably simplifies the analysis of device performance although it must be remembered that it is applicable only to linear systems.

In nuclear medicine and radiology the modulation transfer function (MTF) is frequently quoted instead of the OTF (MacIntyre et al 1969). The MTF is the modulus of the OTF, thus

$$\text{MTF}(f_x, f_y) = |\text{OTF}(f_x, f_y)|$$

$$= [C(f_x, f_y)^2 + S(f_x, f_y)^2]^{\frac{1}{2}}$$

where $C(f_x, f_y) = \iint \text{PSF}(x,y)\cos 2\pi(f_x \cdot x + f_y \cdot y)dxdy$

and $S(f_x, f_y) = \iint \text{PSF}(x,y)\sin 2\pi(f_x \cdot x + f_y \cdot y)dxdy$

The MTF does not specify system performance completely since it neglects phase information and only describes how amplitudes are modulated.

The phase information can be represented by the phase transfer function

$$P(f_x, f_y) = \tan^{-1}\left[\frac{-S(f_x, f_y)}{C(f_x, f_y)}\right]$$

However the OTF and MTF are equivalent if the PSF is both real (as it must be in nuclear medicine and radiology) and even, i.e.

$$PSF\ (-x,\ -y) = PSF\ (x,\ y)$$

Figure 2 shows the MTFs for two gamma cameras, one built in 1972 and the other in 1978. Both MTFs show how high spatial frequencies, representing fine spatial detail, are reproduced less effectively than low frequencies. At any frequency the 1978 system has a higher MTF value than the older system so demonstrating its superior spatial resolution.

Whereas the OTF has many advantages from the analytical aspect of image analysis, it cannot predict how effectively the human observer is able to use the image. One simple measure which has been derived from the OTF and which it is claimed (Schade 1975) correlates well with the subjective impression of image quality is the noise equivalent pass-band (NEP)

$$NEP = \iint |OTF(f_x, f_y)|^2 df_x df_y$$

Wagner (1977) has considered the application of the NEP to radiology.

Various test patterns have been used in both radiology and nuclear medicine to allow spatial resolution to be assessed subjectively. One of the main problems is to devise a perceptual task which will permit a quantitative measure of resolution to be made. In many patterns (Hay 1964, Rollo 1977) resolution is measured in terms of the observer's ability to detect detail of various sizes or to resolve detail in periodic structures such as the bar phantom or Anger phantom (Anger 1973). The former have the advantage of relating resolution to the clinical task of detecting abnormal pathology but there does appear to be a need to reconsider what information these patterns give.

Having decided upon a measure of spatial resolution, what physical factors affect it?

Patient movement is one over which little control can be exercised. The use of short exposures is a possible solution in radiology but is not feasible in nuclear medicine. In nuclear medicine movement compensation has been applied successfully either by using the centre of gravity of the data to realign the images electronically (Hoffer 1972) or by synchronising data acquisition with physiological signals (Parkin 1980). Physiological gating has been particularly useful in cardiology for imaging the heart at different times during the cardiac cycle.

Imaging geometry poses many problems. In radiology the finite size of the anode prevents it being considered as a simple point

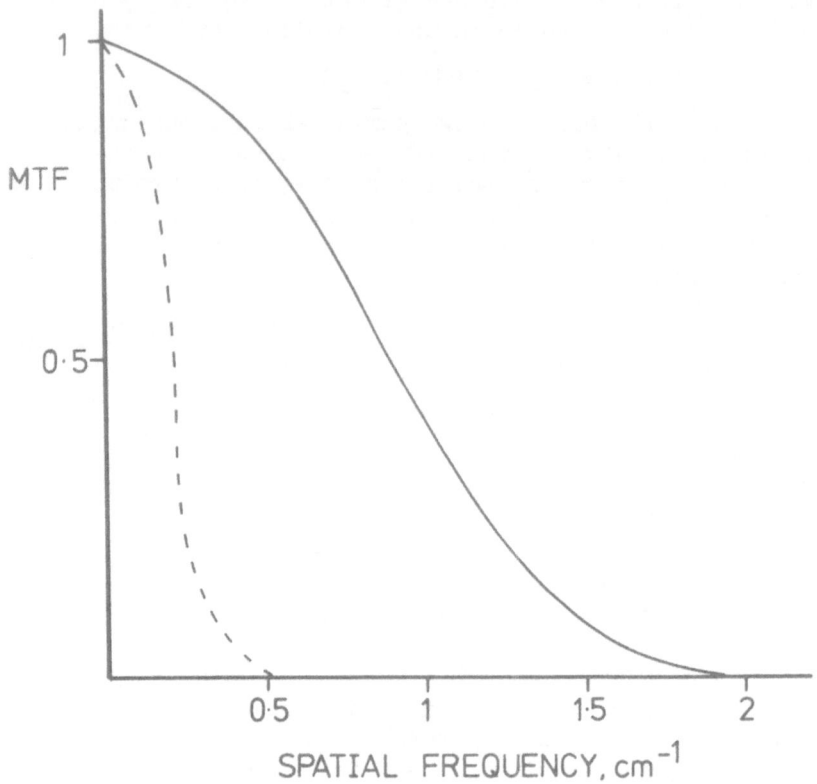

Fig. 2.   Modulation transfer functions for a gamma camera made in
1978 (———) and in 1972 (----).

source of X-rays. If the resulting PSF is asymmetrical then the
OTF may have negative values at some spatial frequencies. The
consequence of this will be a phase shift of those frequencies
associated with negative OTF values so giving spurious resolution
(Wagner et al 1974).

     An analogous problem in nuclear medicine arises from the need
to collimate the gamma-ray flux in order to produce an image. As
gamma rays are emitted isotropically, to form an image it is
necessary to exclude from the detector of the gamma camera all
rays except those travelling in a particular direction. The
commonly used parallel-hole collimator consists of a series of
parallel holes separated by lead septa so only those gammas
travelling along the hole axis, perpendicular to the face of the
detector, will contribute to the image. (Anger 1964). As the
diameter of the holes is large, approximately 1.5 mm, it is inevitable that some obliquely incident gammas will also be detected.
Thus the image of a point source will be blurred having a FWHM at
best of 5 to 7 mm. This blurring becomes worse as the source is

moved further from the collimator.

Scatter of radiation is a major factor in degrading
spatial resolution. Over the range of photon energies of interest
to radiology and nuclear medicine, say between 40 and 300 keV, the
main attenuating process is Compton which involves not only
absorption of energy but also the scattering of X- or gamma photon.

The energy retained by the scattered photon is related to
its incident energy hν and angle θ, through which it is
scattered by the formula

$$h\nu_{out} = h\nu_{in}[1 + \alpha - \alpha \cos \theta]^{-1}$$

$$\text{where } \alpha = \frac{h\nu_{in}}{m_0 c^2}$$

Thus the proportion of the energy retained by the low energy
photons (Fig. 3) is high even for large values of θ

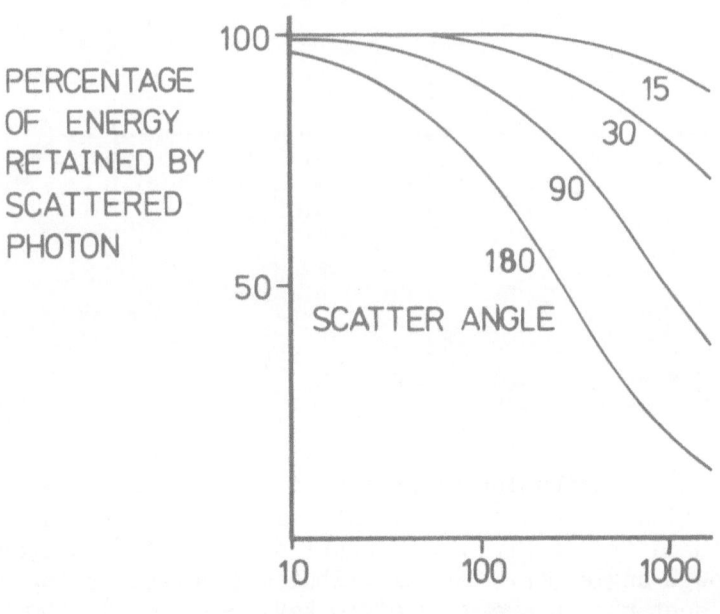

Fig. 3. Fraction of energy retained by a photon after Compton
scattering through various angles.

226

The likelihood of a photon being scattered through a particular angle is given by the differential cross-section which is shown in Fig. 4. Note that at low energies there is a high probability of scattering through large angles.

Obviously the imaging of scattered photons will result in a loss of image sharpness so how can it be minimised? Discrimination against scatter can be made either on the basis that these photons will have lost energy, or that they will have deviated from their expected direction.

The former approach is used in nuclear medicine since the expected, photo-peak, energy of the gamma photons is known and pulse height analysis can be applied to exclude from the image all those photons whose energy is different from the photo-peak energy. Since the accuracy with which the gamma camera can measure the energy of a photon is only about 11%, to ensure that most of the

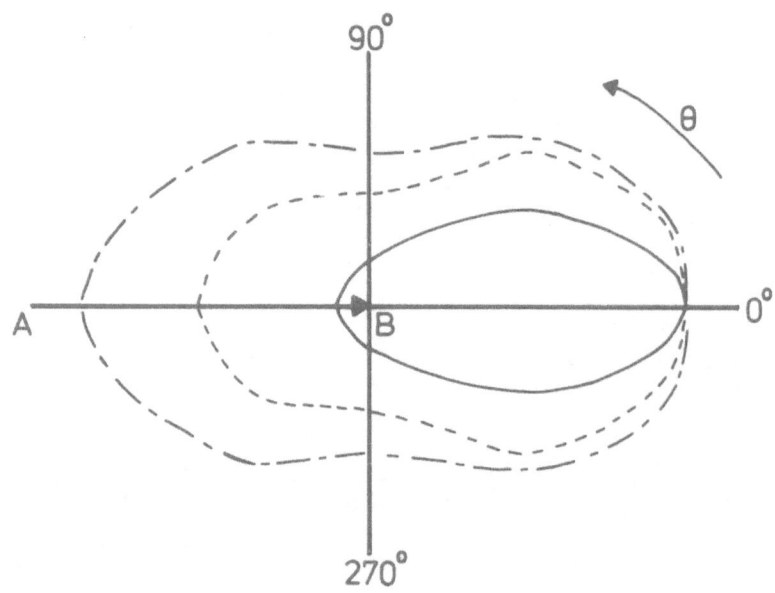

Fig. 4. Spatial distribution of photons incident along the direction AB and scattered at point B. The radial distance from B gives the amount of radiation scattered through angle θ . The distributions corresponding to incident photon energies of 10 keV (— - —), 100 keV (-----) and 1 MeV (———).

unscattered photons are used, it is necessary to accept photons whose energies fall within a range of values usually chosen to be 20% to 30% of the photopeak. Since it has been shown that photons have a high probability of being scattered through large angles with only a small loss of energy, it can be appreciated that pulse height analysis is a relatively inefficient method of excluding scatter. Under typical conditions about 30% of the total number of photons forming the image will have been scattered (Atkins et al, 1977).

Since the X-ray beam in radiology is polyenergetic, pulse height analysis cannot be applied. However the direction of the beam is now well-defined so the change in direction of scattered photons can be used to exclude them. One of the most common techniques is to place a lead grid in front of the X-ray film. In situations where the scatter can be considered as occurring uniformly over the image, it can be shown (Barrett and Swindell, 1981) that the system MTF is reduced by a factor $(1 + S/P)^{-1}$ where S/P is the ratio of scatter to primary photons. This ratio may be as high as 10 for large field sizes whilst with grids it can be reduced to about unity.

We must now consider the efficiency of the detector itself. In the gamma camera this consists of a scintillation crystal viewed by an array of photomultiplier tubes (see page 195). The image is formed in the scintillation crystal by the collimator, as already discussed, but these scintillations must be converted into electronic signals by the photomultipliers if pulse height analysis is to be carried out. This process will also degrade spatial information.

The accuracy with which the photomultiplier tube array can convert the image formed in the scintillation crystal into electronic signals will depend upon the number of photons resulting from a gamma-ray interaction in the crystal. As the number of light photons arriving at a particular photomultiplier tube follows Poisson statistics, consecutive scintillations from the same position in the crystal will produce slightly different electronic signals and will be interpreted as originating from slightly different spatial locations. The blurring of a point of image in the crystal on the final display is known as the intrinsic resolution (FWHM), R_i, of the gamma camera and is of the order of 3mm for gamma rays of 100 keV energy.

The total spatial resolution, R_T, of the camera is, to a good approximation, given by

$$R_T = \left[R_c^2 + R_i^2 \right]^{\frac{1}{2}}$$

where R_c is the collimator resolution.

Other factors affecting camera resolution are discussed elsewhere.

In radiology the detector usually consists of double-sided photographic film sandwiched between two intensifying screens. Deterioration of resolution is mainly caused by diffusion of light in the phosphor of the screens, the film itself being too thin to contribute significantly to this effect. Increasing screen thickness or phosphor grain size will worsen resolution. In these systems cross-over of light photons from, say, the front screen to the rear emulsion, must also be taken into consideration.

3. NOISE

At the low photon densities encountered in nuclear medicine and radiology it is readily apparent that the images are made up of individual quanta. In nuclear medicine the visual effect is dependent upon the size of light spot used to record each detected photon onto photographic film (Fig. 1) so the appearance of the noise can to some extent be controlled.

In radiology noise appearance will depend upon the characteristics of the intensifying screen and film used. Each X-ray photon interacting with the screen will produce a blurred spot image on the film - the point spread function of the screen. If the film-screen combination is inefficient so that a large number of these spots is needed to give a particular film density, then the spots will overlap and produce a blurred but uniform image. However, if a more efficient system is used in which only a few spots are needed to produce the required film density, then individual blurred spot images will be apparent so giving a mottled appearance to the image known as quantum mottle. The efficient system will produce the sharper images but quantum mottle will obscure low contrast detail.

How can noise be described?

The emission of photons from a source follows the Poisson process so if the average rate of detection of photons over the image area is constant, N per unit area, statistical variability in the number found in some area A will have a standard deviation of $\sqrt{NA}$.

For a wide variety of photographic films, the subjective assessment of graininess correlates well with the variance in density σ_D^2, measured by scanning the film with a microdensitometer of aperture A. The Selwyn granularity coefficient S, is defined as

$$S = \sigma_D (2A)^{1/2}$$

As S is found to be independent of A for a wide variety of films it provides a useful measure of granularity (Selwyn 1935) but unfortunately it is less effective where used on images produced with film-screen combinations probably because of the correlations introduced by the overlapping blurred spots. It also has the drawback of not providing an estimate of how effectively the noise will mask a signal.

Simple measures based on the first order statistics of the Poisson process, such as variance, will only give an indication of the magnitude of the noise. It is also necessary to know the spatial structure or texture of the noise. This can be described by second order statistics. The second order probability density function gives the probability that at a point (x, y) the image density, say, lies in the range D_1 to $D_1 + \Delta D_{\dot{\nu}}$ and at (x + a, y + b) it lies between D_2 and $D_2 + \Delta D_2$. Assuming that density variation is described by a Gaussian distribution, this second order probability density function can be described by its first joint moment, the autocorrelation function C (a, b) (Dainty and Shaw 1974b), where

$$C(a,b) = \text{limit of } \frac{1}{2X} \cdot \frac{1}{2Y} \int_{-X}^{X} \int_{-Y}^{Y} D(x,y) \cdot D(x+a,y+b) dx dy$$

as X and Y tend to infinity.

In practice it is more convenient to use the fluctuation, ΔD of the density from its mean rather than absolute density values so

$$C^1(a,b) = \text{limit of } \frac{1}{2X} \cdot \frac{1}{2Y} \int_{-X}^{X} \int_{-Y}^{Y} \Delta D(x,y) \Delta D(x+a,y+b) dx dy$$

as X and Y tend to infinity.

If a and b are zero then $C^1(0, 0)$ is equal to the variance of the density fluctuations, i.e. variance provides a measure of noise magnitude while the shape of the autocorrelation function describes texture.

If this measure of noise is to be used in a linear system theory approach, it is necessary to define it in terms of spatial frequencies. The Fourier transform of the autocorrelation function is known as the Wiener spectrum or noise power spectrum.

$$W(f_x, f_y) = \int\int C^1(a,b) \exp[(-2\pi i (f_x a + f_y b))] dx dy$$

The variance of the density fluctuations is now given by the area under the two dimensional Wiener spectrum.

The autocorrelation function and Wiener spectrum are equivalent ways of describing noise, analogous to the use of PSF or OTF to describe spatial resolution.

If the input noise now passes through a linear filter system, the output Wiener spectrum, Wout, is linked to the input Win, by the relationship

$$W_{out}(f_x,f_y) = W_{in}(f_x,f_y)|OTF(f_x,f_y)|^2$$

where the OTF is the system transfer function.

The Wiener spectrum can be measured by scanning a uniformly exposed piece of film with a microdensitometer and analysing the frequency output. The measurement and application of Wiener spectrum description of noise in radiology is described by Rossman (1963) and in nuclear medicine by Tsui et al (1981).

4. IMAGE QUALITY

The problems of determining subjective image quality experimentally will be considered in a later presentation. Here we will consider mainly mathematical descriptions of quality incorporating the measures of spatial resolution and noise discussed earlier.

De Vries (1943) and Rose (1942) proposed that the limiting factor on detecting detail in photon limited images was the ratio of signal intensity to noise, SNR. This led to the concept of describing the performance of an imaging device in terms of the ratio of SNR at the output to SNR at the input, a ratio known as the detective quantum efficiency, DQE.

For photographic systems the DQE has been expressed in terms of system MTF and output Wiener spectrum as

$$DQE(f_x,f_y) = \frac{\gamma^2 (\log_{10} e)^2 MTF(f_x,f_y)^2}{N \cdot W(f_x,f_y)}$$

where γ is the film gamma and N the number of quanta exposing the film (Shaw 1974).

For radiographic systems using rare earth screens, DQE values of up to 40% have been reported.

If we wish to know to what extent such a measure of performance reflects image quality it is necessary to relate objective measures, such as SNR, with subjective behaviour, that is to say, how the observer utilises the information presented to him.

The De Vries-Rose model proposed that, when faced with a noisy image the observer sampled areas equal in size to that of the signal. Thus for Poisson noise of average N photons per unit area, the noise will have a standard deviation $\sqrt{NA}$, A being the signal area. Rose suggested that the signal would be seen provided it exceeded this noise level by a factor of 4 or 5.

The most obvious difficulty with this approach is in defining A, particularly if the signal does not have sharp edges and if the observer has no prior knowledge of its size. Later (Morgan 1965) it was proposed that A should be related to the MTF of the visual system while others have proposed more sophisticated statistical tests (Mallard and Corfield 1969).

Sharp and Mallard (1974) suggested that the general concept of a statistical test was implausible and the statistics should be used instead to predict the probability of the non-signal (background) area producing by chance a signal resembling the true one. The probability of detecting the correct signal is thus given by

Probability of a correct response = $\sum_{i}$ [Probability of no signal of intensity greater than i being present in the background]

X [Probability that the true signal has an intensity of (i + d) exactly].

However, the problem of predicting the minimum intensity difference, d, needed to discriminate between two signals has not been answered convincingly.

A different statistical approach is offered by decision theory. The task considered is that of discriminating between two possibilities, namely that the observation is derived from a signal I, represented by a series of reading I_i, or from a signal J, represented by a series J_i. If it can be assumed that the noise in the imaging system is Gaussian of variance σ^2, then the likelihood of a particular set of observations, R_i originating from I is

$$L(I) = \prod_i \left[\frac{1}{\sqrt{2\pi}} \exp\left\{ \frac{(R_i - I_i)^2}{2\sigma^2} \right\} \right]$$

and of it originating from J

$$L(J) = \prod_i \left[\frac{1}{\sqrt{2\pi}} \exp\left\{ \frac{(R_i - J_i)^2}{2\sigma^2} \right\} \right]$$

If the two alternatives are equally likely to occur, then the decision should be made in favour of the one with the greater likelihood ratio.

Swets (1961) suggested that human decision-making is based on the likelihood ratio $L(I)/_{L(J)}$ leading to the response that I was present if the ratio exceeds some initial value or criterion. This criterion will depend upon such factors as prior knowledge of signal probabilities. The application of this signal detection theory approach to the subjective measurement of image quality will be discussed in a later presentation.

Harris (1964) and Roetling et al (1968) considered how decision theory would be used to specify quality. They showed that the probability of a correct decision is a function of

$$\frac{1}{2\sigma}\left[\sum_i (I_i - J_i)^2\right]^{\frac{1}{2}}$$

which in frequency space becomes

$$\frac{1}{4\sigma^2}\iint [I'_i(f_x,f_y) - J'_i(f_x,f_y)]^2 |OTF(f_x,f_y)|^2 df_x \, df_y$$

where I^1 and J^1 represent the signals prior to their passage through a system whose transfer function is OTF. It can be shown (Roetling et al 1968) that for a simple case of imaging a line source in a uniform background, the above relationship is proportional to the noise equivalent passband.

Decision theory has also been used by De Belder et al (1971) in analysing radiographic images, his formulation also including the MTF of the visual system.

Another approach has been to apply information theory to the measurement of image quality. Rather than use the information content of the image, it has been suggested that information capacity is more relevant since it refers to the best that can be achieved given optimal coding and utilisation of the data (Gregg 1968).

$$I_{CAP} = \frac{1}{2}\int\limits_{-\infty}^{\infty}\!\!\int \log_2\left[1 + \frac{W_S(f_x,f_y)}{W_N(f_x,f_y)}\right] df_x \, df_y$$

where W_S and W_N are the Wiener spectra of signal and noise respectively. However, Metz et al (1978) have demonstrated that, in general, information capacity does not provide a good measure of quality.

Although much work has been done on the objective specificat-
ion of image quality, it is inevitable that its usefulness will be
limited by the ability to model the visual system and at present
much work remains to be done particularly with regard to more
complex images.

REFERENCES

Anger H.O. (1964) Scintillation camera with multichannel collimat-
    ors. J. Nuc. Med. 5, 515
Anger H.O. (1973) Testing the performance of scintillation cameras.
    USAEC Report LBL-2027
Atkins F.B., Beck R.N., Hoffer P.B. and Palmer D. (1977) Dependence
    of optimum baseline setting on scatter fraction and detector
    response function. In Medical Radionuclide Imaging (IAEA Vienna)
    101
Barrett H.H. and Swindell W.(1981) Radiological Imaging. The theory
    of image formation, detection and processing. Vol. 2, p.646
Dainty J.C. and Shaw R. (1974a) Image Science. Principles, analysis
    and evaluation of photographic type imaging processes. (Academic
    Press, London) p.212
Dainty J.C. and Shaw R. (1974b) ibid. p.220
DeBelder M., Bollen R. and Duville R. (1971) A new approach to the
    evaluation of radiographic systems. J. Photog. Sci. 19, 126
DeVries H. (1943) The quantum character of light and its bearing
    upon the threshold of vision, the differential sensitivity and
    visual acuity of the eye. Physica 10, 553
Gregg E.C. (1968) MTF, information capacity and performance criteria
    of scintiscans. J. Nuc. Med. 9, 116
Harris J.L. (1964) Resolving power and decision theory. J. Opt.Soc.
    Am. 54, 606
Hay G.A. (1964) A physical assessment of the Cinelux electro-
    optical image intensifier in television fluoroscopy. Radiology,
    83, 86
Hoffer P.B., Oppenheim B.E., Sterling M.L. and Yasillo N. (1972)
    A simple device for reducing motion artefacts in gamma camera
    images. Radiology, 103, 199
Keller E.L. and Coltman J.W. (1968) MTF and scintillation limitat-
    ions in gamma-ray imaging. J. Nuc. Med. 9, 537
Longhurst R.S. (1967) Geometrical and physical optics (Longmans,
    London) 165.
MacIntyre W.A., Fedoruk S.O., Harris C.C., Kuhl D.E. and Mallard
    J.R. (1969) Sensitivity and resolution in radioisotope scanning.
    In Medical Radioisotope Scintigraphy Vol. 1 (IAEA, Vienna) 391
Mallard J.R. and Corfield J.R. (1969) A statistical model for the
    visualisation of changes in the count density on radioisotope
    display images. Phys. Med. Biol. 19, 348

Metz H.J., Ruchti S. and Siedel K. (1978). Comparison of image quality and information capacity for different model imaging systems. J. Photogr. Sci. 26, 229

Morgan R.H. (1965) Threshold visual perception and its relation to photon fluctuations and sine wave response. Am. J. Roent. 93, 982

Parkin A. and Unsworth G.D. (1980) Improved gamma camera images of the liver using a physiological gating mechanism. Brit. J. Radiol. 53, 900

Roetling P.G., Trabka E.A. and Kinzly R.E. (1968). Theoretical predictions of image quality. J. Opt. Soc. Am., 58, 342

Rollo F.D. (1977) Evaluating imaging devices. In Nuclear medicine physics, instrumentation and agents. Ed. F.D. Rollo (C.V. Mosby, St Louis) 436

Rose A. (1942) The relative sensitivities of television pick-up tubes, photographic film and the human eye. Proc. IRE, 30, 293

Rossman K. (1963) Spatial fluctuations of X-ray quanta and the recording of radiographic mottle. Am. J. Roent. 90, 863

Schade O.H. (1975) Image Quality: a comparison of photographic and television systems. (RCA Labs, Princeton)

Selwyn E.W.H. (1935) A theory of graininess. Photogr. J. 75, 571

Sharp P. and Mallard J. (1974) A proposed model for the visual detection of signals in radioisotope display images. Phys. Med. Biol. 19, 348

Shaw R. (1979) Some modern aspects of image quality. In The physics of medical imaging: recording systems measurement and techniques. Ed. A.G. Haus (AAPM, New York) 515

Swets J.A. (1961) Detection theory and psychophysics: a review. Psychometrika, 26, 49

Tsui B.W., Beck R.N., Doi K. and Metz C.E. (1981) Analysis of recorded noise in nuclear medicine. Phys. Med. Biol. 26, 883

Tsui B.W., Metz C.E., Atkins F.B., Starr S.J. and Beck R.N. (1978) A comparison of optimum detector spatial resolution in nuclear imaging based on statistical theory and on observer performance. Phys. Med. Biol. 23, 654

Wagner R.F. (1977) Toward a unified view of radiological imaging systems. Part II. Noisy Images. Med. Phys. 4, 279

Wagner R.F., Weaver K.E., Denny E.W. and Bostrom R.G. (1974) Toward a unified view of radiological imaging systems. Part I. Noiseless Images. Med. Phys. 1, 11.

THE PRESENTATION OF PHOTON-LIMITED IMAGES

P.F. SHARP

Department of Biomedical Physics and Bioengineering,
University of Aberdeen, Foresterhill, Aberdeen AB9 2ZD,
Scotland, UK.

1. INTRODUCTION

Given the intrinsically poor quality of X and gamma ray images,
it is of the greatest importance to ensure that the image data is
well presented. Presentation encompasses not only display of the
image but also ways in which it can be manipulated to improve interpretation.

Mainly owing to the lack of necessary computing facilities,
manipulation of radiographic images has not been widely used in
routine hospital work and we shall concentrate on nuclear medicine
images. The reader interested in radiographic imaging is referred
to the review by Trussell (1981).

While it is self evident that the acceptance of image manipulation will depend upon demonstrating its effectiveness, surprisingly
little work has been done on this problem. This is perhaps why,
even in nuclear medicine, many of the proposed techniques have not
been widely accepted. In the second part of this chapter, measurement of subjective image quality will be discussed.

2. IMAGE DISPLAY

Although much effort has been put into production of a single
optimum image format,most problems can be avoided by the use of an
interactive system in which the viewer is free to manipulate, say,
contrast and upper and lower threshold levels, until satisfied that
no more useful image information can be obtained. Nevertheless, it
is probably still prudent to ensure that the image the viewer starts
with is displayed in a nearly optimum format so as to minimise the

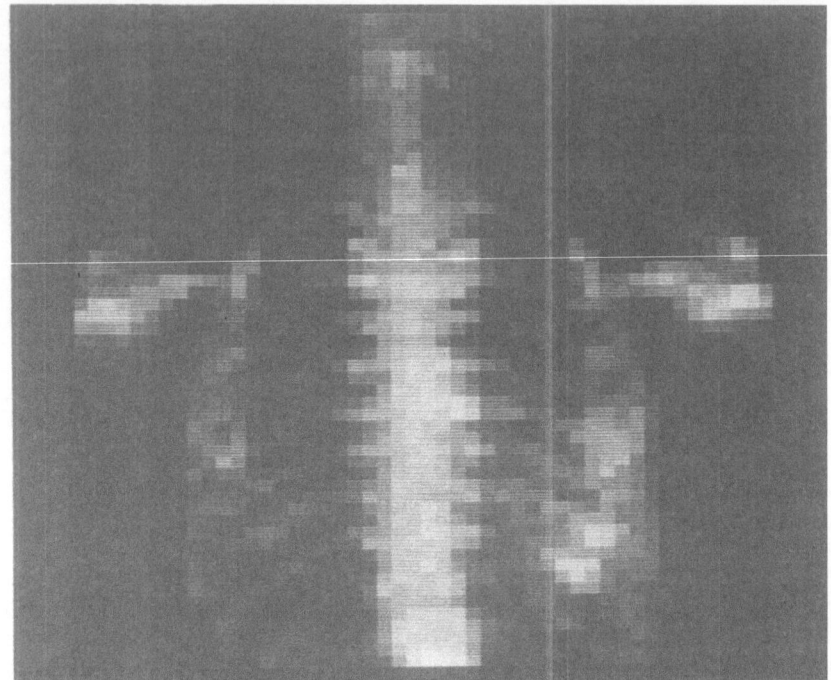

Figure 1

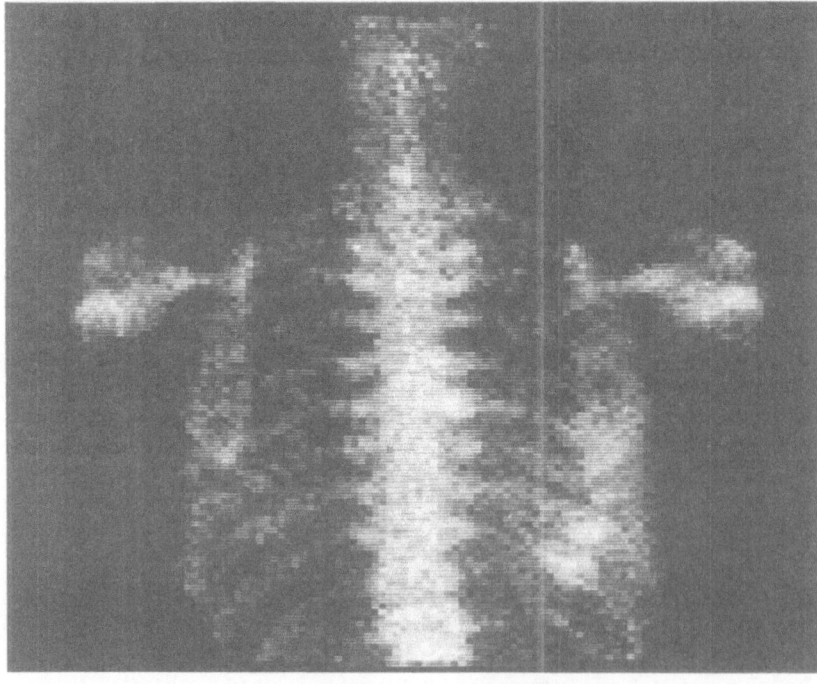

Figure 2

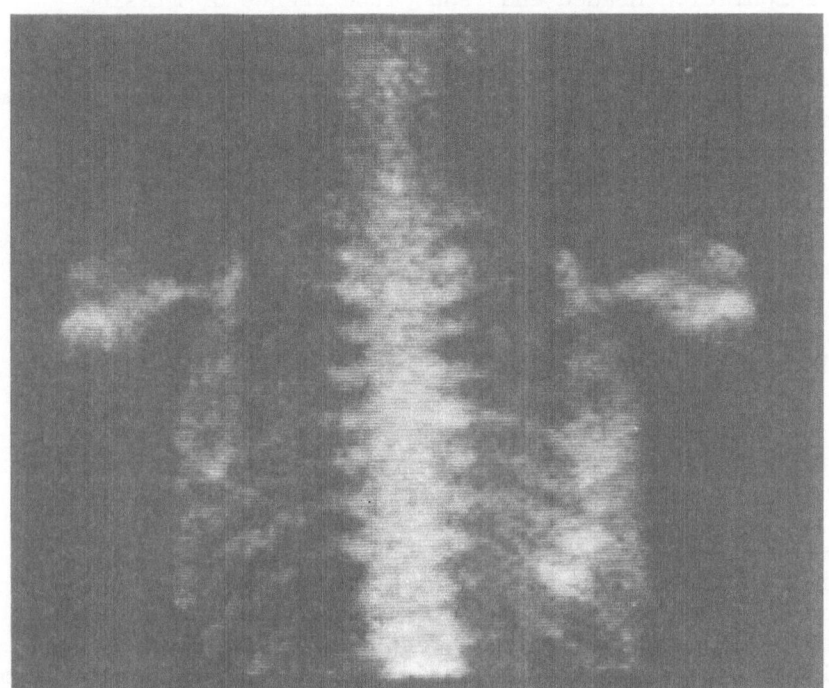

Figure 4

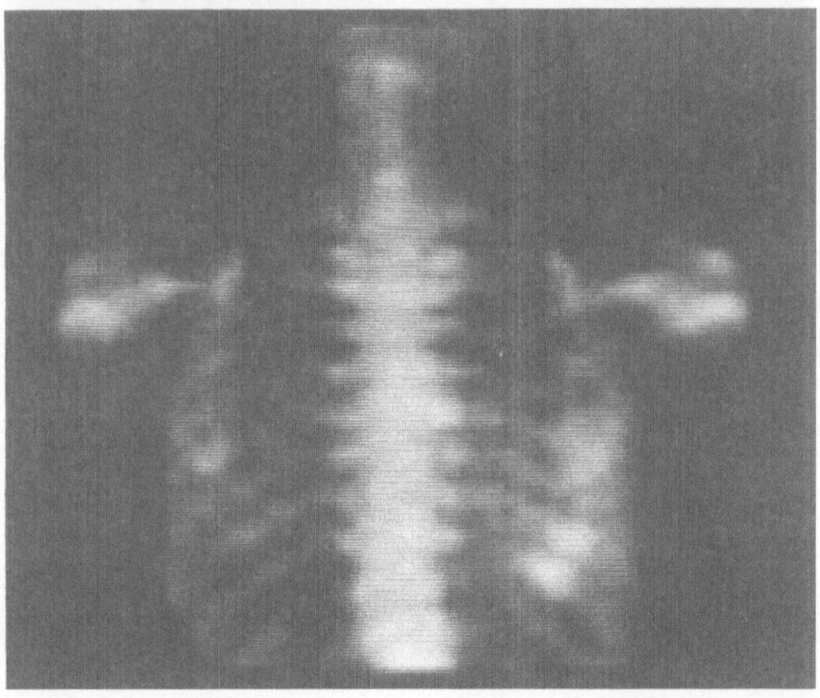

Figure 5

amount of manipulation needed and reduce the chance of missing features of interest.

In radiology, interactive systems are often used for presenting computed tomographic (CT) images but conventional images are invariably displayed on transparency film. In nuclear medicine, where the need to improve image quality is greater, much work has been done on the computer manipulation of data.

Unfortunately, interfacing a display to a computer system immediately produces problems since images must have both their spatial and intensity information digitised. This digitisation into individual picture elements (pixels) can itself degrade image appearance as the viewer is immediately aware of pixel boundaries (Fig. 1). This effect can be minimised by using a large number of very small pixels (Fig. 2), or by interpolating the data, Fig. 3. As can be seen, conventional linear interpolation gives a smoothed appearance to the image owing to the correlation between the interpolated pixel and its neighbours. This can be overcome by random interpolation (Sharp et al 1982), in which random noise is added to interpolated data points (Fig. 4). The intrusiveness of pixellation can also be reduced by image minification as discussed by Pitt elsewhere in this volume. A detailed investigation into the problem of selecting the optimum pixel size has been made by Sharp et al (1982).

FIGURES 1 - 4 see preceding pages

Fig. 1 (Top Left) Posterior view of an abnormal bone scan displayed with a 64 x 64 pixellation

Fig. 2 (Bottom Left) The same image as in Fig. 1 but using a 128 x 128 pixellation

Fig. 3 (Top Right) The image of Fig. 1 after the 64 x 64 pixellated data has been interpolated to 256 x 256

Fig. 4 (Bottom Right) The same image after random interpolation from 128 x 128 to 256 x 256.

As intensity information in nuclear medicine images is already in a discrete form, it is possible to retain all the original data by digitising it into a large number of intensity levels. It is, of course, still necessary to ensure that these levels are visually discriminable and for this purpose some systems have employed colour coding rather than conventional gray-scale coding.

While colour provides an effective way of denoting differences in image intensity, used indiscriminiately it can produce artificial image contours since it deprives the visual system of the power of associability (Cormack and Hutton 1980), i.e. the ability to see adjacent intensity levels as possibly relating to the same image feature. Colour scales have been proposed in which differences in saturation are used to expand the dynamic range of image intensity rather than changes in hue (Milan and Taylor, 1975).

Having decided how display intensity is to be represented, the next problem is to relate this intensity scale to image count density. The simplest approach is to use a linear scale (Fig. 5a) in which each of n different intensity levels represents one nth of the range of count densities in the image. If each intensity level represents a large range of count densities there is the danger that small but nevertheless statistically significant changes in image count density may be missed because they fail to cause a change in display intensity. Alternatively, using a very narrow range of count densities per level may simply enhance image noise.

The statistical level system relates the range of counts per intensity level to the expected statistical (Poisson) image noise. It has been investigated by Sharp and Mallard (1976) who suggested that the range of count density represented by an intensity level should not exceed three standard deviations of the mean count density of that level if significant changes in county density were not to be missed. Of course these results also provide a guide as to the maximum count density range which should be used with linear levels. An image displayed with 2 and 4 standard deviation statistical levels is shown in Figs. 5b and 5c.

Histogram equalisation (Goris et al 1976) provides another way of coding intensities. Count densities are assigned in such a way that each intensity level appears the same number of times in the image (Fig. 5d). A similar coding resulted from an approach based on information theory (Cormack and Hutton 1980).

The effectiveness of different display techniques is still an open question, mainly due to the difficulty of measuring image quality although several studies have been carried out (Sharp and Mallard 1976, 1977; Houston and MacLeod 1977; Houston 1980; Sharp et al 1982).

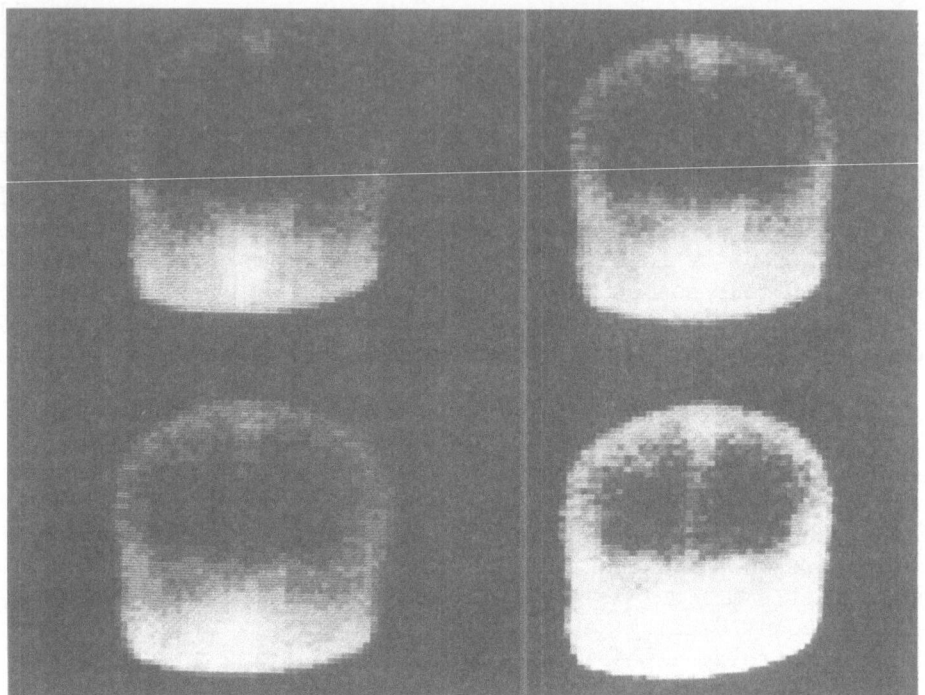

Fig. 5 Anterior view of an abnormal brain scan displayed a) using
linear intensity levels (Top left), b) two standard deviation wide statistical levels (Top right), c) four standard
deviation wide statistical levels (Bottom left) and d)
histogram equalised levels (Bottom right).

3. IMAGE FILTERING

The filtering of image data prior to display is potentially a
very effective way of improving quality. Early attempts, which
were limited to analogue display systems, involved such techniques
as defocussing the displayed image so as to blur out high frequency
noise (MacIntyre and Christie 1966), but they had little success.
With the introduction of computer interfaced display systems, the
sophistication of filtering techniques has increased greatly.

An image filter is most conveniently described in terms of its
effect on the spatial frequency content of the images, the maximum
frequency present in nuclear medicine images being about 1.5 cycles/
cm.

There are two main techniques for implementing the filter process. In the Fourier transform method, the image is first transformed into frequency space, multiplied by the filter weighting coefficients for different frequencies and then transformed back into real space. The second method, the convolution process, is the most commonly used technique and here the filter coefficients, initially defined in frequency space, are transformed into real space and convolved directly with the image. The advantages and disadvantages of the two approaches are discussed fully by Miller and Sampathkumaran (1982).

Most filters can be classified into one of three categories.

(i) Smoothing filters. These aim to reduce the noise in an image while leaving spatial resolution unchanged. Since noise is most evident at high frequencies, the basic smoothing filter is a band-pass one in which all frequencies above some upper limit are removed. In practice the sharp cut-off associated with this is not realisable and a more gradual attenuation of high frequencies is used as seen in the frequency plots in Fig. 6. Figure 7 demonstrates the effect of these filters on an abnormal brain image and shows how decreasing the cut-off frequency reduces noise but also sacrifices image sharpness.

Other widely used smoothing filters in nuclear medicine are the nine point filter (Todd-Pokropek 1980) which shows a steady drop in response with increasing frequency, and the matched filter (Turin 1960) which is designed to maximise the signal to noise ratio.

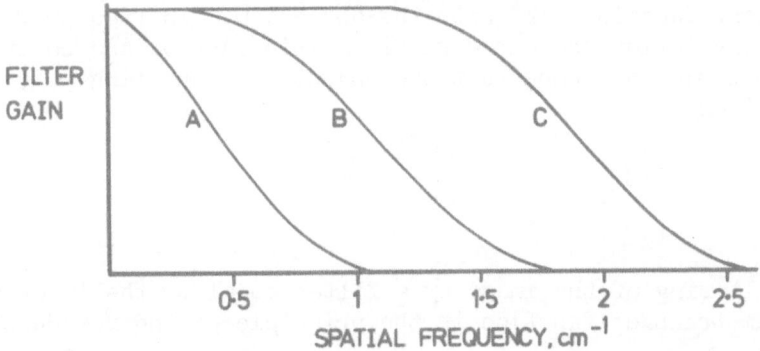

Fig. 6 Three examples of band-pass smoothing filters in frequency space.

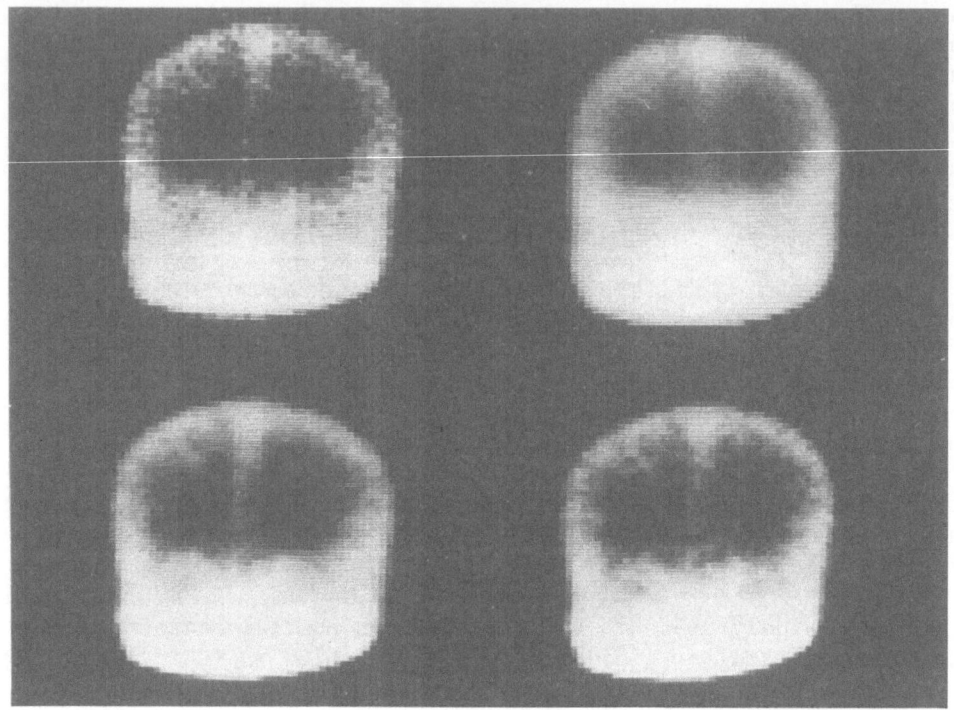

Fig. 7   The same abnormal brain scan as in Fig. 5 shown before
filtering (top left) and after processing with the filters
in Fig. 6; filter A (top right), filter B (bottom left)
filter C (bottom right)

(ii) Refocussing filters. The degradation of spatial information by the imaging device is described by the convolution of the
device transfer function, TF, with the object, O. In frequency
space this convolution becomes a simple multiplication and so it
should be possible to reconstruct the object from the image, I,
by deconvolution:

$$O(f_x, f_y) = \frac{I(f_x, f_y)}{TF(f_x, f_y)}$$

This filtering of the image by a filter equal to the inverse
of the system transfer function is the principle behind refocussing
filters.

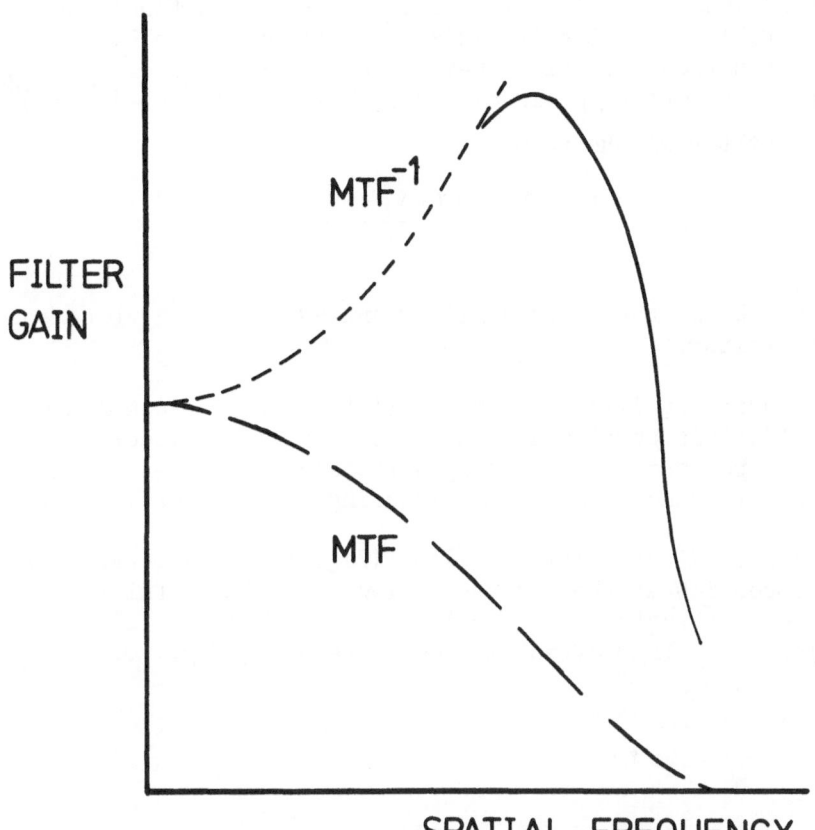

Fig. 8 The refocussing filter shown by the finely broken line
(------) as the inverse of the system MTF (– – – –) and
with the roll-off (———————) necessary to avoid noise
amplification at high spatial frequencies.

As can be seen in Fig. 8, the modulation transfer function of
a camera falls rapidly to near zero at high spatial frequencies.
Thus the refocussing, or inverse, filter will give greatest weight
to these frequencies with the unfortunate consequence that image
noise will be amplified. As a practical compromise, filters are
rolled off at some limiting frequency to avoid undue noise amplification.

In the filter proposed by Tanaka and Iinuma (1970)

$$F(f_x, f_y) = \frac{1}{TF(f_x, f_y)} \text{ for } f_x, f_y < f_c$$

$$= \frac{TF(f_x, f_y)}{TF(f_c, f_c)} \text{ for } f_x, f_y > f_c$$

where f_c is a cut-off frequency chosen to avoid undue "ringing" in
the processed image. Typically a value of about 0.4/FWHM, where
FWHM is the full width at half maximum height of the collimator's
line spread function, is used. Metz (1969) proposed a class of
filters in which $1/TF(f)$ is expanded in terms of $(1 - |TF(f)|^2)$

so giving a filter of the form:

$$1 - \frac{[1 - |1 - TF(f)|^2]^{n+1}}{TF(f)}$$

which for $n = \emptyset$ is the matched filter and as n becomes
the inverse filter.

(iii) Enhancing filters. While all filters aim to enhance
image quality, this particular class specifically excludes those
whose primary purpose is to rectify deficiencies in the imaging
process e.g. by reducing noise or restoring spatial resolution.

The simple differentiating filter (Fig. 9), by its emphasis
on high frequencies at the expense of low ones, is useful for
sharpening edge structures and so improving the perceptibility of
adjacent lesions. This effect is demonstrated in Fig. 10.

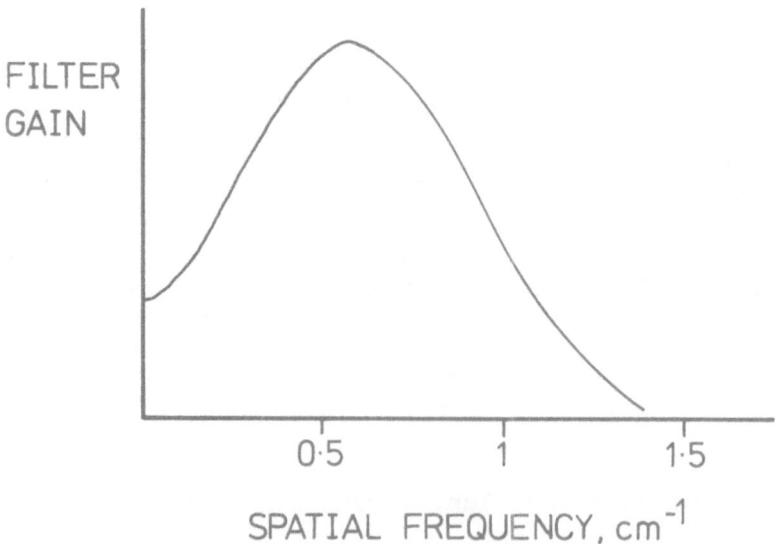

Fig. 9 A differentiating filter in frequency space.

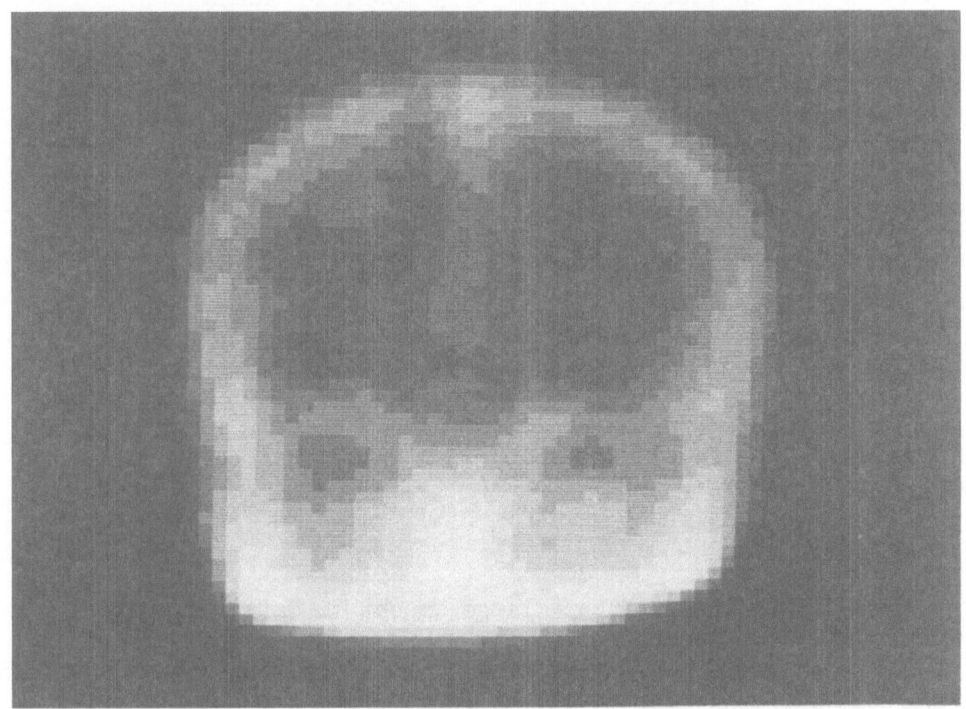

Fig. 10 The abnormal brain image of Fig. 5 after the differentiating filter of Fig. 9 has been used. Note how edges have
been sharpened.

Unsharp masking, in which a highly smoothed version of the
image is subtracted from a lightly smoothed one so as to remove the
low frequency component, has a similar effect to the differentiating filter.

The convexity/concavity filter (Neil and Hutchinson 1971) aims
to enhance regions of count density having a convex or concave
profile, these being associated with "hot" or "cold" lesions.

Several workers have tried to define the pattern of uptake in
a normal image since abnormal features could then be identified as
producing statistically significant changes from the norm. One of
the most successful methods has used principal component analysis
(Barber 1976).

Despite the variety of filters available, there has been a
marked reluctance to accept them for routine clinical work. Two
factors may have led to this. First, the reluctance of clinicians

to rely only on data which has been "manipulated" and secondly, the unfortunate impression, often given, that for a filter to be of value it must be applicable to all imaging problems, whereas filters are often useful only for very specific tasks.

A more profitable approach may be to present the clinician with the original unprocessed image together with the same image processed with a selection of filters chosen to deal with particular problems of interpretation. The viewer's confidence about features appearing on the filtered images can then be increased by reference to the unprocessed image and vice versa. In an inter-comparison of filtering techniques (Houston et al 1979), observers were presented with brain images filtered with a differentiating filter and a convexity seeking filter as well as the original data. The differentiating filter was chosen to enhance lesions situated close to the edge of normal anatomical structures, such as the superior sagittal sinus, while the convexity filter should emphasise lesions present in the region of almost uniform low uptake over the cerebral tissue. In the comparison with other techniques this method performed very well.

4. FUNCTIONAL IMAGING

Dynamic studies produce a large amount of data and that relating to relative changes in the radiopharmaceutical distribution can be expressed in quantitative form. The most common approach is to use time-activity curves but in reducing the data to a few curves it is necessary for each one to refer to a relatively large image area and so there will be loss of spatial information.

Functional or parametric imaging (Kaihara et al 1969; MacIntyre 1970) provides a way of retaining spatial information while also obtaining a quantitative measure of the temporal variation of the radiopharmaceutical distribution. It involves the selection of parameters which will describe the time-activity curve. Images are then produced by replacing the count density information in a pixel by the value of one of the parameters describing the time-activity curve for this pixel, one image being made for each chosen parameter. For example it has been applied to lung ventilation studies by using the half-time of the gas washout phase from the lungs as the parameter (Burdine et al 1972). Further applications have been in renal studies (Wiener et al 1974) cerebral blood flow (Lassen et al 1978), cardiology (Pavel 1981) and hepato-biliary imaging (Basic et al 1983)

The greatest problem is in the selection of suitable parameters as they must not only describe the curve shape but also be sensitive indicators of disease. Ideally they would be based on knowledge of physiology underlying the handling of the radiopharmaceutical, for example in terms of a compartmental analysis of the system, but such

knowledge is all too frequently absent. Principal component
analysis can provide an efficient way of describing curve shape
and may possibly allow some physiological interpretation to be
placed on the parametric images (Barber 1980; Oppenheim and
Appledorn 1981; Houston et al 1982).

5. MEASURING IMAGE QUALITY

While many techniques have been suggested for improving image
quality, their effectiveness is rarely evaluated beyond showing
examples of an image before and after processing. This is, to a
large extent, due to the difficulty of assessing image quality.

Measures of quality can be divided into three main categories.
First there are the purely objective measures of device performance,
such as MTF, which are aimed mainly at describing the physical
aspect of device performance.

Secondly, there are the semi-objective measures such as
detective quantum efficiency which have taken into account the
perceptual process while not actually requiring that subjective
measurements be made.

Finally there are the purely subjective quality indices
derived from the results of perceptual measurements. It is this
last category that we shall now consider.

Image quality can be defined, in its most general sense, as a
measure of the effectiveness with which the image can be used for
its intended purpose. The most basic perceptual task is to detect
a signal i.e. the details of interest, in the presence of noise.
Noise includes all image features irrelevant to the perceptual task
such as quantum mottle, anatomical noise e.g. interference from
overlying body structures, and visual noise, i.e. physiological
and psychological inconsistencies in response.

Three techniques have been widely used to measure this
discriminability between signal and noise; the method of constant
stimulus, signal detection theory and ranking techniques.

5.1 Method of constant stimulus

Consider the simple visual task found in nuclear medicine of
detecting an abnormality, a small area of increased count density,
in a background of constant mean count density but having random
noise caused by statistical (Poisson) fluctuations in count
density (Fig. 11). Because of these random variations in count
density, the ease with which an abnormality can be seen, i.e. the
visual stimulus it produces, will vary from image to image even
though its contrast remains unchanged. In Fig. 12 this variation

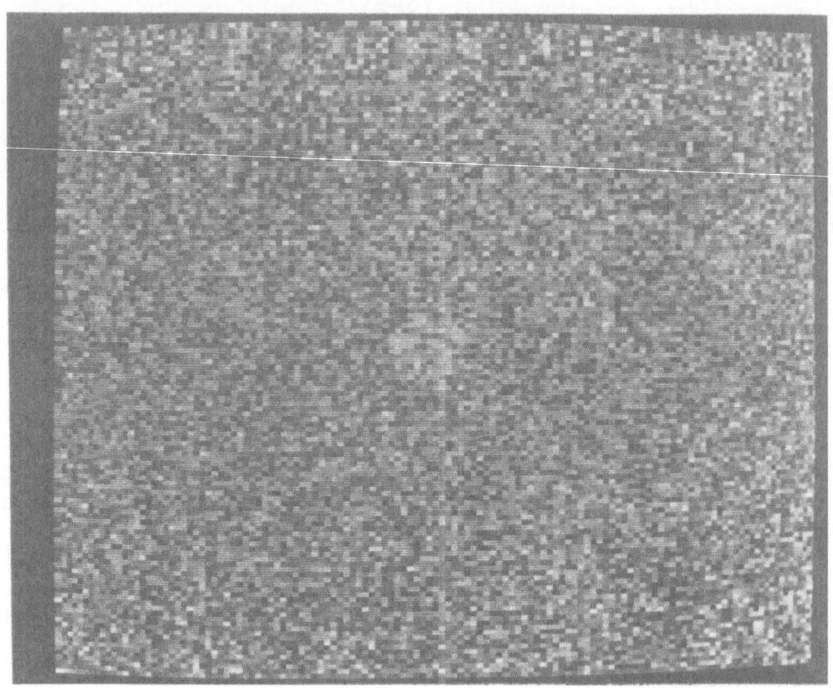

Fig. 11 Test pattern showing a solitary circular "abnormality" of
increased count density in the centre of a background of
constant mean count density but containing statistical
noise.

in appearance (visual stimulus) of an abnormality of a given
contrast is denoted by the bell-shaped probability curves.

It is assumed that the observer uses a visual threshold, T,
and reports that an abnormality is present only if the stimulus
exceeds this value. The contrast of an abnormality does not
therefore determine whether it will be visible but rather the
probability that it will be seen. This probability, known as the
observer's true positive visual response, is equal to the area of
the probability curve falling to the right of T.

To measure visual response, the observer must be presented
with several images of the test-pattern all showing the abnormality
at the same contrast. The percentage of these images in which the
observer sees the abnormality is the true positive response rate
for this contrast. The curve showing true positive response rate

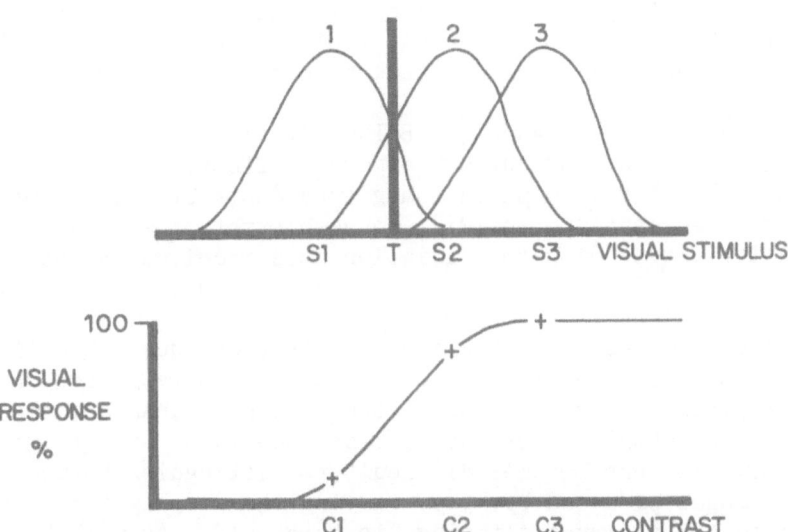

Fig. 12   The production of a visual response curve. Owing to
          statistical noise the abnormality, such as that in the
          test-pattern in Fig. 11, does not always have the same
          appearance, i.e. produce the same visual response, for a
          given contrast. The visual stimulus corresponding to
          abnormalities of contrast (C_1, C_2, C_3) is described by
          probability distributions of mean   values S_1, S_2 and
          S_3 respectively.

          An observer asked to detect the abnormality will adopt
          some visual threshold T such that if the visual stimulus
          falls below this value he will report that he cannot see
          the abnormality. As contrast is increased the average
          rate of abnormality detection, the visual response, also
          increases giving the ogival shaped visual response curve
          shown in the lower part of the figure.

as a function of abnormality contrast is known as the visual
response curve or psychometric function. This is the lower curve
in Fig. 12.

     There are obvious advantages in selecting a particular response
rate at which the abnormality can be said to have been detected.
While the 100% response might seem an obvious choice, it is not
satisfactory as the response curve approaches this figure very
slowly. The 50% response rate is usually chosen and has the

advantage that the contrast producing it coincides with the position of the visual threshold.

In nuclear medicine this approach has been used to measure how image quality is influenced by choice of display modality (Sharp and Mallard 1976, 1977; Sharp et al, 1982) and how it is affected by radiopharmaceutical behaviour (Lakshmanan et al 1978).

De Belder and colleagues (De Belder et al 1975) have used a similar technique to find out how exposure affects the quality of radiographs. In X-ray computed tomography contrast-detail-dose curves have been used (Cohen 1979) to demonstrate how detail detectability varies with the radiation dose received by the patient.

There are two major criticisms of this technique. The first is whether it is reasonable to assume that an observer uses one particular visual threshold, for if the threshold changes, then so will that value obtained for the contrast needed to detect the abnormality. It thus becomes difficult to distinguish between contrast changes produced by, say, varying the display format and those due to the observer altering his threshold. As will be discussed in the next section, the visual threshold does indeed vary and to stabilise it, the observer must be told what decision criterion to use. For example, the observer may be told to give a positive response only if absolutely certain that there is an abnormality in the image.

Secondly, the method of constant stimulus requires that the experimentalist can control the contrast of the abnormality in the test pattern. Obviously this often limits the technique to experiments with simulated images.

5.2 Signal detection theory

In signal detection theory, images are considered to be in one of two groups; those which are in some way abnormal, usually referred to as the "signal + noise" group, and those which are normal, the "noise" group.

The method of constant stimulus only deals in terms of true positive responses, that area of the "signal + noise" probability curve exceeding the threshold, T, but there are three other types of responses (Fig. 13). The false negative response is equal to that area of the "signal + noise" curve falling below the threshold, and gives the percentage of times on which an abnormal image is reported as normal. In a normal image noise may be mistaken for an abnormality and this leads to the false positive response, the area of the "noise" curve exceeding the threshold. Finally there is the true negative response, the area of "noise" curve not exceeding the

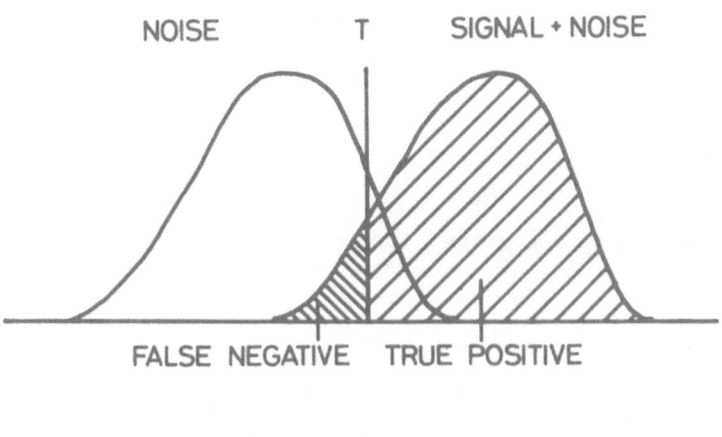

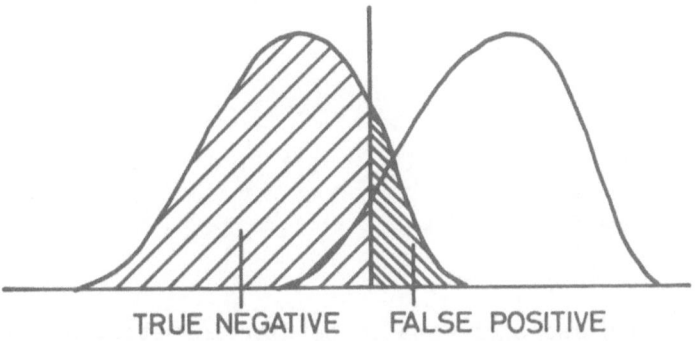

Fig. 13 Types of visual response which may be produced in the
        discrimination of signal from noise.

threshold, in which an image is correctly identified as being
normal.

These four types of visual response are not independent since;

True positive + False Negative = 1

False positive + True Negative = 1

Usually the true positive and false positive responses are chosen.

The quality of images can be measured by the apparent separation between the "noise" and "signal + noise" distributions. The
more widely apart, the better the quality since it is easier to
discriminate between normal and abnormal images. This perceived
separation between two probability curves can be measured by
inducing the observer to vary his visual threshold and noting how

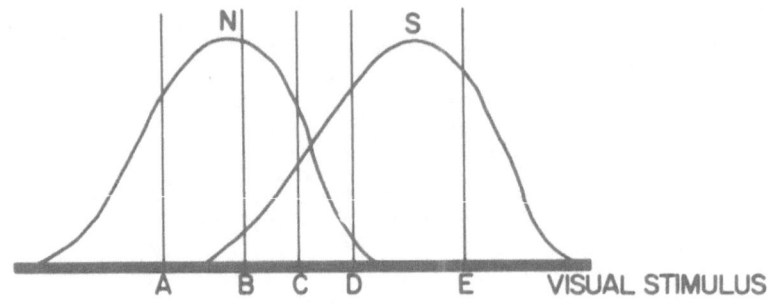

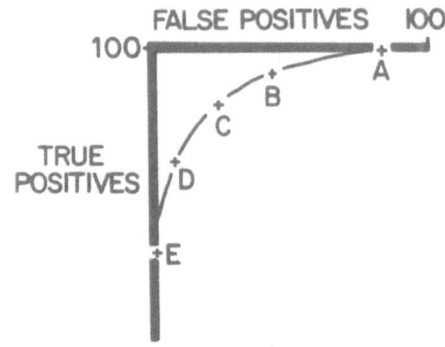

Fig. 14 The two classes of data to be discriminated, noise (N)
and signal plus noise (S), are represented by bell-shaped
probability curves. When the observer adopts different
visual thresholds, A to E, then the corresponding true and
false positive response rates will change. For example,
threshold A implies that the observer is very lax in making
a decision about the presence of a signal (abnormality)
and while correctly identifying all abnormalities (100%
true positive rate) does so at the cost of producing a
high false positive rate (noise incorrectly identified as
being an abnormality). A strict decision criterion, such
as E, gives no false positives but also very few true
positives.

The graph showing the variation of true with false positive
responses is known as the receiver operating characteristic
(ROC) curve and measures the observer's ability to
discriminate between the two groups of images.

the true and associated false positive response rates change. This
is demonstrated in Fig. 14.

There are several ways of persuading the observer to alter his
visual threshold, the most efficient probably being the rating
method. (Green and Swets 1976; Goodenough et al 1974). The plot of
true positive against false positive response is known as the
receiver operating characteristic (ROC) curve. An improvement in
image quality will be chosen by a shift in the position of this
curve towards the top left hand corner, i.e. by a reduction in the
false positive rate associated with a given true positive rate.

This technique has been applied to a variety of problems in
radiology (Lusted 1969; Goodenough et al 1972, 1974; Metz et al
1973, 1976). In nuclear medicine it has been used to evaluate
image filtering (Kuhl et al 1972; Metz and Goodenough 1973; Herath
and Sharp 1976; Houston and MacLeod 1977) and display formats
(Houston 1980; Sharp et al 1982). An intercomparison of X-ray
computed tomography and nuclear medicine has been performed by
Swets and colleagues(1979).

Signal detection theory obviously avoid the need to define
one particular visual threshold since it relies upon varying the
threshold. It is useful in situations in which images can readily
be classified as normal or abnormal but the intensity of the
abnormality is neither definable nor controllable. However, there
is also the disadvantage that one must be careful to ensure that
the images used are representative of the two classes. In clinical
trials it is easy to produce a sample of abnormal images biassed
towards those in which follow-up and hence a definitive diagnosis
is easiest to make.

The positions of the ROC curves indicate whether one
technique is better than another, but it is difficult to produce
an index to quantify the difference. (Goodenough et al 1973).

In many cases it is not sufficient simply to identify an
image as abnormal as it is also necessary to indicate the position
of the abnormality in the image. This means that there are now two
types of false positive response, the second being when an image is
correctly identified as being abnormal but the wrong position for
the abnormality is given. This response is known as a true positive
incorrect location (Starr et al 1975). It is now necessary to plot
the ROC curve in three dimensions, the additional axis showing
these true positive incorrect location responses, and so the
analysis of data is further complicated.

However, the general approach of signal detection theory is
very useful, particularly as it can be used to introduce cost-

benefit analysis into the evaluation of imaging techniques (Metz et al 1976).

5.3 Ranking

The two previous methods are applicable to tasks requiring a binary response from the observer, the abnormality is either seen or not seen, and this limits the range of perceptual tasks which can be investigated.

It is often easy for an experienced viewer to say whether or not an image is of good quality yet difficult for him to define the criteria used to make this judgement. Ranking tries to make use of this intuitive appreciation of quality.

The observer is presented with a set of images in which some parameter suspected of influencing quality has been varied in a systematic fashion and asked to arrange the images in order of preference, the highest quality first, the next best second, and so on. By examining the ranking order produced by several observers, it is possible to decide whether they agree on what constitutes a good image.

We have used this approach in nuclear medicine (Sharp et al 1982) to look at the effect of digitisation on image quality. In radiology Cohen and colleagues (1976) have investigated how various film screen combinations affect the quality of tomographic images.

The main criticism of this approach is that the meaning of "preference" is vague, and the technique relies on the experience and even the prejudices of observers.

This can be avoided by using more specific criteria for ranking in a comparison of five film/screen systems for mammography (Sickles et al 1977) the images were ranked according to various physical criteria, such as film density, and anatomical criteria, such as the clarity of breast calcifications.

Another possible application of ranking is in testing how well objective measures of quality agree with subjective assessments. For example, we have looked at how well various measures of gamma-camera non-uniformity agreed with the observer's judgement of image uniformity (Sharp and Marshall 1981). Rank order correlation coefficients, such as Spearman's, can be used to test the agreement between the ranking of images according to values of the objective parameter and ranking based on subjective impression.

Ranking has the advantage of requiring only an ordering of the images, but since there is no numerical measure it is not possible to say how much better one display format, for example, is than

another. One way to rectify this weakness is to assign a score to
the assessment of quality of various image parameters. Summing
these values then gives an overall score for the quality of the
image. For example, Vucich (1979) used this idea to evaluate the
quality of chest radiographs. The problem is to justify the
scoring system and the assumption that the net effect of the
different parameters can be measured simply by summing individual
scores.

6. CONCLUSION

Whilst the application of processing techniques has undoubtedly proved to be of great value in simplifying interpretation of the
large amount of data produced in dynamic studies, its effect on
improving image appearance is less obvious. To a large extent
this problem is one of demonstrating that a processing technique
has actually enhanced the diagnostic quality of the image, and as
this may involve lengthy trials with clinical data, few investig-
ators have attempted it.

The acceptance of medical image processing may have to await
the introduction of systems with comprehensive interactive
capabilities so that the investigator has much greater freedom to
choose the techniques most suited to a particular problem.

With the increasing power of cheap data processing systems
the application to radiographic images is imminent. As noise will
be far less of a problem than with radionuclide images, many of
the difficulties encountered in this field will be avoided, and as
a wealth of spatial detail is to be found in radiographs many
interesting challenges should be posed for image processing.

REFERENCES

Barber D.C. (1976) Digital computer processing of brain scans
  using principal components. Phys. Med. Biol. 21, 722
Barber D.C. (1980) The use of principal components in the
  quantitative analysis of gamma camera dynamic studies. Phys. Med.
  Biol. 25, 283
Basic M., Sharp P.F. and Dendy P.P. (1983) A functional imaging
  routine for cholescintigraphy. Phys. Med. Biol. (to be published)
Burdine J.A., Murphy P.H., Alagarsamy V., et al (1972) Functional
  pulmonary imaging. J. Nuc. Med. 13, 933
Cohen G. (1979) Contrast-detail-dose analysis of six different
  computed tomographic scanners. J. Comput. Assist.Tomog. 3, 197.
Cohen G., Barnes J.O. and Peria P.M. (1976) The effects of film/
  screen combination on tomographic image quality. Radiology 129,
  515

Cormack J. and Hutton B. (1980) Minimisation of data transfer
losses in the display of digitised scintigraphic images. Phys.
Med. Biol. 25, 271

De Belder M., Bollen R. and Van Esch R. (1975) A new evaluation
method of radiographic systems. SPIE Medical X-ray Photo-Optical
Systems Evaluation 56, 54

Goodenough R.J., Rossman K. and Lusted L.B. (1972) Radiographic
applications of signal detection theory. Radiology 165, 199

Goodenough R.J., Metz C.E. and Lusted L.B. (1973) Caveat on the use
of parameter d' for evaluation of observer performance. Radiology,
106, 565

Goodenough R.J., Rossman K. and Lusted L.B. (1974) Radiographic
applications of receiver operating characteristic (ROC) curves.
Radiology 110, 89

Goris M.L., Scheibe P.O. and Kriss J.P. (1976) A method to optimise
the use of a grey-shade scale in nuclear medicine images. Comps.
and Biomed. Res. 9, 571

Green D.M. and Swets J.A. (1976) Signal detection theory and
psychophysics (Wiley) p. 99

Herath K.B. and Sharp P.F. (1974) Effect of "matched filter"
smoothing as measured by receiver operating characteristic curve.
Phys. Med. Biol. 21, 442

Houston A.S. (1980) A comparison of four standard scintigraphic
TV displays: concise communication. J. Nuc. Med. 21, 512

Houston A.S., Elliott A.T. and Stone D.L. (1982) Factorial phase
imaging: a new concept in the analysis of first-pass cardiac
studies. Phys. Med. Biol. 27, 1269

Houston A.S. and MacLeod M.A. (1977) An intercomparison of computer
assisted data processing and display methods in radioisotope
scintigraphy using mathematical tumours. Phys. Med. Biol. 22, 1097

Houston A.S., Sharp P.F., Tofts P.S. and Duffey B.L. (1979) A
multicentre comparison of computer assisted image processing and
display methods in scintigraphy. Phys. Med. Biol. 24, 547

Kaihara S., Natarjan T.K., Maynard C.D. and Wagner H.N. (1969)
Construction of a functional image from spatially localised rate
constants obtained from serial camera and rectilinear scanner
data. Radiology 93, 1345

Kuhl D.E., Sanders T.D., Edwards R.Q. and Makler P.T. Jr.(1972)
Failure to improve observer performance with scan smoothing,
J. Nuc. Med. 13, 752

Lakshmanan A.V., Sharp P. and Mallard J. (1978) The influence of
size and radiopharmaceutical concentration ratio on the detection of abnormalities in clinical radionuclide imaging. Brit. J.
Radiol., 51, 986

Lassen N.A., Ingvar D.H. and Skinhøj E. (1978) Brain function and
blood flow. Scientific American 239, 50

Lusted L.B. (1969) Perception of the roentgen image: applications
of signal detectability theory. Rad. Clins. N. Amer. VII p.435

MacIntyre W.J. and Christie J.H. (1966) The use of data blending
to reduce statistical fluctuations in radioisotope scanning.
Radiology 86, 141

MacIntyre W.J., Inkley S.R., Roth E. et al (1970) Spatial recording
of disappearance constants of Xe-133 washout from the lungs.
J. Lab. Clin. Med. 76, 701

Metz C.E. (1969) A mathematical investigation of radioisotope scan
image processing. Ph.D. Thesis, University of Pennsylvania.

Metz C.E. and Goodenough D.J. (1973) On failure to improve observer
performance with scan smoothing: a rebuttal. J. Nuc. Med. 14, 873

Metz C.E., Goodenough D.J. and Rossman K. (1973) Evaluation of ROC
data in terms of information theory with applications in radiography. Radiology 109, 297

Metz C.E., Starr S.J. and Lusted L.B. (1976) Quantitative evaluation of visual detection performance in medicine: ROC analysis
and determination of diagnostic benefit in Medical Images:
formation, perception and measurement. Ed. G.A. Hay, Institute
of Physics and Wiley, p. 220.

Milan J. and Taylor K.J.W. (1975) The application of the temperature scale to ultrasound imaging. J. Clin. Ultrasound 3, 171

Miller T.R. and Sampathkumaran S. (1982) Digital filtering in
nuclear medicine. J. Nuc. Med. 23, 64

Neil G.D. and Hutchison F. (1971) Computer detection and display of
focal lesions in scintigrams. Brit. J. Radiol. 44, 962

Oppenheim B.E. and Appledorn C.R. (1981) Optimal functional serial
imaging using iodine-123-hippuran and factor analysis. In
Medical Radionuclide Imaging 1980 Vol. II (IAEA Vienna) p.351

Pavel D., Byrom E., Swiryn S., Meyer-Pavel C and Rosen S (1981)
Normal and abnormal electrical activation of the heart. In
Medical Radionuclide Imaging 1980 Vol. II (IAEA Vienna) p.253

Sharp P.F., Chesser R.B. and Mallard J.R. (1982) The influence of
picture element size on the quality of clinical radionuclide
images Phys. Med. Biol. 27, 913

Sharp P.F. and Mallard J.R. (1976) The measurement of the perform
ance of the display system of a radioisotope imaging display:
the multi-element band display. Brit. J. Radiol. 49, 270

Sharp P.F. and Mallard J.R. (1977) Measurement of the performance
of the gamma oscilloscope display. Brit. J. Radiol. 50, 822

Sharp P.F. and Marshall I. (1981) The usefulness of indices
measuring gamma camera non-uniformity. Phys. Med. Biol. 26, 149

Sickles E.A., Genant H.K. and Doi K. (1977) Comparison of laborat-
ory and clinical evaluations of mammographic screen-film systems.
In Applications of Optical Instrumentation in Medicine VI. Eds
J.E. Gray and W.R. Hendee. (SPIE Boston) p.30

Starr S.J., Metz C.E., Lusted L.B. and Goodenough D.J.(1975) Visual
detection and localisation of radiographic images. Radiology 116,
533

Swets J.A., Pickett R.M., Whitehead S.F. et al (1979) Assessment
of diagnostic technologies. Science 205, 753

Tanaka E. and Iinuma T.A. (1970) Approaches to optimal data
  processing in radioisotope imaging. Phys. Med. Biol. 15, 683-694
Todd-Pokropek A. (1980) Image processing in nuclear medicine
  IEEE Trans. Nuc. Sci. NS-27, 1080
Trussell H.J. (1981) Processing of X-ray images. Proc. IEEE, 69,
  615
Turin G.L. (1960) An introduction to matched filters. IRE Trans.
  Inf. Theory 6, 311
Vucich J.J. (1979) The role of anatomic criteria in the evaluation
  of radiographic images in The physics of medical imaging: recording
  systems, measurements and techniques. Ed. A.G. Haus (AAPM, New
  York) p. 573
Wiener S.N., Borkat F.R., and Floyd R.M. (1974) Functional imaging.
  A method of analysis and display using regional rate constants.
  J. Nuc. Med. 15, 65-68.

THE INFLUENCE OF DIGITISATION ON THE QUALITY OF RADIONUCLIDE IMAGES

W.R. PITT

Department of Biomedical Physics and Bioengineering,
University of Aberdeen, Foresterhill, Aberdeen AB9 2ZD,
Scotland, UK.

1. INTRODUCTION

The availability of relatively cheap computing facilities has
led to widespread use of digitised images in nuclear medicine. The
digitised data may be used subsequently for quantitative analysis,
or may be displayed for visual interpretation by a clinician. The
act of digitising an analogue image imposes a spatially discrete
sampling matrix onto the data and it is important to determine the
size of matrix best suited to the purpose for which the data are to
be used.

2. PHYSICAL CONSIDERATIONS FOR QUANTITATIVE ANALYSIS

When conventional TV images are digitised, it is usual to
employ the smallest picture element (pixel) available in order to
preserve fine spatial detail. Radionuclide images, however, are
quantum limited. Their count density is low - between 10^2 and 10^3
per square centimetre - partly because of restrictions on the amount
of radioactivity which may be administered to a patient, and partly
because the efficiency of the detector is poor. Under these
circumstances the smallest pixel available is not necessarily the
best choice.

The variability in the number of counts in each pixel follows
Poisson statistics, and therefore the standard deviation in the
number of counts equals the square root of that number. The
result of this Poisson uncertainty is that small pixels, which
contain fewer counts than large ones, will have higher relative
variability in their data, so the more finely digitised image will

appear noisier. A simplified example of this mechanism is shown in
Fig. 1. Thus large pixels are required for statistical reliability.

Account must also be taken, however, of the need for adequate
spatial resolution. Nyquist's Theorem states that in order to
preserve all spatial detail in an image, it should be digitised onto
a matrix whose sampling frequency is at least twice the highest
spatial frequency in the data. The spatial resolution of a gamma
camera is poor - a typical point spread function has a full width
at half maximum counts (FWHM) of 3 to 4 mm under the best possible
imaging conditions. When scattering material, e.g. soft tissue, is
present, the FWHM is degraded to 10 mm or more. Reasonably large
pixels will therefore still satisfy the requirement for adequate
spatial resolution.

We have used a simple model to determine the best size of pixel
for quantitative image analysis. It is based on Fourier theory and
will be published elsewhere. The model combines the requirements
of statistical confidence with those of spatial fidelity and predicts

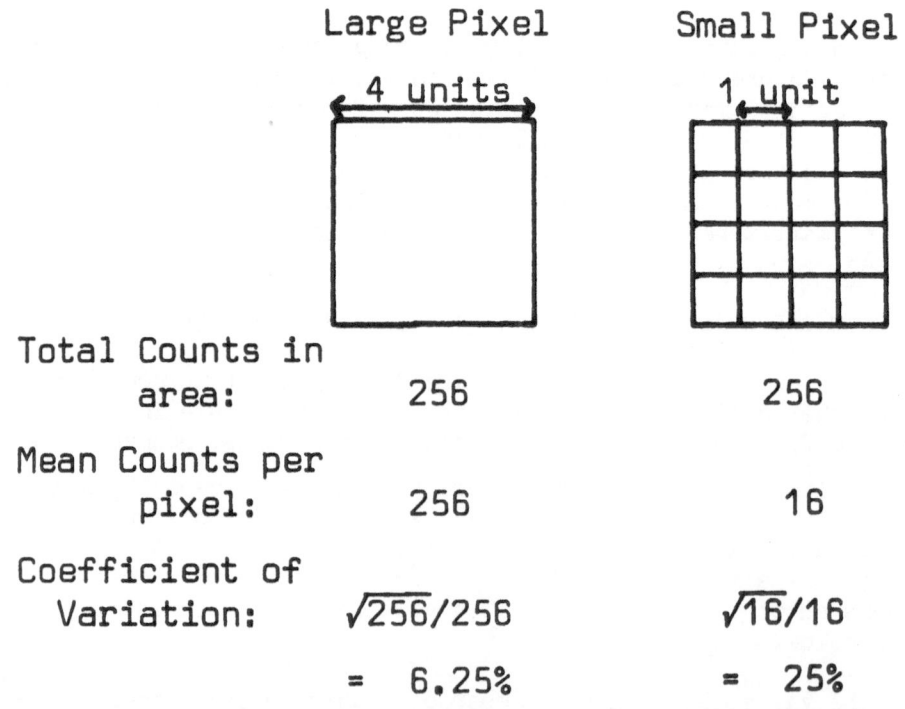

	Large Pixel	Small Pixel
	4 units	1 unit
Total Counts in area:	256	256
Mean Counts per pixel:	256	16
Coefficient of Variation:	$\sqrt{256}/256$	$\sqrt{16}/16$
	= 6.25%	= 25%

Fig. 1. Demonstration of the effect of pixel size on image noise
levels. The coefficient of variation (standard deviation
divided by mean) is lower for the large pixel.

that, for data with count densities typical of those produced in nuclear medicine, a pixel with a side length representing 3.5 mm on the gamma camera face is best for numerical calculations. This corresponds to a 128 X 128 matrix of pixels for our gamma camera system.

3. PERCEPTUAL CONSIDERATIONS FOR VISUAL INTERPRETATION

The above discussion has concerned only physical criteria. If an image is to be interpreted by a human observer rather than by computer, it is necessary to take account of the observer's responses to the image. The pixel size chosen above is not necessarily optimum for this task. Sharp et al (1982) have shown that a large pixel is adequate for simple perceptual tasks, but when more complex data are presented in this way, oberservers are acutely aware that the images are coarsely digitised and find the pixels themselves visually obtrusive. Finely digitised images are preferred to coarsely digitised ones, even though the former appear noisier.

One way to overcome the distracting nature of large pixels is to digitise initially onto a coarse matrix and then interpolate onto a finer matrix. The interpolated images retain the statistical precision of the original large pixels, but they appear excessively smoothed and observers also find this objectionable (Sharp et al 1982). The problem, therefore, is to overcome the adverse reactions of observers to large pixels without using very fine digitisation matrices.

The edges of pixels are composed of high spatial frequencies. Although Fourier analysis shows that these frequencies have a low energy content, it is possible that their regularity enhances their detection by the visual system. This effect may be overcome partially by the use of a low-pass filter. Such blurring is easily achieved by defocussing the eye, or squinting at the image.

It has been suggested by Harmon and Julesz (1973) that the frequency components principally responsible for disrupting the visual interpretation of a digitised image are those which lie within two octaves (1) of the spectrum of the signal. According to their "critical band masking" model, it is the noise frequencies produced by the pixels within this range of the signal that mask some of the information in the image rather than the very high frequencies present in the pixel edges. Whichever model is correct, it would be useful if some simple and convenient way could be found to enhance the spatial frequencies present in the real data and simultaneously to attenuate those frequencies above this range.

1) An octave is the interval between two frequencies which differ by a factor of two.

262

The response of the human visual system varies greatly with spatial frequency. Campbell and Robson (1968) showed the existence of channels in the eye-brain system which are selectively tuned to specific spatial frequencies. They carried out a series of experiments to measure the detectability of sinusoidal, square, rectangular, and saw-toothed patterns, and the results agreed with predictions from Fourier theory. These channels are shown in Fig. 2. The envelope of these channels represents an effective "modulation transfer function" (MTF) for the human visual system.

Figure 2 shows that an observer's response is greatest at a spatial frequency of a few cycles per degree. It should be possible therefore to adjust the viewing conditions of a coarsely digitised image such that the signal frequencies lie around the peak of the "MTF" and are enhanced, while the visual impact of the higher frequency pixel components is reduced. This requires the position of the visual "MTF" to be varied with respect to the spatial frequencies in the image. This variation may be achieved by altering the visual angle of the image, either by changing the size of the

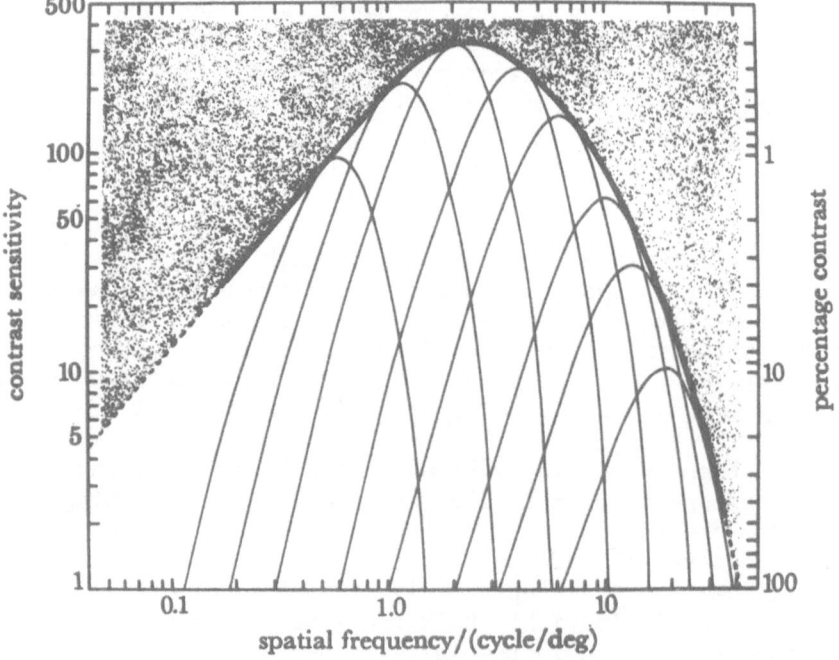

Fig. 2 Response of the human visual system. The light curves represent channels in the human visual system, each of which is sensitive to a narrow range of spatial frequencies. The dark curve, which is the envelope of the individual channels, is the effective "MTF" of the eye. From Campbell, 1980.

image or by viewing a fixed image from different distances. The
latter is the more practical method for experimental purposes, and
is commonly carried out in clinical practice.

Warren and Pandya (1982) have investigated the effect of
viewing distance on the detectability of a specific abnormality in
X-ray CT images. They used very small pixels (0.25 mm in length)
and a disc-shaped target 9 mm diameter. Using a simple measure of
the certainty with which observers identified the presence of the
target, they found that the detection rate increased with viewing
distance at first, but then began to fall. The optimum viewing
distance for that particular task was between 4 and 5 metres.

The pixels used by Warren and Pandya were very small with
linear dimensions less than 3% of their target diameter. The
spectrum of frequencies produced by these pixels is therefore much
higher than that of their target. According to the critical band
masking theory of Harmon and Julesz, therefore, these pixels should
have had little effect on the perception of their target. As
discussed above, radionuclide images require pixels whose sizes
approach those of the signal, so the separation between the spatial
frequencies of the signal and those introduced by the digitisation
matrix is much less distinct, and any critical band masking will be
severe. The effective position of the visual "MTF" when an image
on our system is viewed from different distances, in shown in
Fig. 3. It can be seen from this figure that, despite the indis-
tinct boundary between signal and pixel noise, increasing the
viewing distance to 4 or 8 metres should have a preferential
enhancing effect on the signal, whereas the "critical band" of the
pixels themselves would be most obvious when viewing from 1 metre.

4. EXPERIMENTAL INVESTIGATION OF THE EFFECT OF VIEWING DISTANCE
   ON IMAGE INTERPRETATION

An observer's response to an image is a complex combination
of subjective and objective characteristics. Therefore we carried
out three different sets of experiments, which will be fully
reported elsewhere, to investigate the effects of varying pixel
size and viewing distance with a view to determining the optimum
combination. For all the experiments a monochrome TV display with
128 grey levels was used, and the images were square, with a side
length of 32 cm.

4.1 Observer preference for clinical images

The first experiment investigated a purely subjective response -
it could almost be described as "observer prejudice". Radionuclide
bone images were presented on three digitisation matrices: 64 X 64,
128 X 128, and 256 X 256 (see Figs. 4 - 6). At each of four
viewing distances (1, 4, 8, and 12 metres) observers were asked

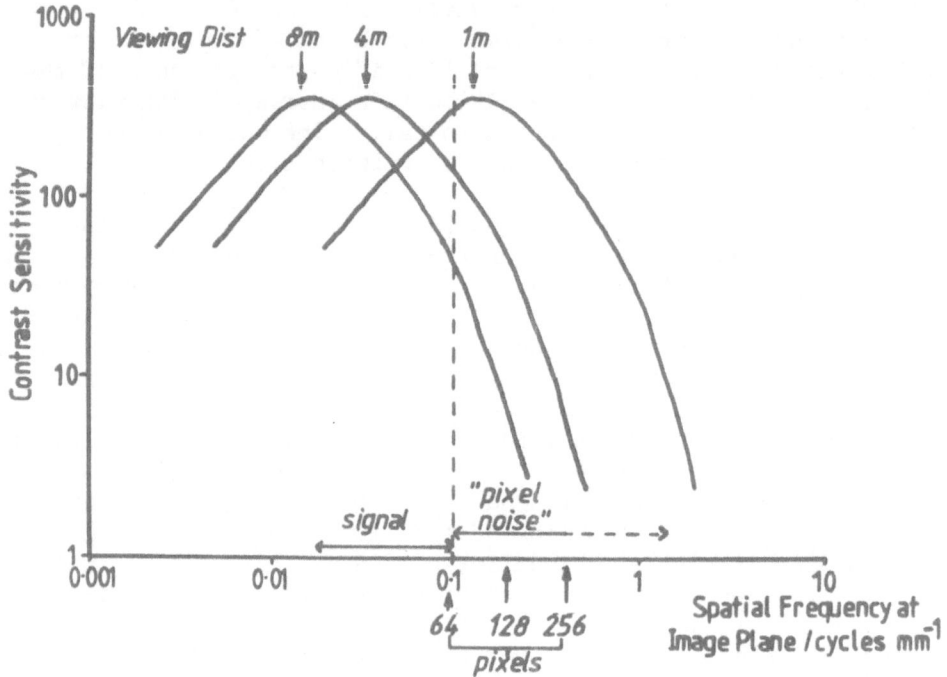

Fig. 3 Positions of the visual "MTF" from Figure 2 with respect to
detail in an image when the image is viewed from different
distances. It can also be seen that many of the frequency
components of the pixels lie in the "critical band" lying
two octaves above the signal frequencies. This method of
data presentation is taken from Warren and Pandya, 1982.

simply to rank the three digitised forms of each image in order of
preference. Bone images were used as test patterns because they
contain the finest spatial detail found in nuclear medicine.

The results showed that the 64 X 64 matrix was disliked even
when viewed from 12 metres, but observers showed no significant
preference for 256 X 256 over 128 X 128 for viewing distances of 4
metres or more.

4.2 Abnormality detection against a uniform background

The fact that observers prefer a particular type of image

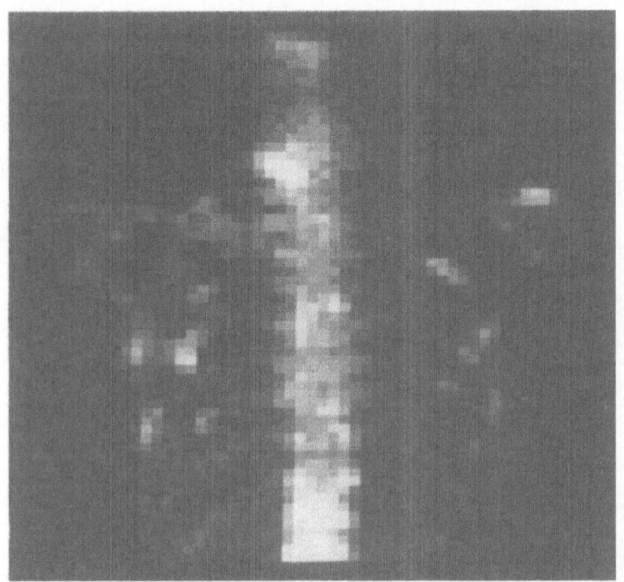

Fig. 4 Abnormal radionuclide bone image digitised onto 64 X 64
matrix. The localised areas of high count density represent
uptake of radiopharmaceutical in areas of tumour growth.

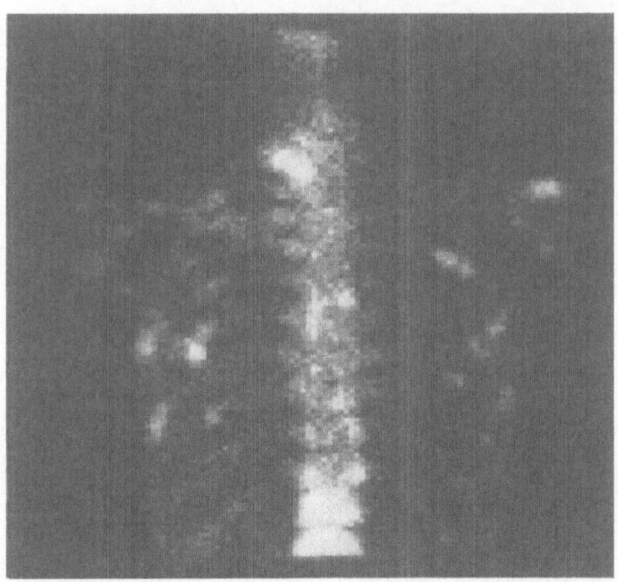

Fig. 5 Data in figure 4 digitised onto a 128 X 128 matrix

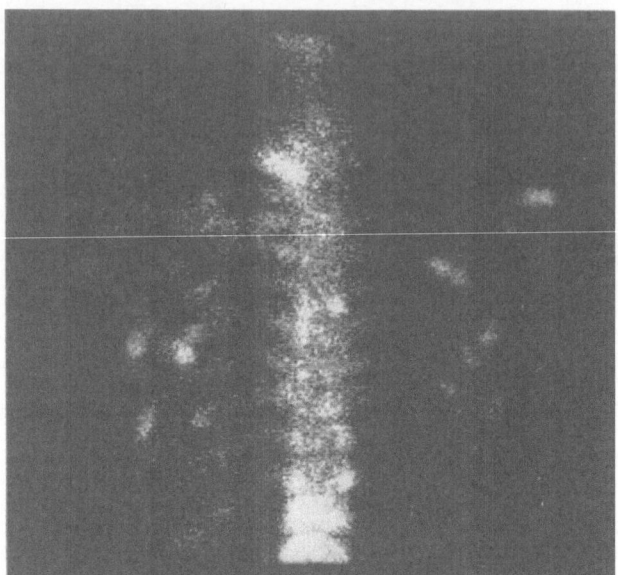

Fig. 6 Data in figure 4 digitised onto a 256 X 256 matrix.

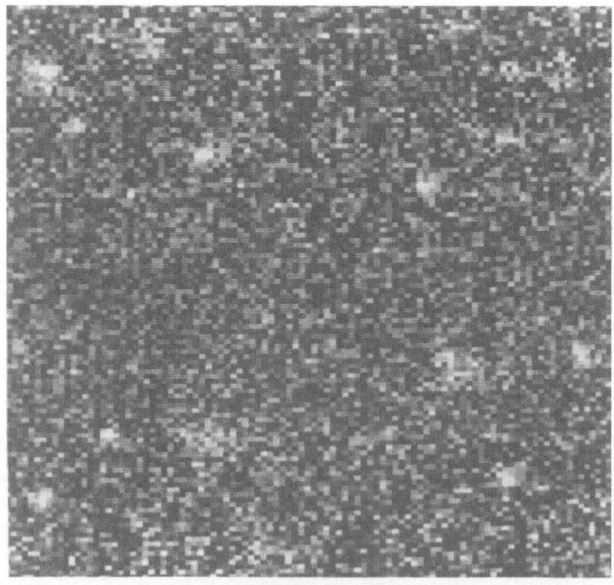

Fig. 7 Test pattern of noisy background containing several hot
spots. The data are digitised onto a 128 X 128 matrix.

presentation does not mean that this is the form which yields most diagnostic information. Thus the second experiment attempted to measure image quality more objectively using the method of constant stimulus. A simple test pattern was generated in which regions of increased count density (hot spots) of Gaussian profile and known contrast were superimposed onto a uniform background containing Poisson noise. The sizes of the abnormalities were typical of those encountered in nuclear medicine. An example of the test pattern is shown in Fig. 7. The data were presented on the same three digitisation matrices at three viewing distances (1, 4, and 8 metres) and six observers were asked to mark any area of the image that contained a hot spot. One of the problems with the method of constant stimulus is that it assumes the decision threshold adopted by the observer will not change during the course of an experiment. In practice, this consistency is only achieved by setting a very strict decision criterion, so each observer was required to be <u>sure</u> that a hot spot was present before marking it. Another advantage of this requirement was that false positive responses were maintained at a negligible level.

For each of the three matrices, detection rates improved as the viewing distance was increased up to 8 metres. Furthermore, results from the 128 X 128 display viewed from 8 metres were always as good as, and often better than, those from any other combination of pixel size and viewing distance.

4.3 Abnormality detection in clinical images

The previous experiment related to detection of targets in a background which was uniform except for the presence of statistical noise, and where any definite increase in count density represented an abnormality. In clinical images, however, this is not the case because normal anatomical or physiological features produce a varying background. This is particularly true of bone images which are highly structured. An observer must be able to distinguish between normal variations in count density as shown in Fig. 8, and those which represent pathological changes. The third experiment, therefore, investigated whether the conclusions from the experiment with the abstract test pattern were still valid for clinical images.

The data used in this experiment were normal bone images to which realistic artificial abnormalities had been added under computer control. Figure 9 shows some high contrast hot spots which have been superimposed onto the normal data of Fig. 8. The experimental procedure outlined in section 4.2 was repeated, again using a strict decision criterion.

The results of this experiment showed that the 128 X 128 matrix was adequate even for complex clinical images. The best viewing distance varied between 4 and 8 metres. Note that the same effect

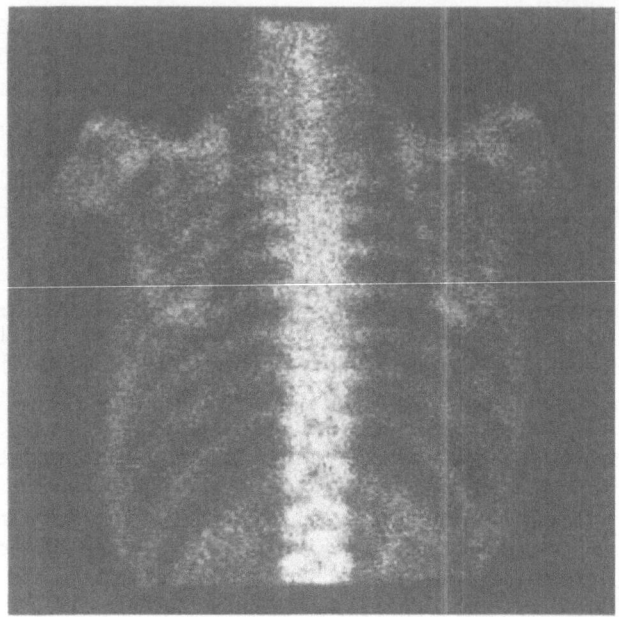

Fig. 8   Normal radionuclide bone image digitised onto a 256 X 256 matrix.

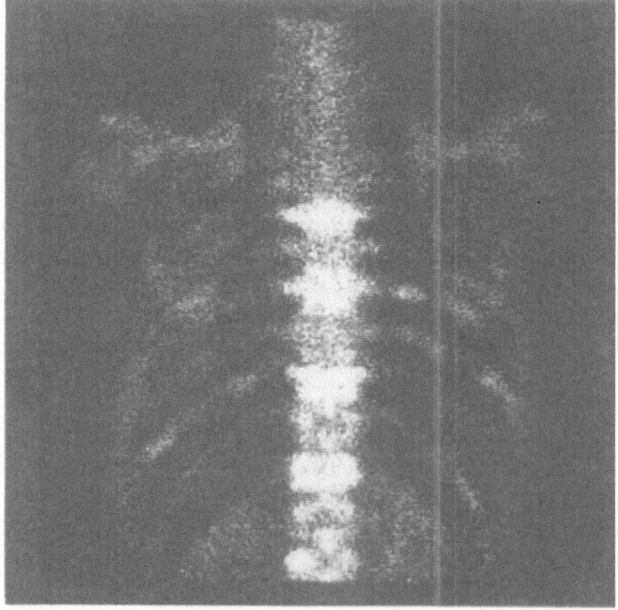

Fig. 9 Data from figure 8 after artificial "hot spots" have been added.

may be achieved by reducing the size of the image by a factor of 4, i.e. to an image with sides 8 cm long, and viewing from both 1 metre and 2 metres.

5. CONCLUSIONS

The choice of optimum pixel size for radionuclide images is a compromise between the requirements of spatial resolution and statistical reliability. For quantitative analysis it appears that reasonably large pixels are the best choice.

When images are presented to observers, however, there is a strong preference for finer matrices. This preference is overcome by minifying the more coarsely digitised images (or by viewing them from a greater distance). When suitably minified, the coarsely digitised images yield abnormality detection rates which are always as good as, and often better than, their more finely digitised counterparts.

REFERENCES

Campbell F.W., and Robson J.G. (1968) Application of Fourier analysis to the visibility of gratings. J. Physiol. 197, 551-566
Campbell F.W. (1980) The physics of visual perception. Phil.Trans. R. Soc. London, B290, 5-9
Harmon L.D. and Julesz B. (1973) Masking in visual recognition: Effects of two-dimensional filtered noise. Science, 180, 1194-1197
Sharp P.F., Chesser R.B., and Mallard J.R. (1982). The influence of picture element size on the quality of clinical radionuclide images. Phys. Med. Biol. 27, 913-926
Warren R.C. and Pandya Y.V. (1982) Effect of window width and viewing distance in CT display. Brit. J. Radiol. 55, 72-74

DIAGNOSTIC PROCESSING AND ANALYSIS OF MEDICAL IMAGES

J.P.J. de VALK

Laboratory of Medical Physics and Biophysics, University
of Nijmegen, c/o Geert Grooteplein Noord 21, 6525 EZ
Nijmegen, The Netherlands.

1. INTRODUCTION

The possibilities of medical imaging for diagnostic purposes
have increased considerably since the introduction of computers
(Hay 1976, Fu and Pavlidis 1979, Huang 1981). Both software and
hardware make important contributions to the production, processing
or enhancement of images, thereby determining their quality.

There is a legitimate need for methods of evaluation that
relate to the true diagnostic value of the medical images themselves.
Performance studies, however, demand the availability of reliable
prior diagnostic knowledge. Experiments testing the diagnostic
quality of images must be based on a maximal amount of diagnostic
information from all sources available. On the one hand, medical
images, especially those produced artificially by computer reconstruction, may contain more information than the human eye can
perceive. On the other hand, humans are excellent pattern recognis-
ers, barely matched by computers in a variety of visual tasks.

Information extracted from images by computation combined with
visual judgements may enhance diagnostic capability. A short treatment of medical imaging and subsequent image interpretation processes will be presented, considering first the digital processing
and analytical methods commonly used, then reviewing four case
studies briefly for the purpose of illustration.

2. MEDICAL IMAGING AND IMAGE INTERPRETATION

2.1 Digitising

Medical images now arise in many ways including radiography, computerised tomography, scintigraphy, ultrasound, nuclear magnetic resonance, thermography, angiography, lymphography, and cell images. Our approach to image analysis is illustrated in Fig. 1.

If the data have not already been digitised, it is usually necessary to do so. The sampling and digitising process requires prior information about resolution and signal to noise ratios. The number of quantisation levels must be chosen optimally with respect to display requirements, thus determining the number of bits per pixel. The sampling grid and technique (static/dynamic) are usually fixed by the hardware used.

The process of digitisation of the medical image can be seen schematically in Fig. 2. Once the medical image has been digitised further processing by computer may start.

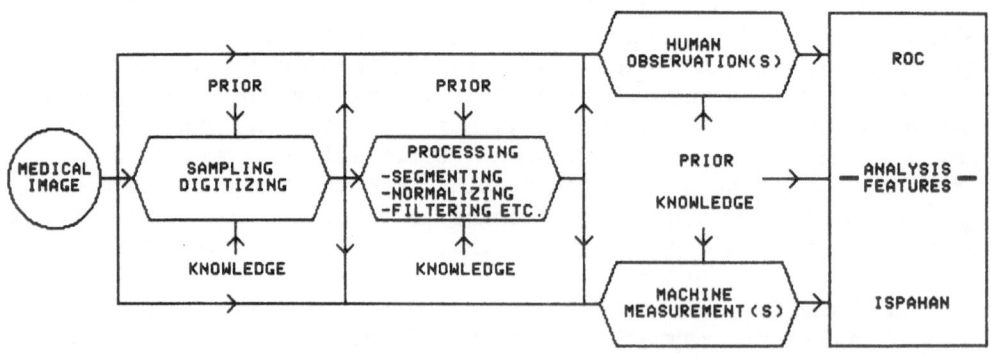

Fig. 1   The analysis of medical images (ROC: Receiver Operating Characteristics; ISPAHAN: Interactive Statistical Pattern Recognition and Analysis.)

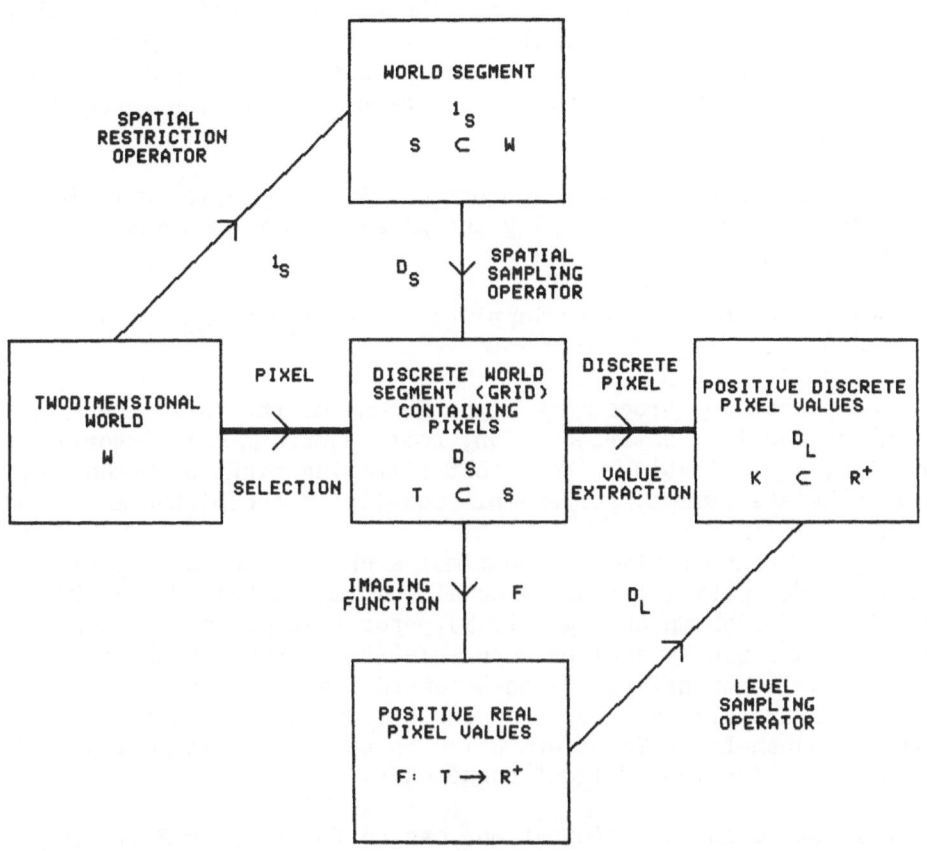

Fig. 2 The digitising of an image

2.2 Processing

There are many possible ways to process the image as shown in the literature (Rosenfeld 1976, Pratt 1981, Cannon 1981). They can roughly be divided into the following categories:

2.2.1 Spatial filtering.
a. Linear; one-pixel based. The new pixel value is only based on the old pixel value (linear transform). Examples are linear intensity transformation (e.g. minimum-maximum rescaling - see Fig. 3 "lt" images) and monochromatic pseudo-colouring.

b. Non-linear; one pixel based. Examples are non-linear intensity transformation (e.g. gamma correction - see Fig. 3 "ps" images) and multichromatic pseudo-colouring.

Categories a and b may be combined using look-up tables.

    c. Linear; multi-pixel based. The new pixel value is based on the old pixel value plus a number of neighbour values (linear combination). An example is convolution with two-dimensional filter point spread function (e.g. Gaussian low-pass filtering - see Fig. 3 "gf" images).

    d. Non-linear; multi-pixel based. An example is averaging within adjacent windows, keeping the pixel number constant (see Fig. 3 ("pa" images).

Note that a and b (intensity mapping) can be considered as a subset of c and d.

    The purpose of processing might be one of the following:- optimal perception; data-reduction; noise suppression; distortion reduction; normalisation. Note that filtering might be based upon pixel value statistics, for example equalising the histogram.

2.2.2 Space Transformation. Sometimes a change from the spatial domain to the spatial frequency domain can be useful (Wintz 1972, Huang 1975, Beauchamp and Yuen 1979). For this purpose orthogonal transforms are widely applied such as:- (Fast) Fourier transform, discrete Cosine transform, Walsh-Hadamard transform and Haar transform; which are all based upon decomposition into "fixed" components or the Karhunen-Loeve Transform which is based upon image statistics and decomposition into "eigen" components.

    All one to one transformations can be followed by filtering in the spatial frequency domain and inverse transformation. The main reason for transformation is to simplify processing in the new domain, resulting in data reduction and easier evaluation. After processing, the medical image must be further inspected and analysed.

2.3 Analysis

    Two approaches may be used, either alone or combined. The first is a qualitative human-based analysis, dealing with for example perception, search patterns, visual psychophysics, observer performance, ROC analysis, signal detection theory. (Marr 1976, Senders et al 1978, Metz 1978, Kundal 1979, Swets 1979, Eijkman 1980, Julesz and Schumer 1981).

    The second is a more quantitative machine-based analysis dealing with random field models, Markovian approach, spatial statistics, local/global calculations, Co-occurrence/texture approach, discriminant/factor analysis (Couley and Lohnes 1971,

Zucker and Terzopoulos 1980, Stevens 1981, Modestino et al 1981).

Used in combination the two analyses can lead to computer (colour) graphics, multi-dimensional imaging, shape and orientation characteristics, feature extraction, pattern recognition, classification, diagnostic information, decisions, and evaluation (Dudo and Hart 1973, Tou 1980, Levine and Shaheen 1981, Witkin 1981, Whitted 1982, Kittler et al 1982). Perhaps the ultimate goal is human/ artificial intelligence!

The lists above are far from complete and show some overlap. They are used to illustrate the breadth of the field under research and its possibilities. Multi-disciplinary co-operation seems to be an absolute necessity.

3. PRACTICAL EXAMPLES

To illustrate the approach as described briefly above, some examples will be given of practical cases under study in Nijmegen.

3.1 Quality assessment of classification of differently stained cervical smears (de Valk et al 1981a)

Signal detection theory and psychophysical measures derived from it can be used for diagnostic quality assessment. Measures such as "detectability" and "diagnostic output" find their applications here and should be related to prior diagnostic knowledge.

The final goal in the case under study was to obtain a quantitative and clear comparison of the classification results achieved with two different cell staining techniques applied to cervical smears.

One may expect different classification results with different preparative procedures and therefore answers were sought to the following questions:

    i) Is there any significant intra-observer difference?
   ii) Is there any significant inter-observer difference?
  iii) Is there any significant difference between results obtained
       by the two staining methods?

Psychophysical methods have been successfuly applied to cell image data and indicate that better classification results can be obtained with one of the staining methods. The results suggest acceptability of the applied psychophysical techniques over a wide range of medical diagnostic problems.

3.2 Evaluation of scintigraphic heart image processing with psychophysical measures (de Valk et al 1981b)

In this example, experiments have been carried out with 61 digitised scintigraphic myocardial images. These were Tl-201 images, recorded in LAO-30 degree position. Images have been low-pass filtered and obtained high emphasis by a Laplacian operator directly after recording in the clinic. They were stored in a 64 x 64 pixel raster format (see Fig. 3).

The final medical diagnosis of each image was known and has been taken by us as the correct diagnosis. The set of 61 images consisted of 27 normal and 34 diseased heart images, showing left ventricular wall and septum. Pathologies were ischaemic areas or infarcts showing up as areas of less intensity in left ventricular wall or septum or in both. The clinical images were considered optimal for diagnostic aims.

It was our intention to investigate the use of different quality factors for description of diagnostic properties. To achieve this, the original digitised images have been compared with degraded versions (deteriorated by filtering) and different measures of diagnostic merit were studied. The chosen processing technique consisted of a simple pixel value averaging within squares of different size, or low-pass two-dimensional filtering using Gaussian filters with different width.

An antilog function was used for all series to compensate for the logarithmic brightness sensitivity of the human visual system. After a learning period, subjects judged all pictures for the possible presence of pathologies. The judging consisted of rating responses according to one of five possible categories. From the judgement data and prior class memberships the detectability and diagnostic output were calculated.

In the diagnostic output the confidence with which both correct and incorrect judgements are made is taken into account. This is done because an uncertain judgement does not contribute much to the medical decision process, despite the fact that it could be a justifiable judgement from an optimal detection point of view.

In all cases we noticed effects according to our expectations, although image quality in a diagnostic sense seems to be relatively insensitive to small variations of the parameters under study, if a picture has an acceptable quality in the first instance.

The main conclusion was that the effect of the image processing techniques under consideration could be described very well using signal detection theory.

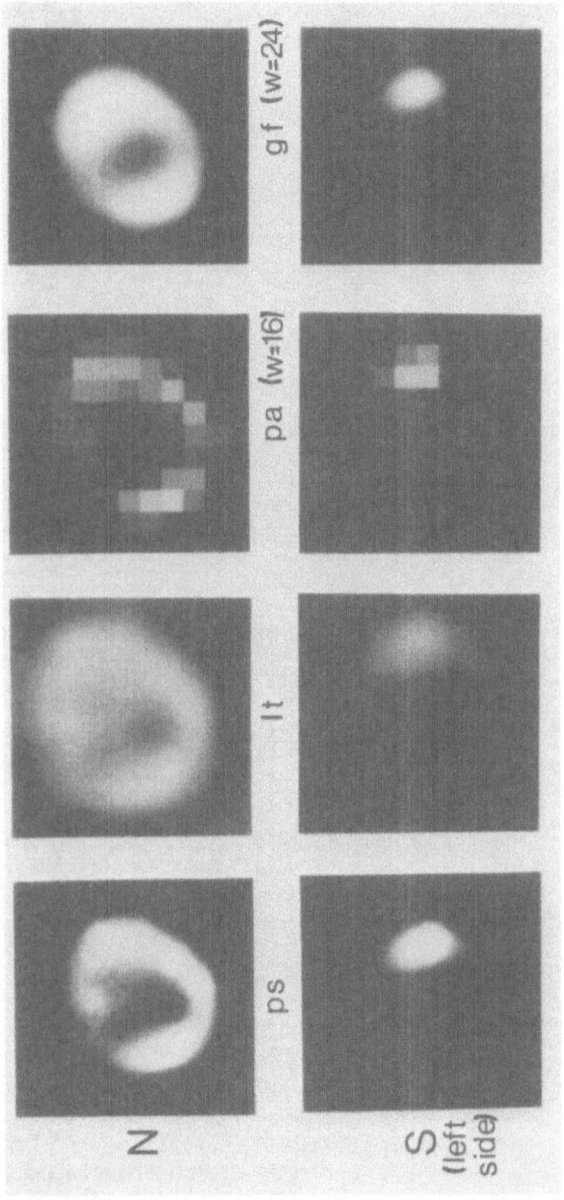

Fig. 3 Examples of scintigram pictures before and after a particular image processing.
(N: No pathology; S: diseased; p.s.: primary series; l.t: primary series with
linear table; p.a.: pixels averaged; g.f.: Gaussian filtered; w: filter/pixel
width. For details see de Valk et al 1981b.)

278

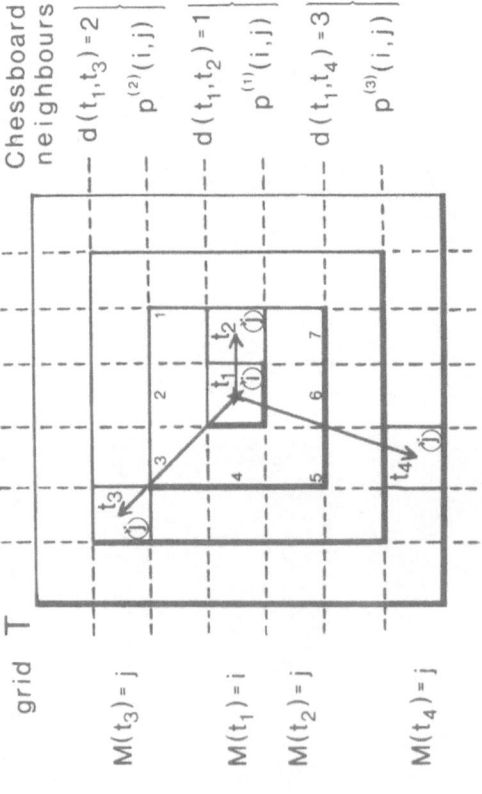

Fig. 4 Spatial relations of pixels in a digitised image. M: discrete image matrix, T: discrete image pixel grid, t: two-dimensional discrete co-ordinate pair, M(t): value of pixel t in image matrix M, $d(t_k, t_l)$: chessboard distance of t_k and t_l, Ng: number of grey levels in image M

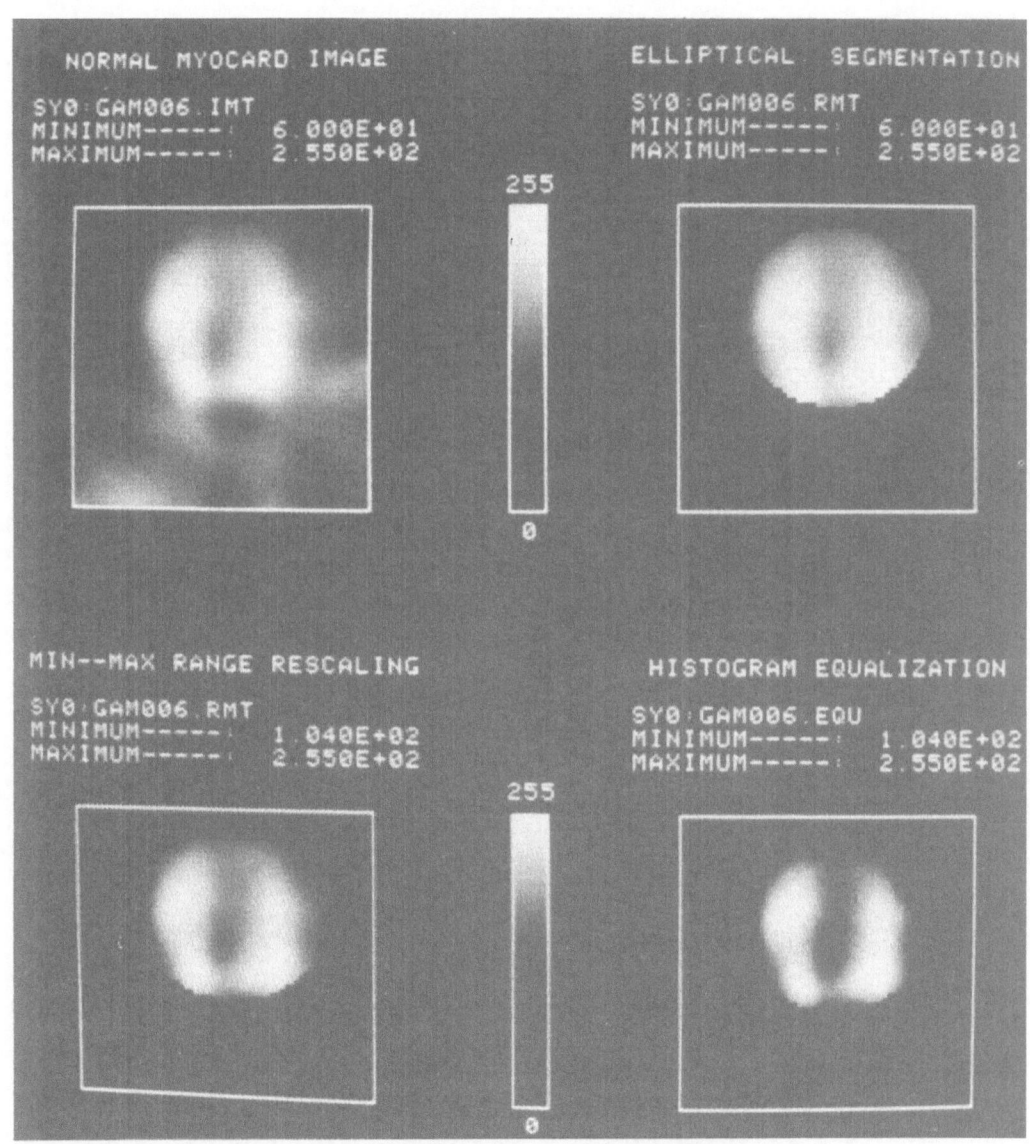

Fig. 5 Normal myocardial image before and after processing.

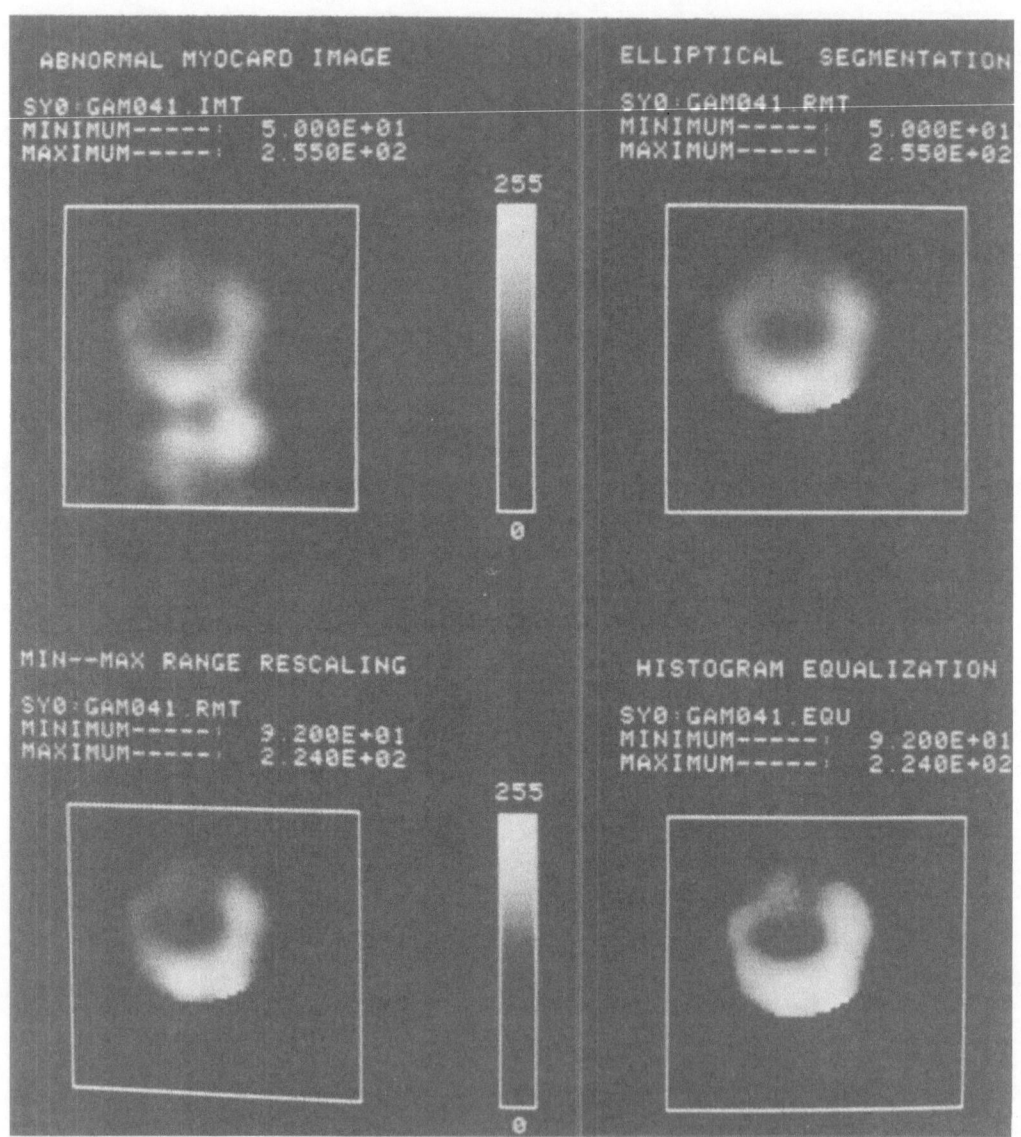

Fig. 6 Abnormal myocardial image before and after processing.
The pathology can be seen in the left part of the horseshoe
shaped myocardium, indicating an ischaemic area or infarct
in the septum, showing up as an area of lower intensity.

3.3 Improved diagnosis of medical images by combination of computer measurements and visual judgements (de Valk and Eijkman 1982)

We compared classification performance of gamma-camera images judged by human observers with the performance of co-occurrence parameters extracted by computation. (Fig. 4)

Our assumption was that information indicating pathology is bounded locally and that global spatial information can be considered to be of minor importance in these images. One can expect human observers to be quite sensitive to shape and symmetry parameters in particular.

In addition, however, information carried by local statistical variables might also contain valuable information for which the human visual system is less sensitive. Combination of human observations and computer calculations would then probably lead to better diagnostic results.

One of the major problems to be solved is the determination of the optimal combination in the sense that the number of false diagnoses is minimised. In the experiments, the digitised images were first segmented such that the myocardium was isolated from its background by an inter-actively matched ellipse. Inside the elliptic segment normalisation of grey levels was carried out, followed by histogram equalisation. Examples of normal and abnormal myocardial images before and after processing are shown in Figs. 5 and 6.

Next, three co-occurrence matrices were calculated for different annular radii. These matrices were the basis for calculation of ten selected statistical parameters, such as energy, entropy and matrix peak value. (Haralick 1979, de Valk and Eijkman 1982). Correlations have been computed between human response values, between linear combinations of the ten selected statistical parameters derived from the three co-occurrence matrices belonging to the three different annular radii and finally between human responses on the one side and linear parameter combinations on the other. Human responses correlated highly with each other as did the linear co-occurrence parameter combinations, but there was much less correlation between human responses and parameter combinations.

For optimal linear combinations the well known Fisher linear discriminant method was used in all cases. This would be an argument to look for a linear (Fisher) discriminant composed of the two kinds of data in our study.

The final diagnostic results could be significantly improved by combining human response values with co-occurrence computer output. The total number of false diagnoses decreased by over 30% for

each of the four observers. Less capable observers profited most
from computer aid and, in that sense, computer aid reduced the
rather large differences between observers. The study of optimal
weighting procedures for human and machine judgements seem just-
ified in final classification tasks. In such procedures, weighting
factors can be updated continuously thus making feasible flexible
co-operation in future tasks between man and computer.

3.4 The interpretation of lung radiographs

The use of chest radiographs for diagnostic purposes is well
known (Starr et al 1975, Kundel et al 1980, Carmody et al 1980,
Swensson et al 1982). For lung diseases such as pneumoconioses,
there are widely used classification schemes (see Fig. 7)

The features proposed and used in these schemes, however, can
be sub-optimal, or ambiguous, which in its turn might lead to mis-
representations and classification erros (Rossiter 1982). This
calls for a study of the features themselves, as well as a search
for an optimal training method for future medical classification
experts who will use the schemes.

Based upon the ILO 1971 reference chest radiographs, we have
set up an experiment, consisting of a training and testing session,
to collect psychophysical data giving answers to questions concerning the efficiency of the features in the ILO scheme (de Valk and
Eijkman 1983). The images are displayed in cyan colour, while

FEATURE	SHORT CLASSIFICATION	EXTENDED CLASSIFICATION
No pneumoconiosis	0	(Rounded 0/-, 0/0, 0/1
PNEUMOCONIOSIS		(Irregular
SMALL OPACITIES		
Rounded		
Profusion*	1, 2, 3	1/0, 1/1, 1/2; 2/1, 2/2, 2/3; 3/2, 3/3, 3/4
Type	p, q(m), r(n)	p, q(m), r(n)
Extent	—	zones: Right, Left; Upper, Middle, Lower
Irregular		
Profusion*	1, 2, 3	1/0, 1/1, 1/2; 2/1, 2/2, 2/3; 3/2, 3/3, 3/4
Type	s, t, u	s, t, u
Extent	—	zones: Right, Left; Upper, Middle, Lower
LARGE OPACITIES		
Size	A, B, C	A, B, C
Type	—	wd (well defined), id (ill defined)
PLEURAL THICKENING		
Costophrenic angle	—	Right, Left
Walls and Diaphragm		
Site		Right, Left
Width	pl	a, b, c
Extent		1, 2
DIAPHRAGM OUTLINE	—	Ill defined: Right, Left
CARDIAC OUTLINE	—	Ill defined: 1, 2, 3
PLEURAL CALCIFICATION		
Site	plc	Walls, Diaphragm, Other; Right, Left
Extent		Length: 1, 2, 3

Fig. 7 Part of the ILO U/C International Classification of
Radiographs of Pneumoconioses 1971.

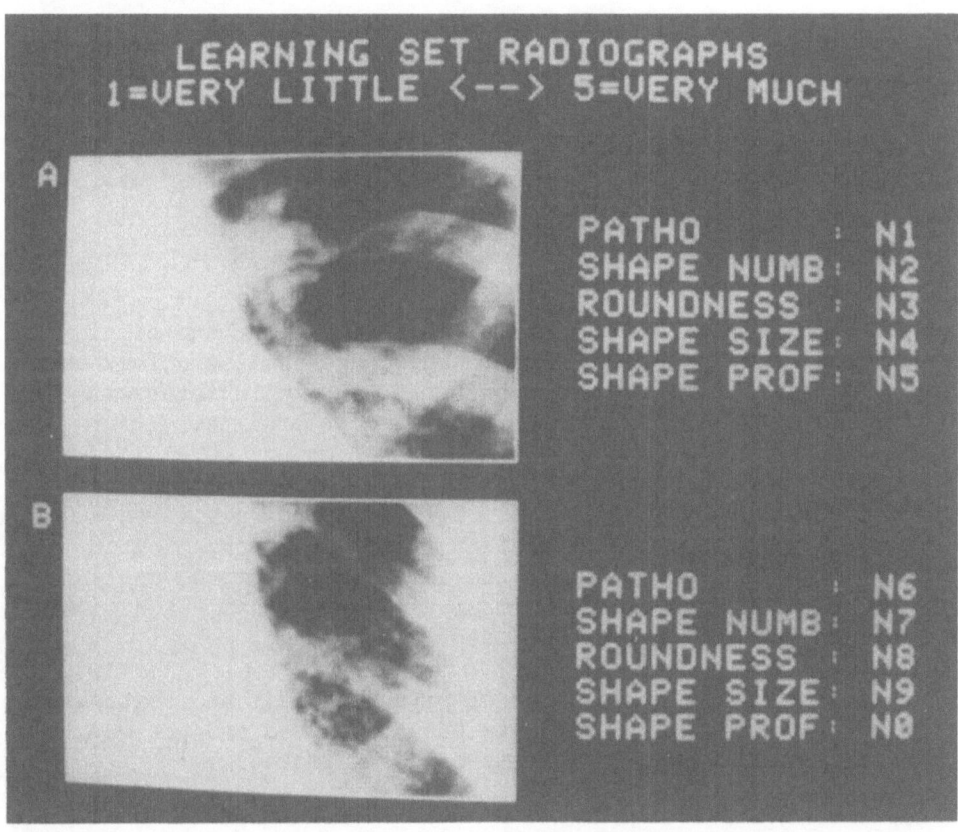

Fig. 8 A training pair of radiograph zones, demonstrating partic-
        ular features.

PATHO: The general impression of pathology of the image rated
            according to its seriousness
SHAPE NUMB: A judgement on the number of observed different shapes
            of pathology indicating opacities
ROUNDNESS: A judgment on the prevailing shape of opacities in the
            image: round or irregular
SHAPE SIZE:A rating of the size of the most frequently occurring
            opacities
SHAPE PROF: A rating of the profusion of the most frequently occurr-
            ing opacities
N1 to NØ: Ratings to be given by the human observers employing a 5
            category rating scale from 1: very little to 5: very much
A: Example of a chest radiograph zone without opacities indicating
   the pathology under study (pneumoconiosis)
B: Example of a chest radiograph zone showing pathological opacities.

gamma-correction is applied in the software of the video display. (See Fig. 8)

All observers were asked to judge six sections in each image corresponding to the commonly used medical sub-division. Features were evaluated and ranked.

Other factors that may be studied include the influence of contrast, brightness, number of discretisation levels as well as eye movements and guided search.

4. DISCUSSION

The analysis of medical images is a very broad and complex area with its own problems. Modern knowledge and technology open many possibilities to overcome these problems, with interactive and interdisciplinary approaches showing most promise.

Quality assessment and diagnostic evaluation are important topics for research. Images are usually samples, digitised and processed (filtering, transformation) prior to observation or computation. All stages before the final (interactive) analysis must be selected with care and inspected thoroughly. Each partic- ular diagnostic imaging problem needs its own approach.

Examples of the application of psychophysical and statistical techniques to practical cases have been discussed and the results indicate the usefulness of the approaches advocated.

REFERENCES

Beauchamp K.G. and Yuen C.K. (1979) Digital methods for signal analysis. Allen and Unwin, London.
Cannon T.M. and Hunt B.R. (1981) Image processing by computer. Scientific American, 10. pp.136 - 145
Carmody D.P., Nodine C.F. and Kundel H.L.(1980) Global and segmented search for lung nodules of different edge gradients. Invest. Radiol. Vol. 15, 3, pp. 224-233
Cooley W.W. and Lohnes P.R. (1971) Multivariate data analysis. Wiley, New York
De Valk J.P.J., Eijkman E.G.J. and Vintcent J. (1981a) Detectability index, likelihood ratio and diagnostic output as psychophysical measures for evaluation of medical image processing. Medical & Biological Engineering & Computing, 19, pp. 597-604
De Valk J.P.J., Oud P.S. and Vintcent J. (1981b) Detectability and diagnostic output for quality assessment of classification of differently stained cervical cells. IEEE Conference Record on 2nd Internat. Conf. on Visual Psychophysics and Medical Imaging, pp. 75-80

De Valk J.P.J. and Eijkman E.G.J.(1982) Diagnosis improvement of
medical images by combination of computer measurements and visual
judgements. Proc. 1st IEEE Computer Society Int. Symp. on Medical
Imaging and Image Interpretation, ISMIII '82 Berlin, pp. 224-230

DeValk J.P.J. and Eijkman E.G.J. (1983) Feasibility of features
for visual classification of lung radiographs. Accepted Proc.
MEDINFO '83 4th World Conference on Medical Informatics, Amsterdam,
August 22-27 1983.

Duda R.O. and Hart P.E. (1973) Pattern classification and scene
analysis. Wiley, New York.

Eijkman E.G.J. (1980) Psychophysics In Handbook of Psychonomics I,
North Holland, Amsterdam. pp. 303-363

Fu K.S. and Pavlidis T. Eds. (1979) Biomedical Pattern Recognition
and Image Processing. Verlag Chemie, Weinheim.

Green D.M. and Swets J.A. (1966) Signal detection theory and psychophysics. Wiley, New York

Haralick R.M. (1979) Statistical and structural approaches to
texture. Proceedings of the IEEE, Vol. 67, 5, pp.786-804

Hay G. Ed. (1976) Medical images: formation, perception and measurement, Wiley, New York.

Huang T.S. Ed. (1975) Picture processing and digital filtering.
Springer, Berlin.

Huang H.K. (1981) Biomedical image processing. CRC Critical Reviews
in Bio-engineering. 5, pp. 185 - 271

Julesz B. and Schumer R.A. (1981) Early visual perception. Annual
Review of Psychology, 32, pp. 575 - 627

Kittler J., Fu K.S. and Pau L.F. (Eds) (1982) Pattern recognition
theory and applications. Reidel, Dordrecht

Kundel H.L. (1979) Images, image quality and observer performance.
Radiology, 132, pp. 265 - 271

Kundel H.L., Nodine C.F. and Carmody D. (1980) Visual scanning
pattern recognition and decision-making in pulmonary nodule
detection. Investigative Radiology, Vol. 15, 3, pp. 175-181

Levine M.D. and Shaheen S.I. (1981) A modular computer vision system
for picture segmentation and interpretation. IEEE Transactions on
pattern analysis and machine intelligence. Vol. PAMI-3, 5, pp.
540-556

Marr D. (1976) Early processing of visual information. Philosophical
Transactions of the Royal Society, B.275, pp. 483 - 524

Metz C.E.: (1978) Basic principles of ROC-analysis. Seminars in
Nuclear Medicine, 8, pp. 283-298

Modestino J.W., Fries R.W. and Vickers A.L. (1981) Texture discrimination based upon an assumed stochastic texture model. Transactions on pattern analysis and machine intelligence. Vol.PAMI-3,
5, pp. 557-580

Pratt W.K. (1978) Digital Image Processing. Wiley, New York

Rosenfeld A. (Ed.)Digital Picture Analysis. Springer, New York,1976

Rossiter C.E. (1972) Initial repeatability trials of the UICC/ Cincinnati Classification of the Radiographic appearances of pneumoconoises. Brit. J. of Industr. Med. 29, pp. 407-419

Senders J.W., Fisher D.F. and Monty, R.A. (Eds) (1978) Eye movements and the higher psychological functions. Wiley, New York.

Starr S.J., Metz C.E., Lusted L.B. and Goodenough D.J. (1975) Visual detection and localisation of radiographic images. Radiol. 116, pp. 533-538

Stevens K.A. (1981) The information content of texture gradients. Biological Cybernetics, 42, pp. 95-105

Swensson R.G., Hessel S.J. and Herman P.G. (1982) Radiographic interpretation with and without search. Invest. Radiol. Vol. 17, 2, pp. 145-151

Swets J.A. (1979) ROC-analysis applied to the evaluation of medical imaging techniques. Invest. Radiol. 14, pp. 109-121

Tou J.T. (1980) Pictorial feature extraction and recognition via image modelling. Computer Graphics and Image Processing, 12, pp. 376 - 406

Whitted T. (1982) Some recent advances in computer graphics. Science Vol. 215, pp. 767-774

Wintz P.A. (1972) Transform picture coding. Proceedings of the IEEE, Vol. 60, 7, pp. 809-820

Witkin A.P. (1981) Recovering surface shape and orientation from texture. Artifical Intelligence, 17, pp. 17 - 45

Zucker S.W. and Terzopoulos D. (1980) Finding structure in Co-occurrence matrices for texture analysis. Computer Graphics and Image Processing, 12, pp. 286 - 308

ACKNOWLEDGEMENTS

The author wishes to thank Professor E.G.J. Eijkman, Dr A.M.H.J. Aertsen, Dr H.F.P. v.d. Boogaard and Mrs E. Piersma.

THE PRINCIPLES AND PRACTICE OF RADIOPHARMACEUTICAL PRODUCTION

Per O BREMER

Institute for Energy Technology, Box 40,
N-2007 Kjeller, Norway.

1. INTRODUCTION

Nuclear medicine has progressed a long way from the nineteen-forties, when radioactive iodine was given for the first time to patients, and a Geiger counter was held directly over the thyroid gland to measure the iodine uptake. The great value of nuclear technology in medicine lies in the possibility of painless, non-invasive and non-destructive investigations that can give information both on body structure and regional body function. The principle of nuclear diagnosis is to use the property of many radioactive nuclides to emit gamma rays in the process of decaying. These rays penetrate the tissues of the body, and measurements can be made on the surface with various degrees of spatial resolution.

The scintillation detector was first introduced for imaging in the rectilinear scanner in 1951. In this instrument crystals of a scintillation material, such as sodium iodide activated by thallium, were moved mechanically back and forth over the area to be imaged. The output from the crystals was fed to a printing device, whose head was synchronised with the movement of the detector. In this way it was possible to make a crude image of the distribution of radioactivity in the patient line by line.

Such scanners are still in use (e.g. for thyroid scanning), but the work horse of the nuclear medicine department is now the Anger gamma camera. In these cameras large, flat scintillation crystals are used, and behind the crystal at a short distance is placed an array of photomultiplier tubes, with fields of view overlapping the whole area of the crystal. The output of these tubes is fed into a computer, which identifies the point on the crystal

from which the scintillation occurred. This information is
registered on an oscilloscope screen. The picture can be recorded
on film or tape, and the data stored for further processing.

The rapid development of new equipment in nuclear medicine
has led to the use of many new radionuclides. In the beginning it
was necessary to use isotopes with high energy gamma rays, as the
detection system was not so sophisticated. The first gamma cameras
were designed with thicker crystals that responded well to a wide
range of gamma ray energies. The crystals had poor spatial
resolution for low energy gamma rays because of the low number of
light photons produced in the crystals. With the explosive development of new technetium-99m labelled radiopharmaceuticals, which
give a relatively low absorbed dose of radiation to patients and
hospital staff, the manufacturers of gamma cameras have directed
production towards cameras designed primarily for this radionuclide.
Cameras can still be used for iodine-131 and other radionuclides
with high energies, but the efficiency is reduced.

The trend today is to reduce absorbed doses as far as
possible by using radionuclides with low radiation energy and short
half-lives. The present selection of radionuclides in use is
therefore very different from those used ten years ago. The use of
new radionuclides, development of new radiopharmaceuticals and the
new detection system have together opened up a whole new range of
investigational techniques and have made others obsolete. Other
factors have also changed the direction of nuclear medicine. When
radionuclides were first introduced to medicine to treat thyroid
diseases, many thought this was just the first step towards a new
approach with a multitude of radiopharmaceuticals that could seek
out and destroy diseased tissue within the human body. However,
developments have not fulfilled hopes for this type of radiotherapy,
and diagnosis is today the most important part of nuclear medicine.
The localization of radionuclides in specific organs or body structures makes possible several different types of diagnosis, such as
determination of size and position of an organ, determination of
blood and plasma volume and detection of tumours, metastases,
abscesses, etc.

Nuclear medicine has now come to a cross-roads, due to the
competition from new investigational techniques, such as ultrasound, transmission computerized tomography (CT scanning), nuclear
magnetic resonance and digital subtraction angiography. The
importance of static imaging will decrease, and the main area will
be dynamic imaging. With the new gamma cameras it is possible to
observe the flow of radionuclides through and into organ systems.
This dynamic process can be recorded, and special areas of interest
can be chosen for further analysis. Count rate variations as a
function of time can be calculated and give detailed information on
the process in question.

The classical functional study in nuclear medicine is determination of kidney function with iodine-131 orthoiodohippurate (hippuran). This technique allows determination of the function of the two kidneys separately, which is a great advantage compared to other methods available. The test is still one of the most frequently performed in a nuclear medicine department, and is also used to check the result of kidney transplantations, where it is important to know the function of the kidney soon after the operation. Several other methods for dynamic imaging are now available for the determination of liver and bile duct patencies blood flow in organ systems, cardiac and pulmonary function, etc. In highly advanced studies it has also been shown by dynamic imaging how the different centres of the brain are put into operation by external stimulations and by thinking.

2. RADIOPHARMACEUTICALS

We can divide these products into three categories:

(i) diagnostic radiopharmaceuticals
(ii) therapeutic radiopharmaceuticals
(iii) research radiopharmaceuticals

More than 90% of radiopharmaceuticals are used for diagnostic purposes. They are administered to patients to differentiate normal from abnormal anatomy, physiology and biochemistry. The therapeutic radiopharmaceuticals are administered to patients for the purpose of delivering radiation to body tissues internally to destroy cancer or other abnormal tissue. Research radiopharmaceuticals are often regular drugs labelled with small quantities of radioactive tracer, and are used to study biodistribution, kinetics and metabolism of the drug.

From a pharmacist's point of view the radiopharmaceuticals make up an extremely varied and inhomogeneous group, with many dosage forms that are seldom found among non-radioactive drugs. The majority of the products are solutions or suspensions for intravenous injection, but the following list shows a great variety in the most common dosage forms:

1. Radioactive gases for inhalation
2. Radioactive gases for injection
3. Oral solutions and solutions for injection
4. Colloidal solutions
5. Suspensions of radioactive labelled particles
6. Capsules
7. Radionuclide generators and inactive preparation kits

2.1 Radioactive gases for inhalation

The gases most used are radioactive noble gases like xenon-133, xenon-127, krypton-85 and krypton-81m. They are supplied as precalibrated doses in a definite volume, and the gases are often diluted with air. These gases are most often used for ventilation studies, in which the patient breathes the gas in a special, closed system. The closed system reduces the amount of gas needed for the study and permits collection of waste gas, so contamination of the laboratory with radioactivity can be avoided. The gamma camera will detect areas in the lungs where the uptake of gas is reduced, or where there is no uptake at all. The ultra-shortlived radionuclide krypton-81m, with a half-life of only 13 seconds, can be supplied to hospitals in the form of a radionuclide generator system, as the parent rubidium-81 has a half-life of 4.5 hours. The patient is breathing gas that is coming directly from the generator, as the generator is eluted with nitrogen or air and then mixed with air in a special breathing mask.

2.2 Radioactive gases for injection

The same radioactive noble gases can also be used to study lung perfusion. For these studies the gases are available in sterile saline solution in multidose or single-dose vials. After intravenous injection the gamma camera image will show how the radioactive gas spreads to the various parts of the lungs, and how the gas later will disappear from lung tissue. For this investigation a new krypton-81m generator has been designed since the gas must be given to the patient as continuous infusion directly from the generator. Such a concept demands very high standards in design, production and use with special emphasis on the aseptic aspects of the procedure.

By combining ventilation and perfusion studies, it is possible to get a detailed picture of the state of the various regions of the lungs and to evaluate the nature of damage.

2.3 Oral solutions and solutions for injection

This group of relatively simple radiopharmaceuticals should not be expected to give many problems in manufacture or use, but in this group we encounter problems created by differences in chemical composition for radiopharmaceuticals and non-radioactive drugs. Several unexpected problems are due to the low concentration of active molecules in the product. Irradiation in reactors and cyclotrons transforms only fractions of the target material to the desired radionuclide. When solutions and later dilutions of these stock solutions are made, the concentration of radioactive material will be so low that the normal macrochemical laws seem not to be valid.

Stability problems in radiopharmaceuticals are often caused by small amounts of impurities that can change the chemical form of the radionuclide by acting as an oxidizing or reducing agent. Such impurities can be introduced to the product in several ways, for instance from the raw materials, water or other solutions used in the production, from glass-ware or glass vials or from rubber stoppers. Stoppers seem to be a particular source of problems, and their choice and treatment must be made with great care.

High radioactive concentrations and high specific activity may also affect the stability of the product, as they can lead to self-decomposition by internal radiation effects and to induction of radiolysis in the solution. Some of these problems can be avoided by supplying radiopharmaceuticals in more dilute form. However, dilution can be the origin of new stability problems, as the oxygen in the solvent may start oxidation processes in the preparation. For each product it is therefore necessary to find the right balance between these factors.

Parenteral solutions often contain preservatives against microbial contamination. When these are added to radiopharmaceuticals for injection, they can be destroyed by radiolysis, and only a limited number of such preservatives can be used in the preparation of radiopharmaceuticals. However, some preservatives, such as benzyl alcohol have a scavenging effect, which will reduce the problem of radiolysis. This agent is frequently used, as it is added to the product both as a preservative and a stabilizer.

The pH solutions can be changed by radiolysis. This can be a problem for the chemical stability of the product, but most important is that such pH changes can create really hazardous situations. Solutions of iodine will liberate fumes of radioactive iodine at pH below 7.0. These can be a great health risk for personnel. Addition of buffers to the product will avoid this problem, but it is important to remember that the buffer can interfere unfavourably with the product, and to keep in mind that buffers can also be changed by radiolysis.

Manufacture of radiopharmaceuticals for injection has special problems, as the methods of production are limited. Sterile filtration is only used to a small degree, as this technique often gives a high radiation dose to the operator. Many of the products cannot be sterile filtered, because they contain particles or colloids, or they have a high viscosity that will prolong the filtration time and lead to adsorption to the filters. Preparation of injections is therefore most often done by aseptic working techniques or by heat sterilization of the product in the final container.

2.4 Colloidal solutions

Products in this group generally have a particle size distribution between 10 and 500 nm, and only products with particle sizes in the lower range of these limits can be called true colloid systems. The homogeneous phase, the solvent, is usually water, with addition of stabilizers and other adjuvants. The classical product in this group is colloidal gold-198, produced by reduction of a solution of gold-198 with ascorbic acid. This will form a lyophobic sol, which is thermodynamically unstable. To stabilize the colloid, gelatine is added to the reaction mixture. Gelatine forms a lyophilic monomolecular layer around the gold particles and thus acts as a protective colloid. Gold colloid has been used for two main indications in nuclear medicine. It was first used to study the liver and spleen, as the colloidal particles are taken up by the Kupffer's cells (RES cells). Ninety per cent of these cells are located in the liver and the rest in the spleen and bone marrow. These scans can give significant clinical information about metastatic disease and other pathological conditions in these organs. Most conditions are detected as "cold spots" due to the replacement of the normal RES cells by disease processes. Unfortunately gold colloid gives unduly high radiation doses to the patients, as the radionuclide emits both beta and gamma rays, and the product has therefore gradually been replaced by other radiopharmaceuticals, such as technetium-99m sulphur colloid and technetium-99m tin colloid.

Gold colloid is still used for therapeutic purposes, for example by injecting large amounts of activity into the peritoneal cavity. Because of their size, colloid particles are engulfed by floating macrophages and then fixed to the cavity wall by tissue macrophages in the wall. The beta rays provide local therapy since they do not penetrate more than four mm into the surrounding tissue, while the gamma rays allow determination of the spatial distribution of radionuclide in the cavity. The therapeutic action will limit the growth of metastatic disease and halt fluid collection. The use of gold colloid in therapy leads to problems of radiation protection because of the larger doses administered to patients. After injection the patient is a considerable radiation risk and must be isolated from other patients. Restrictions must be imposed on visiting and special precautions must be taken for the care of patients by personnel. Colloids containing pure beta emitters would reduce this problem, and such products are commercially available with phosphorus-32.

2.5 Suspensions of radioactive labelled particles

These products can be sub-divided into four groups:

a) Macroaggregated particles
b) Microspheres
c) Microaggregated particles
d) Millimicrospheres

The difference between the two first products and the two last is, as the names indicate, a difference in particle size. This difference decides the diagnostic use of these products. There is no exact division between the two groups, but as a guideline macro-aggregated particles and microspheres have a mean particle diameter greater than 1 μm.

Aggregates are normally produced by denaturing a solution of serum albumin. The mean particle size depends on the conditions of the production process, such as temperature, rate of stirring, liquid medium, etc. The first aggregated particles were supplied to the hospitals ready labelled with iodine-131. Today the pharmaceutical is supplied as a freeze-dried preparation kit containing prefabricated albumin particles. The particles also contain a reducing agent, which will take part in the formation of the complex of albumin particles and technetium-99m, when an eluate from a technetium-99m generator is added to the vial.

Microaggregates and millimocrospheres are used to study the lymphatic system and the RES system, while microspheres and macro-aggregated albumin (MAA) are used for lung studies. For lung studies the particle size should be between 10 and 100 μm, ideally with a mean diameter of approximately 30 - 40 μm. After injection the particles are transported with the blood to the lungs and are distributed in proportion to pulmonary blood-flow. The smallest capillaries in the vascular network of the lungs will be blocked by the particles, as the capillaries have a diameter of 7 - 10 μm. The number of particles injected should be so low that only a minor fraction of the smallest blood vessels will be closed. This is such a small part of the total number that the study will not be danger-ous to the patient. The particles are biodegradable, being metabolised by the body after a few hours, and normal blood flow is restored to the lung tissue. This method is used to study pulmonary embolism and lung perfusion.

It is important that the particle size is within the correct range. If the particles are smaller than 10 μm, they will be taken up by the Kupffer's cells in the RES system, which will make it difficult to interpret the study. If the particles are too large, they will block some of the greater blood vessels, and can thereby seriously damage blood flow to large areas of the lungs. This can be very dangerous for weak and ill patients.

The main disadvantage with MAA is that their shape and size is not well defined. The spread around the mean diameter can vary, and the mean particle size varies from producer to producer. A search for a physically stable, uniformly sized biodegradable particle led to the development of microspheres. As the name indicates, they are small, spherical particles, and they are also made by denaturing albumin. They can be made with a very narrow particle size distribution, as their greater mechanical resistance allows sorting according to size by sieving after production. One batch can therefore be divided into fractions with different mean particle sizes. Microspheres are also supplied in freeze-dried form for labelling with technetium-99m. One disadvantage with this product is a slower metabolism for the microsphere than for MAA, which prolongs the time to restore the normal blood flow. The rate of biodegradation is determined by the amount of heat used to prepare the microspheres. At present the price of microspheres is considerably higher than the price of MAA.

In the human body we are limited to materials that are non-toxic, non-antigenic and biodegradable to produce the micro-embolization. In medical research another type of microsphere can be applied. This is a tracer microsphere made of an organic matrix which contains a solution of the radionuclide. After carbonization of the organic material, the solution will be sealed inside the microsphere. These tracer microspheres can be produced in a wide range of particle sizes and with many different radionuclides.

2.6 Capsules

Radiopharmaceuticals in liquid form for oral use are, to a certain degree, administered as capsules. This is primarily done to reduce the risk of spillage and contamination. The solution is dispensed in ordinary gelatine capsules, filled with a carrier substance, such as dried sodium phosphate. The volume of radioactive solution in each capsule is normally so small that there is little danger of the capsule wall dissolving because of the liquid present.

2.7 Radionuclide generators

2.7.1 General properties. Radionuclides with short physical half-lives allow administration of large doses without increase in radiation exposure above accepted levels. With high photon fluxes better images with more information can be obtained in a shorter time. Economic limitations and transport difficulties would generally make it impossible to supply these nuclides to medical centres that are distant from the site of production, but for some nuclides the problems have been solved by development of radionuclide generators.

The generators are constructed on the principle of the decay-growth relationship between a long-lived parent radionuclide and

its short-lived daughter radionuclide. The activity of the
daughter nuclide grows as the result of decay of the parent nuclide
until an equilibrium is reached. At this equilibrium the daughter
nuclide appears to decay with the same half-life as the parent.

The daughter nuclide is used in the preparation of radiopharmaceuticals, and must therefore be separated from the parent. If
the system is to be useful, the half-life of the parent nuclide
must be long relative to the transport time. To obtain separation
of the nuclides, the daughter nuclide must have different chemical
properties from the parent, and this difference is usually obtained
by the difference of one in the atomic number of the two nuclides.
The separation is usually by chromatography, where the parent
nuclide is adsorbed to a column material in a small glass or plastic
column, while the daughter will not bind to the column material,
and can therefore be eluted by a suitable solvent. After elution
the daughter activity starts to grow again until equilibrium is
reached. Elution of activity can be made repeatedly.

Several such radionuclide generator systems are in use in
nuclear medicine today, such as

$$Sn\text{-}113 \longrightarrow In\text{-}113\ m \longrightarrow$$

$$T_{\frac{1}{2}} = 117\ days \qquad T_{\frac{1}{2}} = 100\ minutes$$

$$Ge\text{-}68 \longrightarrow Ga\text{-}68 \longrightarrow$$

$$T_{\frac{1}{2}} = 280\ days \qquad T_{\frac{1}{2}} = 68\ minutes$$

$$Rb\text{-}81 \longrightarrow Kr\text{-}81m \longrightarrow$$

$$T_{\frac{1}{2}} = 4.7\ hours \qquad T_{\frac{1}{2}} = 13\ seconds$$

However, one system overshadows all the other in importance,
and will be covered in detail: molybdenum-99/technetium-99m.

The ideal radionuclide generator for parenteral solutions
should have the following characteristics:

(i)   The system supplies a sterile, non-pyrogenic eluate
(ii)   0.9% saline solution can be used as eluent
(iii)   The radionuclide generator may be stored without
        special precautions at room temperature
(iv)   The daughter nuclide has suitable radiation character-
        istics and half-life
(v)   Chemical separation gives minimal amounts of parent
        nuclide in the eluate (no break-through)

(vi) The parent nuclide must have a relatively long half-life, so that the activity of the generator does not decay too quickly

(vii) The eluate should contain the daughter nuclide in a chemical state that makes it suitable for forming complexes with other agents to produce radiopharmaceuticals.

The Mo-99/Tc-99m generator fulfils most of these criteria. The generator is based on the decay of molybdenum-99 to technetium-99 through the intermediate step, the metastable isotope technetium-99m:

$$\text{Mo-99} \longrightarrow \text{Tc-99m} \longrightarrow \text{Tc-99} \longrightarrow \text{Ru-98}$$

$$T_{\frac{1}{2}} = 67 \text{ hours} \quad T_{\frac{1}{2}} = 6 \text{ hours} \quad T_{\frac{1}{2}} = 210\ 000 \text{ years}$$

The decay of Mo-99 will not produce 100% of the metastable isotope Tc-99m, as approximately 15% will decay directly to Tc-99. Mo-99 can be produced by neutron irradiation of Mo-98 in a reactor or by fission of uranium-235. Mo-99 produced by fission may contain a large number of foreign radionuclide impurities, and very elaborate separation processes and strict quality control are necessary to ensure a product of high quality. Today the market is dominated by generators filled with fission Mo-99.

In the generator, Mo-99 is adsorbed to the upper part of a small chromatographic column filled with high grade alumina (Al_2O_3). The molybdenum is fixed to the alumina, and will not be dissolved by 0.9% saline solution. By decay the activity of Tc-99m on the column will grow until a maximum is reached after approximately four Tc-99m half-lives. Tc-99m will not bind to the column material, and the two radionuclides can be separated by passing 0.9% saline solution through the column. When sterile solutions and aseptic techniques are used, the generator will give a ready-to-use Tc-99m solution.

After elution the Tc-99m activity will start to grow, and a new maximum will be reached after 23 hours. Elution may be carried out, if needed, before equilibrium is reached. The amount of Tc-99m available will depend on the time elapsed since the last elution.

The technetium generator allows hospitals to have their most important isotope available every day. Each generator is normally used for one week. The total amount of Tc-99m activity available will decrease each day because of decay of the parent Mo-99, and it is therefore important to plan the studies to be performed each week.

In the eluate Tc-99m and Tc-99 will be found as pertechnetate

ions $Tc^{99m}O_4$ and $Tc^{99}O_4$ which are very soluble in water. The
eluate may contain radionuclidic and radiochemical impurities,
the most common are "break-through" of the parent isotope Mo-99
and aluminium ions or small particles from the alumina in the
column material. Because of the possibility of long-lived
impurities in the eluate the pharmacopoeias set very strict limits
in their monographs for the content of foreign radionuclides in
the eluate.

2.7.2 Malfunction of technetium generators. Since the first
Tc-99m generators were introduced two decades ago, producers have
worked continuously to make systems as reliable and easy to use as
possible. Today most generators have automatic elution, and most
of the earlier problems have been solved. But there are still a
couple of problems that can upset routine work in the nuclear
medicine departments:
   (i) No yield of activity is obtained when the generator is
       eluted
   (ii)Reduced yield when the generator is eluted.

   If no activity at all is obtained when the generator is
eluted, this may be due to substantial internal damage during
transport, which makes it impossible to pass saline through the
column, or the needle of the cannula may be blocked with particles
from the rubber stoppers of the vials. In systems where generator
elution is based on evacuated vials, a leak in these vials will
also make it impossible to perform the elution process.

   Reduced yield has two main causes. Pertechnetate may be
reduced to ionic species of oxidation states that will fix to the
column material. This can be due to the presence of reducing agents
in the column material or strong local radiation at the upper part
of the column. In generators where some of the saline solution is
left on the column, this radiation will lead to the formation of
free radicals. These radicals will act as reducing agents on the
pertechnetate ions, and form species that will not be eluted. Most
modern generators solve this problem by ensuring that all solvent
is removed from the column at each elution. Vials with reduced
vacuum may give partial elution and leave some solution on the
column and thereby reducing yield at the next elution. Low yields
are also obtained when the eluent only reaches a fraction of the
surface of the column material on which the parent nuclide is
absorbed. This can be caused by poor quality column material,
improper initial packing of the bed or disturbance in the bed
during transport, or the formation of channels in the bed during
transport.

3. TECHNETIUM CHEMISTRY

Technetium belongs to the heavy metals, and its chemistry resembles that of manganese and rhenium, which belong to the same group in the periodic table. Technetium is not found naturally, as the name indicates, and all isotopes of the element are radioactive. The pertechnetate ion TcO_4^-, with technetium in the oxidation state +7, is the most stable form in water and air. It corresponds to the permanganate ion MnO_4^-, but is a somewhat weaker oxidizing agent. The pertechnetate ion is very like iodide, I^- in size and charge, and these ions have approximately the same biodistribution.

The pertechnetate ion is not particularly reactive, and in pharmaceutical chemistry the ion is most often reduced to a lower valence state, between +6 and -1, to form complexes. After reduction technetium can be complexed with a number of different ligands, which will change the characteristics and biodistribution of the nuclide. Reduction and complex formation depend on several factors, which are important when new radiopharmaceuticals are designed:

1. The redox potential for the desired state of oxidation must be chosen in relation to the potential of the reducing agent

2. The optimal stoichiometric ratio for reducing agent, ligand and pertechnetate ion must be found

3. The reaction kinetics must be considered

The reducing agents most used are various stannous salts (Sn^{2+}), such as stannous chloride and stannous fluoride. The chemistry of technetium is very difficult to study because of the very low concentrations of pertechnetate ion in the relevant radiopharmaceuticals, but it is possible to set up the following balanced redox reaction:

I. $\quad Sn^{2+} \rightleftharpoons Sn^{4+} + 2e^-$

II. $\quad TcO_4^- + 8H^+ + 3e^- \rightleftharpoons Tc^{4+} + 4H_2O$

and combined

III. $\quad 2TcO_4^- + 16H^+ + 3Sn^{2+} \rightleftharpoons 2Tc^{4+} + 3Sn^{4+} + 8H_2O$

Reduction is reversible, but the reaction is not in equilibrium because Tc^{4+} is bound by complex formation when it is formed. Even though it is possible to set up such a balanced equation for the redox reaction, it is not certain that the solution will contain only technetium in valence state +4 in practice.

Depending on the conditions in the reaction mixture, the whole range of oxidation states may be present. Stoichiometrically only traces of stannous ions should be needed to reduce all pertechnetate ions present in the eluate, but radiopharmaceuticals must contain a considerably larger amount of these salts to produce a stable product after labelling. Stannous chloride has the disadvantage that it readily undergoes hydrolysis in aqueous solution at pH 6 - 7 and forms insoluble colloids. These colloids bind to the reduced pertechnetate and compete with the complexing agents in the labelling process.

Technetium ions have unfilled electron orbits which may be filled with electrons from other chemical compounds, <u>ligands</u>. Among the most well known ligands used in radipharmacy are anions like chloride, fluoride and hydroxyl, molecules like alcohols and water and groups such as amines, phosphates and sulphydryl. When complexes are formed, one or more ligans can be bound to the metal ion. Mixed complexes are formed when the ligands are of different species. The ligands are in dynamic equilibrium, and the end result depends on the concentration of different components, and the strength and type of bonds formed between the components. The time needed for the formation of complexes varies from fractions of a second to several hours. Complexes containing the ligands OH^- of H_2O can split off water molecules and form insoluble oxide, TcO_2,

which is stable in water, and corresponds to MnO_2 in manganese chemistry.

Chelating agents, a group of nucleophilic ligands, are frequently used in technetium radiopharmacy. These molecules have two or more sites for formation of bonds with the metal ion. In this way the molecules can form a network around the heavy metal ion and chelates are considerably more stable than normal complexes.

In a preparation of a Tc-99m labelled compound three different species may be present:

1. Free Tc-99m, pertechnetate that has not been reduced by stannous ions
2. "Hydrolyzed" Tc-99m, such as $Tc^{99m}O_2$

3. "Bound" Tc-99m, which is the desired compound formed by the binding of reduced Tc^{99m} and the chelating agent.

4. DESIGN OF NEW RADIOPHARMACEUTICALS

The first radiochemicals were the results of scientists' eager work to explore new possibilities provided by radioactive tracers in biological investigations. Uptake measurements of radioactive iodine in the thyroid were the start of routine use of radionuclides in medicine. Since then nuclear medicine has advanced rapidly, and the new radiopharmaceuticals introduced today are based on extensive research and development work.

Radiopharmaceuticals must fulfil several criteria that are not generally relevant for non-radioactive drugs. The following list gives some of the characteristics that would be sought in an "ideal radiopharmaceutical".

(i)   Easy availability. The radiopharmaceutical should be easy to produce, inexpensive and readily available for all interested users.

(ii) The radionuclide should have no particle emission. Emission of alpha or beta particles from the nuclide increases radiation damage to the tissue.

(iii) High target-to-non-target ratio

(iv) Short effective half-life

(v)   Low toxicity

The essential principle must be to design a radiopharmaceutical, which is organ or function specific, which is simple to use in routine practice and allows a safe and effective examination. The characteristics of a radiopharmaceutical are decided by the choice of radionuclide and the choice of a suitable chemical structure that can incorporate it. This chemical structure can be a simple chemical form of the radionuclide or a complicated chemical molecule or complex where the radionuclide is incorporated by labelling or synthesis.

A number of the following factors must be evaluated in the development of a new radiopharmaceutical:

4.1 Choice of radionuclide

The important consideration must be to choose a suitable radionuclide that will maximize detectable photon yield and minimize radiation dose absorbed by the patient. Ideally the emission from the decaying nuclide should consist of one principal photon with energy in the optimal range for the detection system. Many radionuclides in current use emit beta particles, for instance iodine-131 products, which add to the radiation dose, but they are used for the photons that these radionuclides emit.

The trend to short-lived radionuclides introduces some technical problems that should be noted:

(i)   Instrument settings may have to be adjusted during a study to ensure a constant count density

(ii) The products will have short shelf-lives

(iii) Short effective biological half-lives will make it impossible to obtain delayed studies, which often are of clinical importance.

The list of radionuclides used in medicine will expand when good positron cameras and compact medical cyclotrons are introduced in more medical centres. This will allow routine use of ultra short-lived isotopes such as carbon-11, nitrogen-13 and oxygen-15, which all decay by positron emission, sending out photons of 511 keV. These radionuclides can be incorporated into pharmaceuticals and biochemicals in which these elements make up a large proportion of the molecules.

4.2 Choice of chemical structure

This structure will decide the physical and chemical properties of the pharmaceutical, the mechanism of localisation in the target area and the dosage form. The choice depends on the final target for the product, and thereby primarily on the method of localizat- ion for the product. If the radionuclide has replaced an atom of the same element in the compound, the process is called isotopic labelling (or native labelling). In a non-isotopic labelled compound (foreign labelled) an atom has been replaced by a radio- nuclide of a different element. These compounds are not isomers, and it should not be taken for granted that their biological and chemical characteristics will be identical.

The list of possible methods of localization is varied:

4.2.1 Capillary blockage. Study of the relative regional perfusion of blood in a capillary bed is possible by blocking a few of the many capillaries in the area with small particles labelled with radioactivity. Mini-embolization will occur in proportion to the local blood flow. A typical example of this type of localization is a lung study with magroaggregated albumin particles or micro- spheres.

4.2.2 Phagocytosis. Radioactive labelled particles can be used to image organs where phagocytosis takes place. These organs contain cells of the reticuloendothelial system (RES cells, Kupffer's cells) which remove opsonin-coated foreign particles from the blood stream.

Radioactive labelled particles of the right size will appear as
foreign material to the body, and will be coated by the plasma
protein opsonin and thereafter taken up by the RES cells. Approximately 90% of these cells are in the liver, the rest are in the
spleen and bone marrow. It has been suggested that the size of
the particles will influence the distribution between organs to a
certain degree. The smaller particles will be deposited in the
bone marrow, the medium sized in the liver and the larger particles
in the spleen.

4.2.3 Cell sequestration. One of the functions of the spleen is
removal of deformed and damaged blood cells from circulation. If
a sample of red blood cells is labelled with a suitable radionuclide
and damaged slightly before reinjection in the patient, these cells
will concentrate in the spleen, and an image of the organ can be
obtained. The methods used for damaging cells must be mild, otherwise a liver image will be obtained, as severely damaged blood
cells will concentrate in the liver rather than in the spleen. One
method much used to damage chromium-51 labelled red blood cells is
by heating.

4.2.4 Active transport. The body has transport systems specific
to individual organs, which can be used to localize radioactive
compounds. The method can be used both to study function and to
obtain organ images. The most typical example is the use of radioactive iodine for thyroid studies. Iodide administered orally or
by injection will be taken up actively in the gland and used to
produce thyroid hormones. Radioactive iodine can be used to study
the steps of thyroid hormone formation, storage and use. Primary
and metastatic malignancies and other lesions of the gland can be
detected by their differential uptake of the radioactive tracer.

Technetium-IDA derivatives (iminodiacetic acid) use the
principle of active transport to study the function of the liver,
as they are actively taken up by the polygonal cells in liver.

Kidneys apply the principles of active transport during blood
filtration. The renal tubules sift the substances in the glomer-
ular filtrate into those to be reabsorbed into the blood and those
to be excreted. Iodohippuric acid (hippuran) is actively secreted
by the renal tubules and passively partially reabsorbed into the
blood. Hippuran labelled with radioactive iodine and similar
substances is used to study the kidney function.

4.2.5 Compartmental localization. The body has several well
separated compartments. When a radioactive tracer is introduced
into a compartment, it will not leave the compartment under normal
circumstances. Thus the tracer can be used to measure parameters
within the compartment, such as flow dynamics, fluid composition
and volume determination. Typical examples of these methods are

determination of plasma volume and red cell mass, and the study of cerebrospinal fluid flow by ytterbium-169 DTPA.

In radionuclide cisternography, radiopharmaceuticals are introduced intrathecally into the CFS pool, and make possible a study of medical problems involving the production, distribution and a reabsorption of this fluid.

Brain tumours and vascular lesions of the brain are typical examples of conditions where the normal blood-brain barrier is changed. Many of the radiopharmaceuticals used in blood-pool studies will be able to cross the barrier at these damaged sites, and thus permit scintigraphic visualization of these areas.

4.2.6 Ion exchange. In ion exchange localization, the ion of the radiopharmaceutical will be deposited in the target area while a native ion will be removed from the site. Two of the nuclear medicine procedures most used today are based on this principle - use of Tc-99m phosphates and phosphonates for bone imaging and use of thallium-201 for study of regional coronary blood flow.

The first step in the localization of bone imaging agents is diffusion from blood into the extracellular fluid. The porous mineralized surface of the bones is surrounded by this fluid, and the Tc-99m complexes fix rapidly to the solid phase of the bone by ion exchange and inclusion in the hydroxyapatite lattice. The concentration of radioactivity in the bone will be highest where there is increased bone activity and blood flow, for instance in areas of healing bone or areas with primary or metastatic tumours.

Thallium-201 is concentrated in the myocardium actively by the sodium/potassium pump. Even though the thallium-201 ion is not a true potassium analogue, its biological distribution is very similar, and the pump will treat Tl^+ ion as if it were a K^+ ion. The use of thallium-201 makes it possible to distinguish between infarcted myocardium, ischaemic myocardium and myocardium with normal blood flow.

4.2.7 Pharmacological localization. The accumulation of pharmacologically active compounds is either due to the physical properties of the compound, such as lipid solubility and size, or the result of molecular binding to specific receptors. Several modern radiopharmaceuticals have been developed with such receptor specificity in mind. Examples are I-131 iodocholesterol and I-131 6-iodomethylnorcholesterol for adrenal imaging and Se-75 selenomethionine for pancreatic imaging. The adrenal cortex is involved with the synthesis and storage of steroids. Steroids and steroid precursors were therefore likely substances in the evaluation of compounds for adrenal imaging. I-131 iodocholesterol was the first really successful product, but later it was found that an impurity in the pharmaceutical, 6-iodomethylnorcholesterol, showed greater adrenal uptake and was more stable with regard to deiodination. The products most used now are thereofre based on this compound.

Receptor-specific substances are often regarded as consisting of two parts. One part of the compound contains the essential site, which is the part of the molecule that will interact with the receptor site. If this is changed significantly, the molecule will not be able to bind the receptor site. The other part of the molecule is modifiable to a certain degree. When a radionuclide is introduced into such a molecule to produce a radiopharmaceutical, incorporation should take place in the modifiable part of the molecule, but it is impossible to ensure that introduction of a foreign label, even in the modifiable part of the molecule, will not change the properties of the compound. It is therefore essential to make the structural changes as small as possible.

The development of this type of radiopharmaceutical is central in research today. For example much work has been done to make complexes of gallium-67 and indium-111 that will concentrate in specific tumour sites. No pharmaceutical presently in use is totally tumour specific, but such products would be very helpful in the search for early stages of primary and metastatic lesions and for therapy. A wide range of anti-tumour agents are available for labelling, and methods have been developed for antibodies, hormones, enzymes and other substances that specifically attach to malignant cells. The introduction in biochemistry of the cloning technology makes it possible to produce in vitro, large amounts of pure antibodies from one single cell. This has resulted in an enormous increase in the scientific work aimed at using such radioactive labelled monoclonal antibodies as tumour markers.

4.3 Biodistribution

For many radiopharmaceuticals the rate of localization is related to the rate of the plasma clearance after administration. Plasma clearance is influenced both by excretion processes and by localization of the radiopharmaceutical in the target area, and most importantly it is affected by the plasma protein binding. To a certain degree most radiopharmaceuticals will bind to plasma proteins, and this binding will depend on factors such as the charge on the radiopharmaceutical molecule, pH, and the nature of the plasma protein.

The half-life for plasma clearnace of Tc-99m tin colloid is short because of rapid extraction of the colloidal particles by Kupffer's cells in the liver. The plasma clearance of Ga-67 is long, because of strong binding of the radionuclide ion to the plasma component transferrin. The plasma clearance rate is easy

to determine by collecting blood samples at different time intervals
after injection and plotting plasma activity versus time after
injection.

Other factors affecting the biodistribution are molecular
size and the water and lipid solubility of the radiopharmaceutical.
Substances of low molecular weight do not generally cross the
intestinal barrier, while molecules of high molecular weight will
not be filtered by the glomeruli in the kidneys. Lipid solubility
is a determining factor because the cell membranes are primarily
composed of phospholipids. Radiochemicals that are insoluble in
lipids will not be able to cross this membrane barrier.

The disappearance of a radiopharmaceutical from a biological
system will depend both on the decay of the radionuclide, characterized by the <u>physical half-life</u>,T_p, and the <u>biological half-life</u>,
T_b. The biological half-life is the time required to eliminate
half of the dose by the regular elimination processes of the
body. The biological disappearance follows an exponential law
similar to that of radionuclide decay if behaviour approximates to
a one-compartment model.

To calculate the amount of radiation absorbed, it is necessary
to know the <u>effective half-life</u>, T_e, of the radiopharmaceutical,
which includes loss of activity due to both physical decay and
biological elimination:

$$\frac{1}{T_e} = \frac{1}{T_p} + \frac{1}{T_b}$$

or

$$T_e = \frac{T_b \, T_p}{T_b + T_p}$$

Radiopharmaceuticals should have a short effective half-life,
ideally not much longer than the time needed to make a particular
study. Radiopharmaceuticals with long effective half-lives give
an unnecessarily high radiation dose to the patient. If T_b or T_p
is very long compared to the other, T_e will be almost
equal to the shorter. In this way a radiopharmaceutial incorporating a radionuclide of very long physical half-life can still be
a good agent if the biological half-life is short.

4.4 <u>Target/non-target ratio</u>

In many diagnostic studies activity uptake in non-target areas
can disturb the interpretation of structural details in the image.
A high target/non-target ratio is therefore important to minimize

this problem. The main goal must be to develop an organ or system-specific radiopharmaceutical, so that localization in secondary sites is avoided. For existing radiopharmaceuticals, it is sometimes possible to improve this ratio by changes in the formulation. Lung uptake has been a problem in studies of the liver by Tc-99m sulphur colloid and Tc-99m tin colloid. This can be reduced by using preparations containing particles of optimal size and a very narrow particle size distribution around the mean value. Another way of increasing the contrast to obtain a good image is to increase elimination of the radiopharmaceutical from the surrounding tissue. This can be done by giving drugs acting together with the radiopharmaceutical. The ratio can also be improved for certain studies by voiding urine before imaging to eliminate interference of radiopharmaceutical excreted to the bladder. The last important factor for obtaining good contrast in the image is determination of the optimal time of imaging. This time is a function of the target/non-target ratio and the absolute uptake in the area.

4.5 Stability in vitro/in vivo

Both in vitro and in vivo stability are important for the quality of the radiopharmaceutical, and these problems do not differ much from those found for non-radioactive drugs. In vitro stability is influenced by physical factors, such as storage temperature, light, pH in the solution, etc. The radiopharmaceutical must also be stable in vivo so that localization in the target area can take place before elimination and break-down processes are too far advanced. However, the ideal is also a product that will disappear as soon as its task is done, and one that will not localize in other areas. Competing materials may be given to load secondary sites, avoiding accumulation of radiopharmaceuticals there, and thus also accelerating excretion of the radiopharmaceutical.

The most important difference from non-radioactive drugs is the change in stability due to the presence of the radionuclide in the active compound. This can result in radiolysis, which can damage the radiopharmaceutical or the additives, or affect the elution of radionuclides from a generator.

4.6 Toxicity

The three potential sources of toxicity for radiopharmaceuticals are:
  (i) Radiation toxicity
 (ii) Chemical toxicity (caused by the radionuclide, the carrier or by excipients)
(iii) Biological toxicity

Radiation toxicity is usually only a problem for radiopharmaceuticals used for therapy. For diagnostic products the radiation dose is well below toxic levels, but it is important to know the estimated radiation doses for each product.

The radionuclide itself is only administered in trace amounts, and toxic effects are not anticipated at this level. The problem of chemical toxicity is therefore mainly associated with the non-radioactive materials in the product either in the form of "carrier materials" of the active ingredients or of excipients, such as the reducing agents in the preparation kits and suspending agents and preservatives in solutions. The risk of a toxic reaction may increase with the time elapsed from production, as decay of the radionuclide will make it necessary to inject a larger volume to give the same dose. In this way the amount of chemical substance administered will increase.

Biological toxicity is due to microbial contamination and presence of pyrogens in the product. These factors are controlled in the quality assurance programme by performing sterility tests and use of a rabbit or limulus ambocyte lysate test to detect pyrogens.

4. 7 Adverse reaction and drug interaction

As only small amounts of radiopharmaceutical material are administered to patients, there has been a low incidence of adverse reactions for this product group. The toxic effects of many compounds are only discovered after the administration over a long time and long-term effects are almost unknown for radiopharmaceuticals, as they usually are administered once or a small number of times to the same patient.

It is not easy to register potentially adverse reactions, as the radiopharmaceuticals are given to ill patients who often receive other medication at the same time. However, any unexpected reaction occurring in a patient who has recently received such a product should be registered as an adverse reaction to the radiopharmaceutical. It is important to have a national register for such reactions, as the low total number of incidents will make it unlikely that any single nuclear medicine facility will have enough experience to identify the potential danger with a particular product.

Many pharmacological agents modify the expected biodistribution of radiopharmaceuticals. This concept of drug-induced change in the distribution was discovered at an early stage in nuclear medicine, when it was found that various drugs could influence the uptake of radioactive iodine in the thyroid. Since then several reviews have been published, and the list of examples show such

308

interactions for most radiopharmaceuticals. More knowledge of
these interactions is needed, as this can help to avoid errors in
diagnosis and can improve diagnostic methods by using a combination
of radiopharmaceuticals and non-radioactive drugs in the same study.

5. PRODUCTION OF RADIOPHARMACEUTICALS

This subject comprises both the production of radionuclides
and the manufacture of radiopharmaceuticals.

Almost all radionculides used in nuclear medicine are
produced artifically in reactors, linear accelerators or cyclotrons.
At present more than 1500 radionuclides have been produced in this
way. The type of radionuclide produced depends on the target
material, the irradiating particle and its energy.

In a nuclear reactor, a self-sustained fission chain-reaction
takes place in a controlled manner. One of the products of the
fission process is the emission of neutrons. Some of the thermal
neutrons are allowed to escape from the reactor elements, and are
used for irradiation of materials. The targets may either be fixed
in special irradiation positions, or they can be brought into the
reactor as mobile targets in a pneumatic system. By reactor irradiation a neutron is added to a stable nucleus. The new nuclide is
often an unstable radioactive isotope of the same element as the
target material. These neutron-rich nuclides will either decay by
emitting gamma rays or by emitting beta or alpha particles. The
latter nuclides are therefore not very interesting for use in
nuclear medicine, but they are often precursors to a gamma emitting
nuclide, that can be used in radiopharmaceuticals. Tellurium-130
will for instance produce tellurium-131 by neutron irradiation,
which willl decay to iodine-131.

The neutron activation process does not turn all the atoms
of the target material into the desired radionculide, and it will
not be possible by chemical means to separate unreacted target
from the radioactive product, as the nuclide is of the same
element, and has the same chemical properties. The unchanged
target material is called carrier. The term specific activity is
used to state the quantity of radioactivity produced per mass unit
of the element present in the target material. With few exceptions
high specific activity is a desirable characteristic of a radiopharmaceutical for diagnostic purposes, as the compound itself will
not overload the biological system under investigation or produce
pharmacological effects. However, high specific activity can
cause radiolysis and internal molecular damage, thereby shortening
the shelf-life of the product.

Fission is the process of splitting a heavy nuclide into two
fragments of approximately similar mass. Fission may occur

spontaneously or be induced by the capture of bombarding particles, primarily neutrons. When a target of a heavy element with atomic number 92 or above is inserted into the reactor core, fission will occur when the target absorbs thermal neutrons. In addition to the fission products, neutrons and energy in the form of gamma rays will be produced. Most fission products have atomic numbers in the range 42 - 56. As they are isotopes of different elements, they can be separated by appropriate chemical methods, such as precipitation, liquid-liquid extraction, ion exchange, chromatography and distillation. The radionuclides produced by fission are generally carrier-free, and isotopes of high specific activity are therefore available for these nuclides. The most common nucleus for fission processes is uranium-235, which is used for the production of important medical nuclides, such as molybdenum-99, iodine-131 and xenon-133.

Another way to make radioactive nuclei is to bombard stable nuclei with charged particles, such as electrons, deuterons, protons, and alpha particles. Under normal conditions such particles are unable to interact with nuclei of other atoms, as they have insufficient energy to penetrate the orbital electrons of the nucleus, but the particles may be given the necessary energy in a particle accelerator. In the linear accelerator a beam of ions is accelerated along a linear path using alternating current as the accelerator. The drawback with these accelerators is their enormous size, and cyclotrons are therefore more used to produce medical radionuclides. In the cyclotron the particles are accelerated in a spiral path under vacuum by means of an electromagnetic field. The particles move along the path with gradually increasing energy. The current and the magnetic fields determine the focus of the particle beam and the energy level to which the particles will be accelerated. For each reaction there is a definite energy threshold that must be surpassed to start the reaction between the target material and the particles. The target material should be as pure as possible to avoid production of undesired nuclides. Extraneous reactions can also occur at the target if the operation of the cyclotron is not under strict control.

Cyclotron produced radionuclides are usually neutron deficient, and decay by positron emission or electron capture. The chemical properties of the target and the product will not be the same, as they are different elements. The produced radionuclide will therefore be carrier-free, but radionuclidic impurities from side reactions may be present. Target processing to remove such impurities is of great importance in the further preparation of radio-pharmaceuticals.

Radiopharmaceutical production has several problems that are unknown in the normal pharmaceutical industry. The procedures involved in manufacture are often very simple chemical operations,

but they are made complicated by the fact that it is just as
important to protect the personnel against the product as it is to
protect the product against the influence of the personnel. To
prevent uncontrolled spread of radioactive contamination in the
laboratories, production must take place in a specially ventilated
enclosure. Enclosures will vary in construction depending on the
amount of radioactivity to be handled and the sort of process that
will take place. Examples are:

(i)   Open fume hood
(ii) Sealed glove box with reduced pressure
(iii) Sealed box with reduced pressure and remote handling
      with tongs
(iv) Manipulator cell

Special shielding is necessary for most processes to keep the
radiation doses received by the operators within the internationally
accepted limits. The shielding materials usually used for protection against beta and gamma radiation are lead, steel, depleted
uranium and high density concrete. The production boxes for pure
beta emitters are often made of transparent plastic materials that
do not require any extra shielding.

Production units for gamma emitters are surrounded by lead
walls at all sides. The thickness of the walls depends on the
amount of radioactivity to be handled inside the enclosure, 5 cm
and 10 cm lead walls being mostly used. The walls are made of
interlocking lead bricks that lend stability and prevent leakage
of radiation between the blocks. The need for transparent heavy
shielding material led to the development of lead glass of high
density. This material is used both as windows in the lead wall
and as local shielding in various production enclosures.

In all working areas where unsealed radioactive sources are
handled the ventilation system must be planned carefully. A well-
functioning system forms the basis of contamination control in the
laboratory. The sealed boxes should be at reduced pressure in
relation to the surrounding premises to ensure that the operator
will not be contaminated by radioactive gas or particles spreading
out in the room from the box.

The majority of radiopharmaceuticals are made for injection,
and even though products go through a final sterilization by auto-
claving, it is important that the quality of air in the laboratory
is of a high standard. Room-air may filter into the box through
small cracks and openings because of the reduced pressure inside,
thereby contaminating the products during preparation.

Production equipment in the enclosures should be as simple and reliable as possible. Chemical processing equipment should be made of non-reactive materials such as borosilicate glass. When process equipment is planned, it should be remembered that all sense of touch is lost when hands are replaced by tongs or manipulators. It may be advantageous to build the equipment as modules that can be mounted and dismounted quickly in the enclosure. The reliability of the production equipment is of utmost importance, as the high activity level in the enclosure will make it very difficult to carry out repairs during normal operation. Maintenance and repairs must often take place in periods when there are halts in the production programme, as the activity level has to be reduced to an acceptable level by cleaning and decay, before service personnel can start their work. One or two spare units make it much easier to run a continuous maintenance programme.

For production of parenteral radiopharmaceuticals that cannot be sterilized in their final container, such as iodinated human serum albumin, it is necessary to apply aseptic working techniques. Such procedures require special working enclosures that may vary considerably in design. We have experience with a system that gives satisfactory working conditions for aseptic preparation of radiopharmaceuticals. The boxes are made of stainless steel and can be sealed off completely. In the roof of the box a large high efficiency particulate air (HEPA) filter is fitted with the flow of air directed vertically down onto a stainless steel-lattice working area. Sterile raw materials and production equipment are placed in the box which is then sealed off and circulation of air in the box begins. During the whole production air will be recirculated through the HEPA filter, and production can take place in an environment with very low microbial and particulate contamination risk. When this type of box was first used, temperatures within the box were excessive. This problem has now been solved by freon cooling of the air before each passage across the HEPA filter.

6. QUALITY CONTROL

The final quality control procedures carried out on a product before release are part of the total quality assurance programme for that particular radiopharmaceutical. However, tests are needed routinely to guarantee a number of criteria, such as radionuclide purity, radiochemical purity, product specificity, sterility and apyrogenicity. Other factors such as toxicity and stability will have been evaluated during the development of the product. In this context we shall concentrate on quality control procedures that are particular to radiopharmaceuticals:

The radionuclide purity of a radiopharmaceutical is defined as the ratio of the activity of the radionuclide concerned to the total radioactivity in the product. It depends on the relative half-lives and the quantities of the desired radionuclide and the contaminants present and is a characteristic that will change with time. The presence of radionuclide impurities in a product may increase the radiation dose to the patient and may interfere with the quality of the scintigraphic images. It is therefore necessary to set standards for the radionuclide purity of a product, and often special limits must be set for the presence of specific radionuclide impurities in the product.

To be able to state the radionuclide purity, the activity and identity of all radionuclides present in the preparation must be known. Determination of gamma emitters is most commonly performed by gamma spectrometry, by means of scintillation or semiconductor detectors and multi-channel analyzers. Alpha emitting impurities are detected by silicon-gold surface detectors, whilst determination of beta emitting radionuclides is performed by liquid scintillation counting. Dose calibrators can be used to provide simple and quick tests for radionuclide purity for certain radiopharmaceuticals by using attenuation shielding. Shields of different materials and thickness will reduce the radiation from the radionuclides present disproportionately because of the difference in photon energies. When the attenuation factors are known, it is possible to calculate the content of an impurity after measurements in the dose calibrator with and without shielding. (This method is for example frequently used to determine the content of Mo-99 in a Tc-99m eluate).

Radiochemical purity is defined as the proportion of the total radioactivity present in the desired chemical form in the radiopharmaceutical. Radiochemical impurities can arise by decomposition of the product resulting from solvent effects, dissolved gases, pH changes, influence of light, radiolysis effects, etc. Radiochemical impurities may reduce the quality of the images due to poor organ localization and higher background activity for the surrounding tissues. Some functional studies can give totally erroneous results if the content of radiochemical impurities in the radiopharmaceutical is too high. A number of analytical methods are used to detect and determine the radiochemical impurities:

   (i)   Paper chromatography
   (ii) Thin layer chromatography
   (iii) Gel chromatography
   (iv) Paper and polyacrylamide-gel electrophoresis
   (v)   Ion exchange
   (vi) Autoradiography

The chromatographic type of analysis is easy to perform and sensitive to impurities, but a major disadvantage with such methods is that one can not be sure that all impurities have separated from the main component or from each other. The solvent system used

must be chosen with care and evaluated for that particular radiopharmaceutical. Systems where the main component moves with the
solvent front or remains at the starting point should be avoided if
possible. Problems may occur because of irreversible adsorption
or decomposition before and during the development of the chromatogram. This is related to the minimal physical amount of radiopharmaceutical present in the sample to be tested. The addition of
carrier may eliminate or reduce these problems. In other cases the
solvent system must be purged with nitrogen and chromatography
performed in a nitrogen atmosphere to avoid the influence of oxygen
during analysis.

It is not sufficient simply to detect the different radiochemical impurities in the product, it is also necessary to quantify
the impurities. This can be done in several ways. When autoradiography and simplified thin-layer chromatography are used, the
chromatograms are cut into portions and the activity in each part
is measured. However, in normal paper and thin-layer chromatography
the use of radiochromatography scanners is most common. The
chromatogram and the detector are moved relative to each other, and
a chart is made of the distribution of radioactivity on the chromatogram. New instruments have now been developed where the whole
chromatogram can be measured at the same time without any movement
of the detector or the chromatogram.

Gel chromatography is very useful for separating proteins of
different molecular weight, but can also be used to separate
different technetium species present in a preparation. Free, bound
and hydrolized technetium can be separated in this way. Note
however that several technetium chelates can bind to this type of
gel, thereby causing problems in the separation of the impurities.

In paper electrophoresis and polyacrylamide-gel electrophoresis, components of the sample move to different positions along
the paper or the gel when an appropriate voltage and buffer is
chosen for the system. The distances will depend on the charge
and ionic mobility of the compounds.

Ion exchange techniques involve passing a sample of radiopharmaceutical through a column of ionic resin and then eluting the
column with a suitable solvent. Separation of the different components in the sample is due to their relative affinities to exchange
with ions of the resin.

The product specificity of radiopharmaceuticals is most
frequently checked by use of animal models. The distribution of
the product is determined by measuring the retained radioactivity
in the organs of interest after dissection of the carcass.

The <u>sterility</u> test is identical to that for other non-radioactive pharmaceuticals. This test controls the whole manufacturing process including the sterilization step, and aims to assure the absence of viable micro-organisms in the product. Standard culture media are used for the test and the method and sampling procedures are described in the various pharmacopoeias. It is admissible to release radiopharmaceuticals for sale before completion of the sterility test because of the short shelf-life of these products, but it is important to remember that no radiopharmaceutical can be regarded as self-sterilizing due to its own radiation.

Sterility tests should be performed by personnel with training in microbiology, sterilization technology and aseptic working techniques in an environment where external micro-organisms cannot be added to the samples during the test procedure. Laminar air flow (LAF) units are used extensively in this type of work for non-radioactive pharmaceuticals, but can only be used for radiopharmaceuticals if modifications are made to avoid possible contamination to the person carrying out the test. Small down-flow LAF units designed to be placed in sealed glove boxes are now commercially available and ideal for this kind of work.

Automated sterility testing has also been attempted for radiopharmaceuticals, since the time required for the test can be shortened considerably, and multiple samples can be examined at the same time. In these instruments the culture medium includes carbon-14 labelled glucose or other labelled substrates. When the sample of radiopharmaceutical is added to the medium and incubated, micro-organisms will produce C-14 carbon dioxide that can be flushed out continuously into an ionization chamber for detection.

<u>Pyrogenic contamination</u> may arise from viable or dead bacteria. The terms pyrogen and endotoxin have been used interchangeably, as endotoxin is the cause of most pyrogenic contamination. Endotoxin is a lipopolysaccharide component of the cell wall of all gram-negative bacteria. It is relatively heat-stable and is a potent producer of fever reactions and haematological changes.

Pyrogen testing is important for parenteral radiopharmaceuticals, both for detection of fever-producing contaminants and as a test of the sterilization, as a positive pyrogen test may indicate that the sterilization procedure has failed. The official method of pyrogen testing is an animal test, where the fever response in rabbits is monitored. Studies comparing human and rabbit response to pyrogens indicate that when pyrogens are injected intravenously on a dose/weight basis, rabbit and humans are equally responsive.

The use of the rabbit test for radiopharmaceuticals is not popular because of the radiation exposure to personnel and problems of radioactive waste in the animal laboratories. For these products the new in vitro method, the Limulus Amebocyte Lysate test is used as an alternative, and this has proved to be a sensitive and convenient method for the detection of endotoxin. The principle of this test is the formation of a solid gel from a solution of a lysate prepared from amebocytes from the blood of Limulus Polyphemus (horseshoe crab) when incubated with a sample of a product containing small quantities of endotoxins.

The time required for this test is short enough to permit testing of short-lived radiopharmaceuticals before administration to patients. The limulus test has a much higher sensitivity for endotoxin than the rabbit test, and is particularly useful in the control of products for intrathecal use. It is possible that this method soon will become the official method for pyrogen testing of radiopharmaceuticals in several of the most used pharmacopeias. For new products the method must be evaluated with care, as some products may inhibit the lysate-endotoxin gel formation.

7. DISTRIBUTION OF RADIOPHARMACEUTICALS

The distribution system will depend very much on local legislation and official quality standards, and such factors as population density, presence of a local radiopharmaceutical centre in the country, type of transportation and transport distances, etc. In some countries products are sent directly from the producers to the hospitals, with the help of shipping agents and smaller distribution centres set up by the manufacturers. Such a system of direct distribution should be based on strict quality standards imposed by government authorities.

The distribution of non-radioactive drugs normally involves both wholesalers and pharmacies as links between the producers and users. For radiopharmaceuticals it is important to make the distribution simple, and the wholesaler link is not commonly found because of the short shelf-life of these products. The pharmacy link seems to become more and more important as more regulations and controls are introduced in this field. These radiopharmacies may be operated by university centres, individual hospitals or as independent health service operations. Commerical radiopharmacies are a new development in the United States and seem to be spreading rapidly. A commerical operation will depend on supplying a large number of nuclear medicine departments in the area. The two major companies in the United States have today respectively 50 and 20 radiopharmacies in operation, and more will open in the future.

The following advantages are claimed for a centralized radiopharmacy:

(i)   Reduced cost per patient dose
(ii)   Laboratory space in the hospitals can be released for
        other purposes than preparation and quality control of
        radiopharmaceuticals
(iii)   Reduction in hospital staff
(iv)   Convenience - several of these radiopharmacies run a
        round-the-clock service. Tc-99m labelled kits are
        delivered in dispensed doses in syringes ready for
        injection
(v)   Reduction of radiation dose to hospital staff
(vi)   Record keeping and billing can be computerized
(vii)   Reduction in the amount of quality control to be
        performed in the hospital
(viii) Waste disposal will be easier, as syringes, empty vials
        and containers can be returned to the radiopharmacy.

Distribution has been arranged differently in other countries.
The three Scandinavian countries, Denmark, Norway and Sweden are
very similar, and their systems for distribution and sale of non-
radioactive drugs have much in common. However, for radiopharmaceuticals, their systems are different because nuclear medicine and
radiopharmacy have developed very differently in these countries.

In Denmark local production of radiopharmaceuticals has been
minimal and the National Energy Technology Institute at Risø
produces routinely only iodine-131 hippuran and some isotopes that
are difficult to import from other suppliers. This production is
kept running mainly for an emergency situation in order to have
operational laboratories and personnel with know-how available.
The country is small with a high population density and has very
good air and rail connections. The supply of radiopharmaceuticals
by importation from abroad is therefore easy to arrange.

Radiopharmaceuticals are included in legislation for drugs in
Denmark which was one of the first countries in the world to impose
registration of radiopharmaceuticals in 1977. Registered products
may be sold directly to the customer by the Danish agents of foreign
producers, in some instances helping to cut the transport time.
Products not registered have to be sold through the Isotope Pharmacy
in Copenhagen. This pharmacy can also supply registered radiopharmaceuticals from all manufacturers. Foreign suppliers may choose
not to have an agent in the country but to sell all their products
through the Isotope Pharmacy. The Isotope Pharmacy is run by the
Danish Health Service, buying products in bulk for dispensing to
individual customers.

Apart from distribution, control and information are the two
most important tasks for the Isotope Pharmacy. Quality control is
at a very advanced level, and the pharmacy now even has a gamma
camera for animal studies used in product control. Pharmaceuticals

within the same group of preparations from different producers are
compared and the results published in report form with price
information to be used as a guide for hospitals. The pharmacy also
provides theoretical and practical information on radiopharmaceuticals to personnel in nuclear medicine departments. The staff of
the Isotope Pharmacy will for example visit these departments and
give information and practical training in aseptic techniques that
are applied in labelling Tc-99m preparation kits and elution of
generators.

In Norway the first radiopharmaceuticals were produced as early
as 1952 at the Institute of Atomic Energy (now named the Institute
for Energy Technology). Gradually a nuclear pharmaceutical
laboratory grew up, and this centre today has responsibility for
all production, control and distribution of these products in the
country. Radiopharmaceuticals have always been subject to the same
legislation as other pharmaceuticals in Norway, but they have been
exempt from registration, and the State Drug Control Centre has not
built up a quality control for this product group. The Nuclear
Pharmaceutical Laboratory at the Institute for Energy Technology
therefore has four main functions:-

(i)   Distribution
(ii)   Production
(iii) Research and development
(iv)   Information

In practice the laboratory functions as a large hospital
pharmacy for all the nuclear medicine facilities in the country.
Approximately 60% of the radiopharmaceuticals are produced in the
laboratories, while the rest, mainly cyclotron products, are
imported from various suppliers. Combined shipments and bulk
dispensing of several products help to reduce the cost for the
customer. As there is no state control, the Nuclear Pharmaceutical
Laboratory has a well-developed quality control programme for both
imported and home-produced products.

In Sweden the situation is very different from the other two
countries. Sweden had a production centre for radiopharmaceuticals
although the range of products was much more limited than in
Norway. Radiopharmaceuticals were, furthermore, not classified as
pharmaceuticals, and the sale and distribution of these products
was directly from the manufacturer through their Swedish agent to
the hospital. There was no quality control system in operation
for these products. A few years ago production ran into economic
difficulties and all production was stopped from 1st January 1981.
Today Sweden imports all radiopharmaceuticals through agents, but
from 1st January 1982 new legislation has been introduced for these
preparations, identifying them as pharmaceuticals. Obligatory
registration will be enforced for all products, but there is no

plan for centralized distribution or quality control.

Registration work is time-consuming and expensive, and a registration fee will be demanded for each product. In Denmark the registration fee is reasonable and should only be paid once. In Sweden the fee proposed is so high that many products cannot carry this expense, and is intended to be an annual charge. This may lead to higher prices of the products, and it is feared that some manufacturers will withdraw certain products from the Swedish market. The Swedish registration system will also include price control.

It would be difficult to say that one of these systems is superior to all others, and should therefore be applied everywhere, but for many countries modified versions of an isotope pharmacy system could be very useful.

REFERENCES FOR FURTHER READING

European Pharmacopoeia. Second Edition 1981
International Atomic Energy Agency (1971) Radioisotope Production
    and Quality Control. (Technical reports series No. 128) Vienna
Rhodes B.A. (1977) Quality control in Nuclear Medicine. C.V. Mosby,
    Company, St Louis
Rhodes B.A. and Croft B.Y. (1978) Basics of radiopharmacy. C.V.
    Mosby Company, St Louis
Saha G.B. (1979) Fundamentals of Nuclear Pharmacy. Springer Verlag,
    New York
Tubis M. and Wolf W. (1976) Radiopharmacy. John Wiley & Son Inc.,
    New York

PLANNING AND RUNNING A HOSPITAL RADIOPHARMACY

Per O. BREMER

Institute for Energy Technology, Box 40, N-2007, Kjeller, Norway.

1. INTRODUCTION

Radiopharmacy is a patient-oriented service that includes the scientific knowledge and professional judgement required to improve and promote health through the use of radioactive drugs for diagnosis and therapy.

The practice of radiopharmacy is composed of the following general areas:
1. The procurement of radiopharmaceuticals
2. Preparation of radiopharmaceuticals
3. Performance of routine quality control procedures
4. Dispensing radiopharmaceuticals
5. Distribution of radiopharmaceuticals
6. Implementation of basic radiation protection procedures and practices
7. Consultation with, and education of, the nuclear medicine community, patients, pharmacists and other health professionals on properties of radiopharmaceuticals
8. Research and development of new products.

An important characteristic of many radiopharmaceuticals compared to non-radioactive drugs is that an essential part of the production or dispensing must take place close to where the product will be used. With the introduction of new nuclides with shorter and shorter half-lives, into radiopharmaceuticals, it is necessary to produce both the nuclide and the radiopharmaceutical at the hospital premises.

The nuclear medicine procedures carried out range from the advanced services of a university hospital to the basic nuclear medicine routines in a small general hospital. This is reflected in the tasks a hospital radiopharmacy will be asked to perform which may vary from simple dispensing of individual patient doses from a commercially supplied ready-to-use radiopharmaceutical, through the preparation of a radiopharmaceutical made from a kit of reagents that has been through a stringent quality control, to production of a preparation that involves complicated chemical or biological synthetic work. The level of service must be planned according to need and available resources, consideration must be given to economy, time and legislation. Thus the level of radio-pharmaceutical service must vary a great deal from country to country and from one hospital to another, and this review will therefore be made on a general basis.

The quality of the products and the service provided by the radiopharmacy are influenced by a number of factors. The principles of "Good Radiopharmacy Practice" (GRP) have defined the basic idea as follows: "The important elements of the safe and efficacious preparation and handling of radiopharmaceuticals are a previously defined manufacturing process carried out and recorded by trained and qualified staff provided with the necessary facilities, including premises, suitable equipment, correct materials and approved procedures".

The two main factors establishing the quality of a radio-pharmaceutical are:
  i) quality standards set for the product by pharmacopoeias or by the manufacturer
  ii) the manufacturing process

In many countries these products are not subject to control by the health authorities, so the local radiopharmacy has a very important task in surveying the quality of commercially supplied radiopharmaceuticals. However, there is now a tendency for government authorities to introduce the same central quality control for radiopharmaceuticals as for non-radioactive drugs. This control will ensure that products supplied to hospitals have been produced according to "Good Manufacturing Practice" (GMP) guidelines and comply with specifications given in the pharmacop-oeias. Several such GMP guidelines have been made for non-radio-active pharmaceuticals, while this work is just starting for radiopharmaceuticals.

The guidelines for GMP must be based on general GMP rules with additions and modifications due to the radioactive nature of the preparations.

GMP is based on many elements: personnel; premises;

equipment; documentation; manufacturing procedures; dispensing procedures; quality control; choice and purchase of products; waste disposal; radiation protection.

The first three items must be resolved at an early stage in planning the radiopharmacy, while the other points must be discussed in detail at the planning stage, but will need modification when put into routine practice.

2. PERSONNEL

There must be sufficient personnel at all levels who are qualified by professional training and experience to carry out the various jobs required. Duties and responsibilities of each job should be defined clearly and described in writing. Persons should be assigned to take over these duties when the responsible person is not present. In larger establishments the responsibilities should be divided between a person responsible for production and a person responsible for quality assurance. This division ensures independent evaluation of a product by a person not involved in production.

With regard to qualifications for personnel involved in production and quality assurance in the radiopharmacy, they should have obtained practical experience and theoretical training in the following subjects:
- i) Pharmacy (including microbiology)
- ii) Chemistry (including radiochemistry)
- iii) Radiation protection
- iv) Radiopharmacology
- v) Radiopharmacy

The other staff members must have a basic knowledge of:
- i) Radiochemistry
- ii) Tracer methodology
- iii) Physiology
- iv) Nuclear medicine
- v) Pharmaceutical techniques required for dispensing and testing products for intravenous use
- vi) Radiation safety techniques for handling radioactive materials

If no radiopharmacist is included in the staff, it is important that pharmaceutical expertise is available to give advice and to help with staff training in the practical aspects of preparation of radiopharmaceuticals and the techniques used in testing for microbiological contamination and pyrogens.

All persons working in the radiopharmacy should be trained in the principles of good manufacturing practice, hygiene and radiation protection. This training must be on-going and the training programmes should be revised frequently and kept up-to-date.

3. PREMISES

Premises must be of suitable design and construction and be maintained in good condition. Design of rooms for preparation of radiopharmaceuticals must aim at protecting the operator against radioactivity and protecting the radiopharmaceutical against contamination from the operator and the environment. For most radiopharmaceuticals the greatest risk of contamination comes from the operator. A high standard of personal cleanliness must therefore be observed, and protective garments must be work in all the rooms where radiopharmaceuticals are handled. Change of footwear should be compulsory, as this improves the hygienic conditions and prevents the spread of radioactive contamination outside the working area. Protective clothing should be worn by everyone entering the working area, including visitors. A check station with radiation monitoring and a physical barrier with change of footwear will help to limit the number of persons entering the restricted area.

The rooms should be arranged in such a way that different types of work are separated, and so that handling of radioactive material does not affect nearby measuring equipment. Within the radiopharmacy the rooms can be classified in categories according to the radiation level. Areas where radioactive work is carried out should be marked with radioactivity warning signs, and should be closely monitored. Production facilities should be well separated from laboratories with microbiological testing or animal experiments. Separate facilities are also needed for handling blood and other patient samples.

Furnishing of rooms should be as described for the general production of radiopharmaceuticals, and can be summarised as follows:

i) Smooth and durable walls and floor coverings for easy cleaning and radioactive decontamination. No junctions or cracks between floor covering and walls and between work surfaces and walls.

ii) A minimum of furniture, equipment and other dust collecting items.

iii) Adequate ventilation to fulfil the requirements for optimum hygienic conditions and for radiation protection.

iv) Non-absorbent materials on working surfaces.

v)   It should not be possible to open the windows in the rooms.

4. EQUIPMENT

The equipment required in the radiopharmacy will depend very much upon the service level of that particular pharmacy. A list of basic requirements is given in the previous chapter.

Written instructions for the use of the equipment and for its maintenance and cleaning, which must be easy, should be available. Cleaning should normally be done as soon as possible after use, but in many cases it is necessary to wait for decay of the radionuclide in order to reduce the radioactive dose to staff. Parts of equipment coming into contact with a separated radiopharmaceutical should be inert.

Measuring and weighing equipment should be checked according to schedules. Sealed sources should be available for frequent calibration of dose calibrators and other instruments for measurement of radionuclides.

Autoclaves and sterilizers should be checked for effectiveness regularly by spore preparations. Time and temperature of each sterilization cycle should be recorded, preferably by automatic recording instruments.

Reduction of radiation dose to personnel is obtained by using tongs and forceps. It is, however, important to note that these utensils are frequently the source of cross-contamination, and should be checked and cleaned regularly.

Special consideration is required when planning equipment and premises for aseptic work. A low level of bacterial and particulate contamination must be achieved to provide the hygienic standard required for this type of work. A work station incorporating a sterile-filtered air supply with laminar air-flow pattern would, if suitably designed, fulfil these requirements (LAF units). In work with radiopharmaceuticals special LAF units with laminar down-flow should be used. These units direct the filtered air down against the working surface and recirculate it. Cabinets with such units are commercially available. A simple but effective and less costly solution for certain operations is to place a specially designed LAF mini-unit in a fumehood for radioactive work or in a glove box. Open LAF units must only be used where there is no risk of airborne radioactive contamination. Filters in LAF units should be tested with a suitable counter and the air velocity checked at specified intervals.

5. DOCUMENTATION

The basis of the quality assurance programme in the radiopharmacy is the documentation work. The aim is to establish a control system that reduces the risk of error which non-written communication may introduce, and make it easy to trace every aspect of an individual preparation from starting materials to waste disposal. Records should provide a trail from manufacturer to patient that can be followed at every step. The documentation system should be designed to include both pharmaceutical and radiation protection aspects. The following matters should be covered:

i)   Inventory of radiopharmaceuticals (with instructions for receipt, handling and storage for each individual product). The inventory should be adjusted for each dispensing operation.

ii) Dispensing and patient administration: A record must be kept for each individual dispensing, making it possible to trace each patient dose back to the relevant stock solution or diluent. Results of activity measurements should be recorded.

iii) Radioactive waste disposal: The routines for disposal should be made according to national and local regulations. Protocols should record storage and removal of radioactive waste from the premises of the radiopharmacy. The mode of disposal should be recorded.

For radiopharmacies producing quantities of radiopharmaceuticals the demand for documentation will be considerably higher. The following principles must be introduced to fulfil the quality assurance requirements of GRP:

5.1 Batch numbers

5.2 Specifications should be made for raw materials, packaging materials and the finished products, giving all relevant information on the criteria for acceptance by the quality control department.

5.3 Master formulas: These should contain a description of the production method, list the specified raw materials and describe the manufacturing and quality control procedures. Such master formulas should be available for each product and for each batch size.

5.4 Batch manufacturing record: This is the single most important part of the documentation for a batch of a radiopharmaceutical. Starting by giving the batch number, this part of the documentation records all events during the production, from information on raw materials including their batch numbers to the results of the quality control. Each step is signed by the operator. Most important, the record must contain the signature of the person authorizing release of the product.

5.5 Dispensing and distribution record: This is a list showing to whom a batch has been supplied, thereby permitting a possible recall in case of a product fault.

5.6 <u>Raw-material record</u>: This should contain all data on the starting materials and the packaging materials used in the preparation of radiopharmaceuticals in the radiopharmacy. Each item should contain a signed release authorization from the person responsible for the quality assurance.

6. MANUFACTURE PROCEDURES

The quality of the finished product depends upon a well-defined manufacturing process and this is ensured when all records described under "Documentation" are available. When new products are evaluated, whether a new preparation, a new process or a change in a currently used process, the whole procedure must be examined. For example, if a change in the production process might influence the stability of the final product, new storage tests are required.

7. DISPENSING PROCEDURES

Dispensing is the most common radiopharmacy procedure and is performed upon written or oral request from the physician for the individual patient. Dispensing of patient doses in syringes or vials are made from:
  i)   Stock solutions
  ii) Eluate from nuclide generators
  iii) A preparation kit made with eluate from a nuclide
       generator

Ready-for-use radiopharmaceuticals may be supplied to the radiopharmacy in pre-dispensed syringes or in single or multidose vials and ampoules. The principles involved in dispensing are very simple, normally involving only one or two transfers between closed containers. As these products are mostly used the same day as they are dispensed, it is not necessary to demand very elaborate areas for the dispensing operation. A good hygienic standard is necessary and radiation protection procedures must be taken into account. The requirements for this protection are determined by the characteristics of the radionuclides dispensed and the activity levels.

Most nuclide generators are designed for automatic elution. The possibility of microbial contamination is low, but it is advisable to use aseptic techniques when eluting the generator. It is good practice to place the generator in a clean room, but placing it in a LAF unit or in a clean room seems unnecessary. Extra radiation shielding is often required to meet radiation protection rules.

Preparation of radiopharmaceuticals from kits by addition of eluate from a generator to the freeze-dried material generally requires only a single-step procedure. Written instructions based on the manufacturer's recommendation should be worked out, giving in detail the necessary operations to be carried out. The finished preparation should be inspected visually to make sure that complete solution or suspension is obtained. As kit procedures are performed in closed systems, and the preparation is used within one working day, requirements for hygienic conditions and radio-active protection are similar to those required for ready-to-use radiopharmaceuticals. It is important not to use higher amounts of radioactivity or dilutions greater than those indicated by the manufacturer.

Before dispensing takes place, a dispensing record sheet with calculations of the individual volumes of radiopharmaceutical to be drawn from stock solutions, should be prepared along with the necessary labels for syringes, vials and lead shields. Individual doses in vials should be labelled in advance to reduce the radiat-ion dose to the operator. Gloves should always be worn during dispensing. Before doses are drawn, vials should be inverted several times to ensure contents are uniformly dispersed in the solution. A visual inspection behind lead-glass shielding should take place either before or after dispensing. Dispensing of particle suspensions requires special attention to obtain the calculated amount in each syringe. Such preparations should not be kept in the syringes for a long time before injection, as particles may adhere to the walls, and particles can absorb lubricant from the syringe.

8. QUALITY CONTROL

Quality control is only part of the quality assurance system. It is mainly concerned with sampling and analytical testing of the products and with the establishment of specifications and document-ations which form the basis for the release of the product. In principle there should be no difference in quality control tests for radiopharmaceuticals and for non-radioactive drugs. However, there may be a need for certain modifications of the rules and regulations because of radioactivity, because of a short half-life or because of the need for special handling and production procedures.

The extent of quality control in a hospital radiopharmacy depends very much on the source of the radiopharmaceutical. We can divide these products into three different types:
   (i) Radiopharmaceuticals from a commercial supplier or a central radiopharmacy unit supplied in ready-for-use individual patient dose or multidose form.

(ii)   Radiopharmaceuticals from generators and approved,
       commercially available, preparation kits or labelling
       of patient samples with such materials.

(iii) Radiopharmaceuticals made from raw materials, from
       nuclide generators, from home-produced preparation kits
       or from labelling procedures with patient samples.

The ready-for-use radiopharmaceuticals will not normally
require any quality control in the pharmacy, as this has been done
by the manufacturer. Visual inspection and control of paper-work
is part of the dispensing procedure. It should always be kept in
mind that one of the quality parameters is constantly changing
during the shelf-life of a pharmaceutical, namely radionuclide
purity. Due to differences in physical half-lifes, the ratio
between the useful isotope and impurities will change with time.
All data on radionuclide purity must therefore be calculated with
regard to the time of administration to the patient.

Radiopharmaceuticals from generators and approved kits will
require further work in the radiopharmacy. The generator should
be checked for changes in properties that can have been affected
by transportation and operation in the hospital, such as:

(i)   Elution yield
(ii)   Break-through of mother isotope in eluate (e.g.
       molybdenum-99 in technetium-99m eluate.) This can be
       performed by using shielding techniques and gamma
       spectrometry or modified dose calibrators.
(iii) Content of column material in eluate.
(iv)   Sterility and apyrogenicity.

Analytical quality control of the routine preparation of kits
should in principle not be needed. However, in many countries
routine checks are performed with regard to radiochemical purity
on each batch prepared. In any case it may be useful to have
quick and simple radiochemical purity tests worked out for this
group of products. They should be applied when:

(i)   Change in generator supply is made (change of manuf-
       acturer or introduction of a generator with a higher
       activity level)
(ii)   New operators are introduced to the work. (Even though
       the same procedures are used, there may be very large
       variations in the results from person to person until
       the right techniques have been acquired.

(iii) Unexpected results are obtained in a patient investigation. With a quick answer from such a test, the
preparation might be excluded as the "trouble-maker",
and the focus for finding the cause of the problem can
be directed to the equipment or to the patient.

These simple checks are mostly performed with paper or thin-
layer chromatography, and the gamma camera can be used to record
the activity distribution of the chromatogram.

For radiopharmaceuticals prepared in the radiopharmacy, a
full quality control system must be established for each product.
This may consist of a complete analytical test of each individual
batch, or it may rely more on the quality assurance system incorporated in the manufacturing process supplemented by checks on
individual batches.

If all aspects of GRP have been fulfilled in the production
of a radiopharmaceutical, it is acceptable to release the pharmaceutical for use before all tests for quality control have been
completed. The short half-life of many radionuclides would make
it impossible to wait for the results from tests on microbial
contamination before use. With the philosophy of GRP this sterility test can be looked upon as a control of the procedures used in
the manufacture of the product.

9. CHOICE OF PRODUCTS AND PURCHASE

Choice of products to be used by the radiopharmacy should be
made jointly by the radiopharmacy and the nuclear medicine department. The three main reasons for the choice of a particular
product should be:

(i)   Quality, including product performance and lack of
      adverse reactions.
(ii) Availability, taking into consideration transport
      problems and regularity in deliveries from the manu-
      facturer.

(iii) Cost (including both product and transport costs).

Some hospitals have special isotope committees and ethical
committees which grant permission to introduce new radionuclides
or new methods of investigation.

For radiopharmaceuticals held in stock in the radiopharmacy
ordering is based on pre-determined stock levels. Other products
are ordered on request. All orders received should be logged in a
record book with information on date of receipt, quantity, manufacturer, lot number and expiry date.

10. WASTE DISPOSAL

The primary principle of radioactive waste disposal is to minimize contamination of the environment. In the radiopharmacy the problem will be centred mainly on the radioactive nature of the waste, but attention should also be paid to disposal of chemical and biological waste. National regulations may determine how waste disposal should be carried out, and it is therefore necessary for the responsible person to be in close contact with local authorities in these matters.

Designated waste storage areas should be readily accessible, but located so that the activity level will not interfere with other work in the radiopharmacy. Waste should be separated according to half-life, radiation intensity and ultimate disposal method. The hospital radiopharmacy will often have to deal with radioactive material returned from other hospital departments, such as syringes, vials and lead containers. This will increase the radiation risk to the staff and increase the risk of radio-active and microbial contamination of clean working areas. The radiopharmacy should therefore be designed so that production and dispensing areas are well separated from areas handling waste.

Radionuclide generators cause special problems, as they often contain long-lived radioactive contaminants. They should be removed from the radiopharmacy as soon as possible after use and placed in a special radioactive store. After a period of decay, they should preferably be returned to the manufacturer for final disposal.

11. RADIATION PROTECTION

Work in the radiopharmacy should be designed to ensure minimum radiation exposure to personnel in accordance with national and international recommendations. The external dose from a given radiation source is determined by three factors:

(i) distance from the source (ii) time of exposure (iii) shielding

Tongs and forceps are used to increase the distance when manipulating radioactive sources. Doubling the working distance reduces exposure by a factor of four for small sources. When the time of exposure is decreased, radiation dose will be decreased in direct proportion. It should not be encouraged to speed up all work in the radiopharmacy to reduce radiation doses since other aspects of the procedure may suffer, but an experienced worker with good working techniques will be able to reduce the time for the job, thereby reducing the dose, compared to an inexperienced worker. Shielding the source will also reduce radiation exposure.

Good radiological protection is obtained by detailed instructions and training of personnel, and by ready availability of
shielding such as lead glass shielding for dispensing operations
and lead syringe shields. A well designed working area and careful
radiation monitoring help to reduce radiation doses. External
radiation is normally monitored by personnel film dosimeters,
pocket electroscope dosimeters or thermoluminescent dosimeters.

Airborne contamination comes either from volatile radionuclides,
such as iodine, from radioactive gases or from dust particles
containing radioactive material. The main cause for internal
contamination is work with iodine. This contamination should be
monitored by measuring iodine uptake in the thyroid.

To avoid accidental contamination with radioactivity, all
working areas should be monitored either directly or by smear
tests at the end of the operation. Paper-work should be separated
as far as possible from handling of radioactivity. Floors and walls
should be checked at regular intervals. At the end of each month
personnel monitoring reports should be available for all staff.
Working procedures should be revised continuously to reduce
radiation exposure to workers in the radiopharmacy. Instructions
should be made for procedures to be carried out in case of
accidents involving radioactive material.

12. CONCLUSION

These points form the basis of planning and running a modern
hospital radiopharmacy. The radiopharmacy also has other tasks
that will require similar planning, such as distribution of radiopharmaceuticals inside and outside the hospital, information and
education on radiopharmaceuticals and research and development
work.

REFERENCES FOR FURTHER READING

Frier M. and Hesslewood S.R. (1980) Quality Assurance of Radio-
    pharmaceuticals. Chapman and Hall Ltd. London
Kristensen K. (1979) The preparation and control of radiopharma-
    ceuticals in hospitals. International Atomic Energy Agency,
    Vienna
Phan T. and Wasnich R. (1981) Practical nuclear pharmacy. Banyan
    Enterprises Ltd. Honululu
Rhodes B.A. and Croft B.Y. (1978) Basics of radiopharmacy.
    C.V. Mosby Company, St Louis
Saha G.B. (1979) Fundamentals of Nuclear Pharmacy. Springer Verlag,
    New York
Tubis M. and Wolf W. (1976) Radiopharmacy. John Wiley & Sons Inc.,
    New York.

NEW ADVANCES IN NUCLEAR CARDIOLOGY

C. CONSTANTINIDES

Director of the Laboratory of Nuclear Imaging, Alexandra
University Hospital, Vas. Sofias Ave., 80, Athens 611,
Greece.

1. INTRODUCTION

Recent advances in nuclear medicine have resulted in the
development of a new field - nuclear cardiology. Because of new
radiopharmaceuticals, new types of equipment and improved techniques,
a new dimension to the evaluation of the cardiac patient is being
provided to the aramentarium of the practising clinician..

In nuclear cardiology we examine the heart as a muscle and as
a pump. As a muscle the heart is examined by using the following
procedures:

(i)   Rest Tl-201 myocardial scintigraphy
(ii) Exercise Tl-201 myocardial scintigraphy
(iii) Scintigraphic studies of the metabolic activity of the
      myocardium by using radiopharmaceuticals labelled with
      positron emitting radionuclides (F-18, C-11, N-13)

As a pump the heart is examined by using the following procedures:

(i)   The first pass isotope angiocardiogram.
(ii) Multiple gated acquisition of the cardiac blood pool
      using Tc-99m labelled erythrocytes

2. THALLIUM-201

Thallium-201 has become established as the imaging agent of
choice for the evaluation of patients with coronary arteriosclerotic
disease (CAD) (Lebowitz et al 1975). It is biologically analogous
to potassium, has a high myocardial extraction efficiency and enters

the myocardium through the cell membrane - bound Na+ - K+ ATpase.
(Kawana et al 1970). Further, the perfusion of Tl through the
myocardium is proportional to the myocardial blood flow.

2.1 Rest and exercise Tl-201 myocardial scintigraphy

Rest and exercise Tl-201 myocardial scintigraphy today comprise
routine procedures for the diagnosis of CAD. Two millicuries of
Tl-201 are administered I.V. at rest or after maximal exercise and
myocardial scintigrams are obtained for three views, anterior, left
anterior oblique (LAO) 45°and LAO 60° by using a gamma camera.

The normal myocardial scintigram is presented as a horse-shoe
configuration (anterior projection) or as a doughnut configuration
(LAO projection). Segmental analysis of the myocardial count
profile is possible today using the computer facility known as
"area of interest" analysis. Normal segmental confidence limits
can be established from data obtained from normals and used to
assess more objectively suspected image defects from patients with
CAD.

2.2. The study of metabolic activity of the heart

The metabolic activity of the heart can be studied today by
using substances like deoxy-glucose or fatty acids labelled with
positrons (F-18 or C-11). In normal conditions the myocardium uses
95% free fatty acids for its metabolic demands and only 5% glucose,
whereas in cases of CAD the glucose demands of the myocardium
increase to 25%.

The metabolic activity of the myocardium can be followed by
using the positron emitting tomographic camera (PET). This
instrument is able to take slice photographs of the myocardium in
vivo in which the distribution is proportional to the blood flow
and therefore areas of ischaemia in the myocardium can be revealed
as areas where the activity of fatty acids decreases but the
activity of glucose increases.

3. RADIONUCLIDE ANGIOCARDIOGRAPHY

Another very important area in nuclear cardiology is the use
of radiolabelled tracers which pass through the cardiac chambers
or stay within the cardiac blood pools. These tracers are imaged
as the bolus makes its first passage through the heart (first pass
isotope angiogram), for the assessment of functional cardiac
anatomy and systemic-to-pulmonary shunting. Alternatively, the
cardiac blood pool can be imaged at equilibrium, several minutes
after the injection, during different phases of the cardiac cycle
for the assessment of regional and total ventricular function
(Mullins et al 1969).

3.1 First Pass Technique

First pass studies can be performed with the heart in any position relative to the detector of the camera. The RAO position is used to see the heart through its long axis. The LAO position is the best for separating the right from the left ventricle. It is important to inject 10-15mCi Tc-99m labelled erythrocytes very rapidly as a discrete bolus into a peripheral arm vein and to follow this immediately by a flushing volume of saline as a "chaser".

The camera and data processing system are triggered by the detection of a preset minimum count rate and events are then recorded continuously by the computer in list mode for 30 seconds. Analogue pictures from the camera can be obtained using the proper imaging device. The first pass technique is used for evaluation of global right and left ventricular performance during rest and exercise. Sequential studies can be performed to evaluate therapeutic interventions. It is also a very simple technique to evaluate cyanotic congenital heart disease with right to left shunts, pulmonary atresia or tricuspid atresia. Intracardiac left to right shunts can be evaluated using suitable data processing systems. Finally, the ejection fraction can be computed by using this technique (Strauss et al 1971).

3.2 Multiple gated radionuclide ventriculography

The first pass technique has some limitations. For example, it requires a special camera such as the multicrystal detector which is able to accumulate high counting rates. Some of the limitations of the first pass method have been overcome by imaging the heart after the tracer has equilibrated in the cardiac blood pool. A special triggering system is used between the camera and an electrocardiograph. This triggers the camera to record certain phases of the cardiac cycle (40 msec from the end diastole to 40 msec from end systole) or the whole cardiac cycle is divided into 10 to 60 gates with the production of corresponding multiple images (multi gated acquisition or MUGA).

The average count rate in each image can be plotted sequentially from the R wave to obtain a time-activity curve representing a ventricular volume curve from which rates of chamber filling and emptying can be calculated and also the ejection fraction (the technique shows a correlation coefficient of 0.93 when compared to biplane contrast angiography.).

3.3 The nuclear stethoscope

One of the greatest limitations of the gamma camera is that this device is not able to examine ventricular function on a beat-to-beat basis. This gap was bridged by the nuclear stethoscope which is a unique tool for the quantitative assessment of left ventricular function (Wagner et al 1975). Further the nuclear stethoscope can extend existing nuclear and ultrasound cardiac imaging modalities. The nuclear stethoscope does not produce images of the heart but portrays a time-dependent image of the rate of filling and emptying of the left ventricle as a whole.

After administration of Tc-99m labelled erythrocytes, the amount of tracer within the heart is measured by a simple crystal scintillation detector, small enough to be held by hand against the patient's chest as the heart fills and empties. The crystal detector is attached by way of intermediate eletronic equipment to a recording device that displays the activity being viewed by the detector in relation to the patient's electrocardiogram.

The device is designed to divide the interval between successive R waves into 48, 96 or 192 intervals. For each of these intervals, the amount of radioactive tracer in the heart is measured and depicted on a television screen in the form of a developing left ventricular time-activity curve.

The ventricular functions obtained from the nuclear stethoscope are heart rate, ejection rate and filling rate, ejection fraction, relative cardiac output, relative stroke volume, relative end-diastolic volume, ejection time and fast filling time.

With the nuclear stethoscope it is possible to monitor left ventricular function for periods of several hours or longer at the patient's bedside.

Diseases of the left ventricle can be divided into three pathophysiological groups:

(i)   Volume overload (aortic or mitral insufficiency or left to right shunts)
(ii) Pressure overload (aortic stenosis, systemic hypertension)
(iii) Diseased myocardium (coronary artery disease, cardiomyopathy)

Differentiation of each of these three causes in a given patient can be made by looking at the changes in the time-activity curve through the nuclear stethoscope.

The clinical uses of the nuclear stethoscope can be summarised as follows:

A. Diagnostic procedures

    (i)   Resting cardiac studies
    (ii) Exercise cardiac studies
    (iii) Myocardial infarction assessment
    (iv) Atrial fibrillation, left ventricular measurements
    (v)   CHF - shock assessment

B. Intervention management procedures

    (i)   Short-term drug response
    (ii) Adriamycin toxicity testing
    (iii) Post surgical rehabilitation studies
    (iv) Presurgical monitoring
    (v)   Adjustable pacemaker assessment
    (vi) Long-term drug assessment
    (vii) Arrhythmia left ventricular function monitoring

      In conclusion, the nuclear stethoscope is a unique new tool for the quantitative assessment of left ventricular function.

REFERENCES

Kawana M., Krizeh H., Porter J., Lathrop K.A., Charleston D., and
Harper P.V. (1970) The use of Tl-199 as a potassium analogue in
scanning. J. Nucl. Med. 11, 333,
Lebowitz E., Green M.W., Fairchild R.,Bradley-Moore P.R., Atkins H.L.
Ansari A.N., Richards P. and Belgrave E. (1975) Thallium-201 for
medical use. J. Nucl. Med. 16, 151
Mullins C.B., Mason D.T., Ashburn W.L. et al (1969). Determination
of ventricular volume by radioisotope angiography. Am. J. Card.
24, 72
Strauss H.W., Zaret B.L., Hurley P.J. et al (1971) A scintiphoto-
graphic method for measuring left ventricular ejection fraction
in man without cardiac catheterization. Am. J. Cardiol. 28, 575
Wagner H.N., Natarajan T.K., Strauss H.W. et al (1975). The
nuclear stethoscope: a bedside device for continuous monitoring
of ventricular function (abstr.) Circulation 52. Suppl.II, 11.

APPLICATIONS OF CONVOLUTION AND DECONVOLUTION METHODS IN NUCLEAR MEDICINE

J.J.P. de LIMA

Laboratory de Radioisotópos, Faculdade de Medicina,
Coimbra, Portugal

1. INTRODUCTION

The process of convolution appears in many aspects of nuclear medicine imaging. For example the output activity-time curve from an organ after a bolus injection is the convolution of an input activity-time curve with the frequency function of transit times of the organ, i.e. its impulse response function. In a two-dimensional situation, the blurring of images in nuclear medicine is partly due to convolution of the object activity distribution with the impulse response function of the camera and collimator system.

2. MATHEMATICAL BASIS

Convolution itself is an important aspect of mathematical physics concerning input-output problems of systems and is expressed by the Fredholm integral equation of the first kind.

$$R(t) = \int_0^t h(t,\tau)I(\tau)d\tau$$

where $h(t,\tau)$ represents an operator which converts an object or input function $I(\tau)$ into a measurable output function $R(t)$. The function $h(t,\tau)$ is known as the kernel of the integral equation and is a characteristic function of the system.

In the particular case where the kernel depends only on the difference between t and τ, this equation takes the form of the convolution integral

$$R(t) = \int_0^t h(t - \tau) I(\tau) d\tau$$

The applicability of the convolution integral depends upon the assumption of linearity in the system under study as well as time shift invariance for the functions involved.

The convolution of functions h (t) and I (t) is represented by the expression

$$R (t) = h (t) * I (t)$$

where the symbol * means convolution. It can be proved that the order of the variables is interchangeable and that in the case of the convolution integral, h (t) is the impulse response function of the system, that is the response of the system to a delta function of unit strength at zero time.

Convolution can be visualised if we decompose the function I (t) into shifted impulse functions at different times and with different strengths and outputs, the summation of which is the output function R (t). The distortion effects resulting from the convolution can be used in a constructive way in filtering procedures such as in the reconstruction process in computerised emission tomography.

In the majority of cases, however, the important problem is the inverse process, i.e. I (t) is wanted when R (t) and h (t) are known, or h (t) is required when R (t) and I (t) are known. In such cases the unknown function is said to be deconvolved from the known output function and the necessary mathematical procedures are referred to as deconvolution. Deconvolution techniques are frequently applied in nuclear medicine studies such as renography, cardiovascular studies, and tomographic reconstruction algorithms.

Convolution or deconvolution with continuous functions can be implemented using the Laplace or the Fourier transformation since by the convolution theorem, convolution in real space is transformed into multiplication in phase space. Thus for the Laplace transformation, $\mathcal{L}$

$$\mathcal{L}[R(t)] = \mathcal{L}[h(t)] \cdot \mathcal{L}[I(t)]$$

A possible deconvolution algorithm when h(t) is required and I (t) and R (t) are known is

$$h(t) = \mathcal{L}^{-1}\left\{\frac{\mathcal{L}[R(t)]}{\mathcal{L}[I(t)]}\right\}$$

which involves calculation of the Laplace transforms of R (t) and
I (t) and of the inverse transform of the quotient of these transforms. Deconvolution, then, involves a division in phase space.

However, continuous deconvolution leads to integral equations
which have general solutions only under constrained conditions or
in rather special cases. Most of the practical situations in
nuclear medicine require a discrete approximation for convolution
and deconvolution.

When the functions involved are sampled functions, the convolution integral becomes

$$R(K) = \sum_{n=0}^{K} h(K-n)I(n) \qquad \ldots \ (1)$$

where R (K), h (K-n) and I(n) are the sampled values of the
functions. This equation can also be written in matrix form as

$$R = H. \ I$$

I is a triangular matrix of (K + 1). (K + 1) elements with line
(n + 1) consisting of the terms I (n), I (n-1), ...I (0)
followed by K-n zeros. H and R are column matrices with K + 1
elements.

H can be calculated if R and I are known, i.e.

$$H = R.I^{-1}$$

An alternative way of considering sampled functions is through
the one-sided Z transforms of the sampled sequence. Defining the
Z transforms of I (t) and h(t) as

$$Z(I) = \sum_{n=0}^{\infty} I(n)z^{-n}$$

$$Z(h) = \sum_{n=0}^{\infty} h(m)z^{-m}$$

and applying Z transforms to equation 1 we have

$$Z(R) = Z(I) \ . \ Z(h)$$

or

$$\sum_{k=0}^{\infty} R(k)z^{-k} = \left(\sum_{n=0}^{\infty} I(n)z^{-n} \right) \left(\sum_{m=0}^{\infty} h(m)z^{-m} \right)$$

Since the Z transformation also transforms convolution into multiplication, convolution now becomes polynomial multiplication.

For the inverse problem we have

$$Z(h) = \frac{Z(R)}{Z(I)}$$

Discrete deconvolution is therefore carried out through poly-nominal division if Z transforms are used. Other methods have also been used in deconvolution such as the fast Fourier transform and numerical function minimisation methods.

When noise affects the data, as is generally the case, decon-volution procedures are strongly perturbed. Deconvolution is an ill-conditioned situation which means that small errors in the two measured functions can produce large errors in the solution.

By using as h (t), I (t) and R (t) known functions with noise added, the perturbation due to noise can be studied. Applying least squares methods to minimise the square of the difference between the original output data and the reconvolved outputs com-puted from the deconvolved functions, it was observed that this minimisation would not necessarily lead to the best deconvolved functions. The quality criterion was based upon δ^2 , the squared difference between the real impulse response function and the deconvolved one. Furthermore, if the sampling interval of the data sequence is decreased, two types of deconvolution problem can be identified. In type A, δ^2 remains finite, converging, and in type B δ^2 increases, diverging. (Hunt 1971; Nimmon et al 1981).

In type A, the effects of noise are not very severe and intro-duce an oscillatory component into the deconvolved curves. To obtain acceptable solutions to the type B problem, the introduction of constraints based upon the characteristics of the specific system involved is required. In this case sophisticated methods of noise filtration are necessary.

3. APPLICATION TO IMAGES

Applications of deconvolution in nuclear medicine have been reported since the fifties in dynamic and metabolic studies (Scholer and Code 1954; Birge et al 1969; Abrams et al 1969).

Renal studies provide one of the main applications where deconvolution work has been applied in detail. A renogram, after background correction, is a representation of how the amount of tracer in the kidney is changing with time. This amount depends, not only on the physiological characteristics of the kidney, but also on the blood concentration of the tracer.

Interest in using reconvolution methods in renography lies in the fact that the renogram curves are strongly affected by the shape

of the blood activity-time curve. This activity-time curve depends
on factors such as the rate of mixing of the radiopharmaceutical
with the blood and the rate of diffusion of the blood to the extra-
vascular space. Also, each individual renogram is dependent on the
function of the other kidney since this function influences blood
activity. In fact, it is easily shown that the renogram of a
particular kidney will appear better or worse according to whether
its partner is working normally or abnormally. This makes the
comparison of renograms from different patients, or even of the
same patient in different stages, rather difficult (De Lima 1980).

Deconvolution of renograms, using a conveniently sampled
activity-time blood function, leads to impulse retention functions
which are equivalent to the response of the kidney to an instantaneous injection into the renal artery, assuming no recirculation.
The impulse retention functions are not affected by those factors
which affect the renograms (Diffey et al 1978).

The deconvolution of renograms is a type A problem which has
been carried out by different authors using different techniques:
e.g. Van Stekelenburg (1973) used Laplace transform and Diffey et
al (1976) and Britton et al (1980) used matrix methods. Either
data smoothing or data bounding before the application of the
deconvolution procedure was used by the various authors.

After deconvolution, the impulse retention function shows an
initial, fast decreasing, segment due to vascular and tissue background, followed by a plateau. The vascular and tissue background
component is generally removed since it does not represent transit
of tracer across the kidney and the resulting impulse retention
function shows a plateau which corresponds to the minimum transit
time followed by a decreasing segment down to zero level.

The height of the plateau of the non-normalised impulse
retention function of a kidney can be shown to be proportional to
its effective renal plasma flow (O'Reilly et al 1979). For every
time, t, the impulse retention function H (t) represents the
fraction of the total injected dose of radiopharmaceutical with
transit times longer than t, i.e.

$$H(t) = \int_{t}^{\infty} h(t)dt$$

where h (t) is the normalised impulse response function or frequency of transit times of the kidney.

The impulse retention functions can either be obtained using
conventional probe renography plus a separate probe to measure the
blood activity, or by using a gamma camera and computer interfaced
display system. In the first case, the renogram and the blood

activity-time curve have first to be digitized and the sampled data is then used as input to a computer. This can be done automatically by interfacing a simple data acquisition system to a microcomputer. In this case either I-131 or I-123 Hippuran or Tc-99m DTPA can be used.

Dynamic imaging with a gamma camera using either I-123 Hippuran or Tc-99m DTPA allows more comprehensive application of deconvolution methods to the data. From the impulse retention function curves, the mean transit time (MTT) of the tracer in the kidney can be calculated. It can be shown that the MTT is the quotient of the area defined by the impulse retention function and the plateau value. MTT data has been shown to permit good discrimination between obstructive and non-obstructive disease except in chronic congenital obstruction (Piepsz et al 1982).

The negative derivative of the impulse retention function is the transit time spectrum or frequency function of transit times of the kidney. The modal structure of this spectrum gives an approximation to the relative value of the intrarenal blood flow in the two nephron populations of the kidneys - the juxtamedullary and the cortical. Nimmon et al (1981) have proposed a correlation method to determine the contribution of the two modes.

In obstructive situations calculating the impulse response functions for the cortical and pelvic areas allows differentiation between the two types of obstruction, with and without nephropathy. For example, increased total MTT with no increase in cortical MTT is in favour of obstructive uropathy.

The quantitative data which can be derived from the deconvoluted impulse retention function makes this technique an attractive tool for renal studies. However, it does increase the processing time.

In dynamic studies, the stability of the response function obtained by deconvolution is dependent on the input function and on the statistical noise associated with the measured data. It has been observed that for the type of functions obtained in first pass studies, the impulse response function is potentially unstable being a type B problem. In recent work where non-invasive measurement of regional cerebral blood flow using Tc-99m red blood cells is reported, the stability of the deconvolution procedure is achieved by using constraints of smoothness, monotonicity and non-negativity through a regularisation method in the Fourier domain. In this technique an intravenous injection is used and activity-time curves are obtained over the aortic arch using a collimated detector and over regions of interest (ROI's) on the brain from a gamma camera vertex view. The mean transit time of the blood through the brain is obtained by deconvolution. Together with this technique, a functional representation of the blood flow in the ROI's has also been used by the same authors to reduce computing time and to give a visual indication of regional blood flow. The method is current-ly being applied in patients with cerebrovascular diseases (Britton et al 1981; Nimmon et al 1981).

Another potentially useful application of deconvolution is in the detection and quantification of left to right cardiac shunts. Following an intravenous injection of Tc-99m, the activity-time curves for the lungs and for a ROI on the middle superior vena cava (input function) are recorded. The variations in input function resulting from the prolonged bolus degrade the accuracy of this technique. Deconvolution allows the calculation of the pulmonary activity-time curve resulting from a standard input i.e. a spike injection in the superior vena cava (Alderson et al 1979, Williams 1979).

During image reconstruction from projections, in emission tomography, algorithms have been used that utilise either convolut-ion or deconvolution as steps to obtain the final image. For example convolution can be used either in mono-dimensional or bidimensional real space to filter either the projections before retroprojection or to filter the retroprojected blurred image. Techniques which perform deconvolution of the blurred images through Fourier transformation have also been proposed (Brooks and Dichiro 1976).

REFERENCES

Abrams M.E., Crawley J.C.W., Green J.R. and Veall N. (1969) A comparative study of digital and analogue computer techniques for deconvolution procedures in clinical tracer studies. Phys. Med. Biol. 14, 225

Alderson P.O., Douglass K.H., Mendenhall K.G. et al (1979) Deconvolution analysis in radionuclide quantitation of left to right cardiac shunts. J. Nucl. Med. 20, 502

Birge S.J., Peck W.A., Berman M. and Whedon G.D. (1969) Study of calcium absorption in man: a kinetic analysis and physiologic model. J. Clin. Invest. 48, 1705

Britton K.E., Nimmon C.C., Whitfield H.N. et al (1980) The evaluation of obstructive nephropathy by means of parenchymal retention functions in "Radionuclide Nephrology," George Thieme, Verlag p. 164

Britton K.E., Granowska M. and Nimmon C.C. (1981) Total and regional cerebral blood flow - a new quantitative non-invasive method for cerebrovascular disease in "Medical Radionuclide' Imaging" 1980 IAEA Vienna Vol. II p.315

Brooks R.A. and Dichiro G. (1976) Principles of computer assisted tomography (CAT) in radiographic and radioisotopic imaging. Phys. Med. Biol. 21, 689

DeLima J.J.P. (1980) Dependence of an individual renogram on the other kidney through the blood activity shown by convolution. Eur. J. Nucl. Med. 5, 469

Diffey B.L., Hall F.M., and Corfield J.R. (1976) The Tc-99m DTPA dynamic renal scan with deconvolution analysis. J. Nucl. Med. 17, 359

Diffey B.L., Hall F.M., Piepsz A. and Erbsmann F. (1978) Renal deconvolution and the poor injection. Eur. J. Nucl. Med. 3, 145

Hunt B.R. (1971) Biased estimation for non-parametric identification of linear systems. Math. Biosci. 10, 215

Nimmon C.C., Lee T.Y., Britton K.E. et al (1981) Practical applications of deconvolution techniques to dynamic studies in"Medical Radionuclide Imaging"1980 IAEA Vienna Vol. I p.367

O'Reilly P.H., Shields R.A. and Testa H.J. (1979) in"Nuclear Medicine in Urology and Nephrology", Butterworths p. 175

Piepsz A., Ham H.R., Erbsmann F. et al (1982) A co-operative study of the value of dynamic renal scanning with deconvolution analysis. Br. J. Radiol. 55, 419-433

Scholer J.F., and Code C.F. (1954) Rate of absorption of water from stomach and small bowel of human beings. Gastroenterology 27, 565

Van Stekelenburg L.H.M. (1978) Hippuran transit times in the kidney - a new approach. Phys. Med. Biol. 23, 291

Williams D.L. (1979) Improvement in quantitative data analysis by numerical deconvolution techniques. J. Nucl. Med. 20, 568

SOME RECENT DEVELOPMENTS IN X-RAY TECHNIQUES

J. GARSOU

Service Universitaire de Contrôle Physique des Radiations,
(University of Liège) Hôpital de Bavière, Bd de la
Constitution 66, 4000 Liège, Belgium.

1. INTRODUCTION

In the field of irradiation, exposure may be expressed in terms
of the number of photons falling on an object, but the biological
effect produced in the object must be related to the energy absorbed,
i.e. the absorbed dose, from the incident radiation.

In radiology also, it was necessary to abandon characteristics
of detectors that related to exposure and to obtain relationships
between these characteristics and the energy really absorbed in the
detectors. One consequence is that the scattered radiation is now
taken more and more into consideration when seeking refinement of
the radiation image or of the image given by the detector.

2. X-RAY PRODUCTION

Since the focal spot is of a finite size, the sharpness of the
image increases with the distance between focus and object. To
increase distance, higher and higher intensities of radiation are
needed, so sources are mainly rotating anodes.

2.1 Electron source

Stability and reproducibility are the most important qualities
required for electron emission. The tungsten filament remains in
favour for its robustness and emission stability in spite of its
relatively low efficiency and appreciable evaporation at the upper
end of the temperature range used.

2.2 Electron focussing

The focussing effect of the electron beam on a focal spot of a desired size on the anode is obtained by inserting the filament in a slot acting as a pair of electrostatic lenses, producing on a focal spot a double image of the filament appearing as two strong lines.

The design of the cathode structure is mostly a question of empiricism and experience, since theoretical analysis is very complex. (see e.g. Moores and Brubacher 1974).

Focal spot measurement by pinhole methods gives results which differ from the nominal size. This technique is felt to be unsatisfactory for assessing imaging performance and a substitute for the pinhole is a slit which affords a number of advantages. For example it overcomes exposure problems in imaging very small focal spots, presents an averaging effect on the detail of intensity distribution, and permits a modulation transfer function to be derived from the information.

However, for predicting imaging performance, the method of determining a value of limiting resolution by means of a resolution pattern in the form of a star is widely accepted. From the limiting resolution figure, the size of a uniform spot with identical limiting resolution can be calculated and the modulation transfer function can be derived.

2.3 Focal spot loading

The high power loading of the focal spot requires a target with high atomic number, high melting point, low vapour pressure, high specific heat capacity and high thermal conductivity. The material of choice for the anode is tungsten alloyed with up to 15% rhenium.

The rating of a rotating anode at short exposure times is a function of the effective focal spot size, target angle (6 degrees has been used successfully with rhenium added in the surface, more than 15 degrees is required for pure tungsten anodes), target diameter (usually 55-125 mm) and target rotational speed (3,000 rpm for "standard" speed, up to 9,000 rpm for high speed and occasionally 18,000 rpm). Longer term rating depends on the specific heat capacity of the bulk material (molybdenum, either pure or alloyed with varying concentrations of carbon) and on cooling characteristics, essentially by radiation, with possible enhancement up to a factor of two by spraying a suitable plasma of mixtures of Al, Cr, and Tl oxides. For a general review see Randmer et al (1980).

2.4 X-ray noise

Extra-focal X-rays are produced by backscattered electrons reattracted by the anode outside the focal spot area. The production of extra-focal X-rays is virtually eliminated if a field free space is provided around a fixed target which is almost hooded. With a radiation cooled rotating anode this is not practical and extra-focal radiation must be reduced by external diaphragms.

Beam quality varies across the beam section from anode to cathode due to the filtering effect of the surface layers of the target within which X-ray production takes place. This variation is more readily detected in the unfiltered beam. It should be added that inherent filtration (80% generally due to the bulb glass) adds some scatter to the radiation.

2.5 Tubes for CT scanning

Either stationary anodes cooled by oil or water circulation with a continuous rating of about 4 kW, or rotating anodes with very high thermal capacity obtained by massive refractory metal discs or by graphite backing (with a liquid cooling system for the tube shield) may be used. With the former anodes, the focal spot is 2.1 x 21 mm, with the latter, it is of conventional size.

As greater stability is required for radiation emission, output modulation of rotating anodes due to slight variation of the target surface around the annular track of the focal spot must be avoided as well as fluctuations of filament emission due to mains frequency.

Pulsing the X-ray beam by grid control, claimed to reduce patient dose, can be used to correct for background effects in the detectors.

3. INTERACTION WITH PATIENT

The resolution of a radiograph is given by the contrast between detail and its surroundings. The spatial frequency in the object at which detailed reproduction is lost is a measure of the resolving power of the detection system.

Resolution is determined by the characteristics of the detector analysing the emerging beam, mainly by its response to the spectral changes produced by a test object, and is affected by the size of the X-ray source, interaction between the radiation and the object producing scatter and movement of the object.

3.1 Assessment of Contrast

If a film or a combination of film + screen is irradiated under a test object of a given line pair frequency v , the signal recorded by a microdensitometer is an optical density D which varies in space at right angles to the bar pattern in a manner depending on the resolution of the system.

The variation in space of the signal can be characterised by the following parameters:

$$\text{Amplitude A} = \frac{D_{max} - D_{min}}{2}$$

$$\text{Mean level } \bar{a} = \frac{D_{max} + D_{min}}{2}$$

$$\text{Modulation M} = \frac{A}{a} = \frac{D_{max} - D_{min}}{D_{max} + D_{min}}$$

Exposure, or better, absorbed dose, can replace D in the region where the response of the film or the film + screen combination is linear.

These equations have recently been investigated by Reece and Roberts (1981) using an experimental arrangement consisting of a lead bar-pattern resolution grid with a line pair frequency varying from 0.25 line pairs/mm to 10 line pairs/mm as test object, various thicknesses of Mix-D, and an X-ray focal spot of 50 μm (in order to ignore geometrical blurring.

Contrast function C_D relative to optical density D at a line-space frequency v was defined by

$$C_D\ (\ v\ ) = \frac{D(v)_{max} - D(v)_{min}}{D(v)_{max} + D(v)_{min}}$$

where $D(v)_{max}$ and $D(v)_{min}$ are maximum and minimum values of optical densities observed above fog level with a microdensitometer at the line-space frequency v in the linear response region of the film-screen combination.

Results showed that for each thickness of Mixed-D, contrast values approached a maximum for each curve as the spatial frequency decreased. The maximum contrast values plotted as a function of tissue thickness are reproduced in Fig. 1.

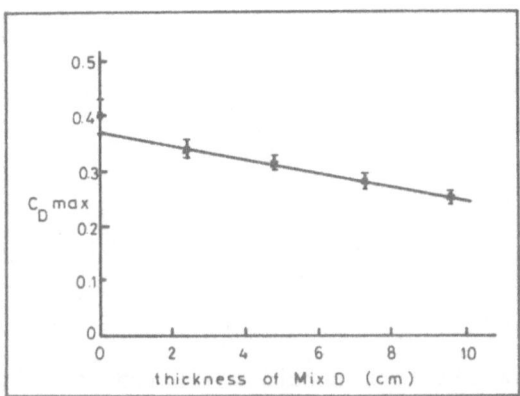

Fig. 1   Maximum contrast as a function of thickness of tissue
         equivalent Mix D (reproduced with permission from Reece
         and Roberts 1981)

3.2   Investigation of spectral changes

The effect of the phantom on the spectrum of X-rays was also
investigated using a Compton scatter technique with an intrinsic
germanium detector (Yaffe et al 1976). As shown in Fig. 2, the
presence of Mix D has a marked effect on the beam spectrum and in
particular on its mean energy.

A contrast function C_S may be used to relate the emergent
spectra from two adjacent areas, one without and two with the lead
contrast

$$C_S = \frac{\psi_1 - \psi_2}{\psi_1 + \psi_2}$$

where the photon energy fluence $\psi = \phi\,(h\nu)\ \times\ h\nu\ \times\ dh\nu$

The calculated value of C_S was shown to decrease as a function
of Mix D thickness in the same way as C_D .

The contrast function C_S can be analysed as a function of the
photon energy using the spectral functions

$$C_S^{\;1} = \frac{\phi_1\,(h\nu) - \phi_2\,(h\nu)}{\phi_1\,(h\nu) + \phi_2\,(h\nu)}$$

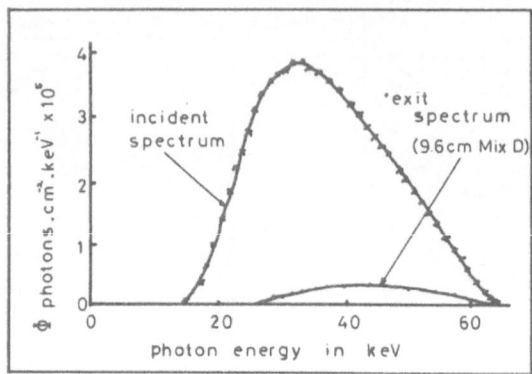

Fig. 2 Effect of Mix D on the beam spectrum (reproduced with permission from Reece and Roberts 1981)

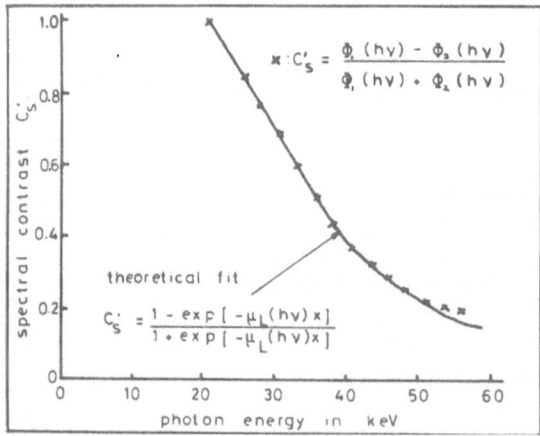

Fig. 3 A plot showing the change in spectral contrast as a function of energy (reproduced with permission from Reece and Roberts 1981)

As shown in Fig. 3 C_S^1 decreases as a function of the photon energy (between 20 and 60 keV) fitting closely a theoretical expression of the form

$$C_S' = \frac{1 - \exp \; -\left[\mu_L \; (h\nu) \; x\right]}{1 + \exp \; -\left[\mu_L \; (h\nu) \; x\right]}$$

where μ_L is the lead attenuation coefficient.

The spectral contrast thus appears to be independent of tissue thickness and depends only on the lead attenuation of the resolution grid.

3.3 Effect of scatter

The production of scattered radiation depends on the X-ray energy, the chemical composition of the tissues and the irradiated volume. Dick et al (1978) have shown that the ratio of scattered fluence to total X-ray fluence transmitted (i.e. the scatter fraction) could be as much as 50% in contact radiography of a 10 cm thick object and a 5 cm diameter field size. For a mean photon energy of 40 keV, Compton scatter and classical Thomson cross-sections are about the same and half the scattered photons lie in the forward hemisphere.

In magnification radiography, with an air gap of at least 25 cm between patient and film, the amount of scatter to the film is reduced and should not impair contrast.

3.4 Response of the image receptor

With both contact and magnification radiography techniques, the imaging system can be chosen so as to enhance the contrast by energy selection. A contrast function of the image receptor C_r where

$$C_r = \frac{\int \phi_1 \; (h\nu) \; \xi_a \; (h\nu) \; dh\nu \; - \int \phi_2 \; (h\nu) \; \xi_a (h\nu) \; dh\nu}{\int \phi_1 \; (h\nu) \; \xi_a \; (h\nu) \; dh\nu \; + \int \phi_2 \; (h\nu) \; \xi_a (h\nu) \; dh\nu}$$

can be derived by finding the useful energy absorbed in the image receptor ξ_a. The value of ξ_a for a pair of calcium tungstate intensifying screens (Ilford High definition) has a maximum of about 19 keV per photon at a photon energy of 25 keV and falls to 9 keV per photon at 10 keV and to 6 keV per photon at 60 keV. For a non-screen film (Kodak Industrex C) ξ_a shows a much flatter response with about 2 keV absorbed per photon at 15 keV going up to 6 keV at 80 keV (without significant increase at the Ag K edge energy due to escape of fluorescent X-rays from the emulsion).

The resulting calculated values of C_r for the image receptor decrease linearly when increasing thicknesses of Mix D are interposed.

4. FILMS AND SCREENS

The detective quantum efficiency (DQE) or comparative noise level is the ratio of the squares of the signal-to-noise ratios for the output from and input to the imaging system.

$$DQE = \frac{(S/N)^2_{out}}{(S/N)^2_{in}}$$

It is well known that image quality is proportional to contrast and sharpness, and to the inverse of noise power. Noise is proportional to contrast, sharpness and to the square of the speed.

4.1 Contrast

Contrast is given by the slope of the characteristic curve. Film-screen combinations used in mammography show a great variety of characteristic curves differing by gamma, shape and speed as measured by the reciprocal exposure to produce unit density. The shape of the characteristic curve provides valuable information on the structure of a film (check for quality control) since it depends upon screen X-ray absorption and light output (intensity and colour), and the film structure and grain size (including development factors).

However, the characteristic curve, expressed in terms of the logarithm of the incident exposure, does not mirror the contrast transfer function or speed, especially as a function of beam quality. Indeed, the number of quanta per unit exposure varies with X-ray energy. Thus it should be more meaningful to express film-screen speed by the X-ray energy absorbed in the screen to produce unit optical density. Furthermore, when a polychromatic X-ray beam is transmitted through a patient, the mean energy may differ from one area to another due to different filtration resulting from the anatomical structures irradiated. A higher filtration effect produces a higher mean energy local beam for which the absorption in the screen is higher and thus the film blackening is higher so image contrast is reduced. This effect is called intrinsic contrast rendition (Stevels 1975). Therefore, radiation contrast is not translated directly into image contrast, but only through the variation with energy of the absorption of screen phosphor. Note that most phosphors show K absorption edges in this photon energy interval.

Finally, the characteristic curve gives no information about transfer of small contrasts. The film gamma in fact affects both signal and noise equally for quantum limited signals corresponding to small contrast differences. The variation of DQE with exposure gives a more realistic description of information transfer. At low X-ray exposure levels, there are not enough light photons for a large number of grains to receive enough photons for development, so the film reproduces most accurately X-ray quantum fluctuations (DQE maximum).

At higher X-ray exposure levels, light photons are in excess, no new information is recorded, and DQE decreases.

Thus important factors are the number of light photons produced per X-ray photon and the average number of light photons required to produce a developable grain. The combination of these two gives the average number of developed film grains per X-ray quantum $\bar{c}$, appearing in the relationship

$$DQE = \frac{\eta}{1 + 1/\bar{c}}$$

where η is the screen absorption and $\bar{c}$ is in the region of 10 - 20 for modern film-screen pairs.

Lassen and Bloch (1978) have demonstrated variation with exposure in low contrast signal-to-noise transfer and an increase in detectability for a particular exposure value.

4.2 Sharpness

The ability of a given imaging system to resolve sufficiently fine detail is usually measured by the point response function or its Fourier transform, the modulation transfer function (MTF).

In film-screen combinations, the MTF is mainly determined by the screen. In the case of $CaWO_4$, it has been found to decrease with increased phosphor thickness and grain size and to depend on phosphor material (Morlotti 1975).

A parameter which provides a convenient, simple measure of sharpness is the noise equivalent passband N_e (Schade 1956) which represents, in two dimensions, the spatial frequency at which an MTF of unit amplitude would have the same volume as the squared transfer function. In one dimension, the area under the squared MTF is considered. It transpires that sharpness, screen thickness and speed are intimately related, and a film-screen combination of a given speed can have a variety of values of N_e.

4.3 <u>Noise</u>

The overall noise properties of a film-screen combination are given by the Wiener spectrum of the density fluctuations, with large values at low spatial frequencies but falling at higher frequencies due to the blurring action of the screen. The ultimate high frequency value represents the "white noise" due to film granularity. It can be shown that the noise properties of quantum limited imaging systems are universally related to the image forming exposure quanta per resolution element, and this is of particular importance in digital systems.

The noise is governed by the number of exposure quanta forming an image and not by the exposure. The number of exposure quanta, q, is related to the film-screen speed S by the relationship

$$1/q \propto S$$ where S is the product of screen absorption, conversion efficiency, light escape and film sensitivity.

Several quantum mottle indices have been defined (see for example Rao and Fatouros 1979) and from such indices it can be seen that screens with similar sharpness have vastly different speeds, but similar noise properties. Screens with similar noise properties have a variety of sharpness and speed combinations. Finally, combinations of fast and slow films with slow and fast screens respectively show very different noise characteristics.

Of fundamental importance in the interplay of sharpness, noise and speed, are the relative contributions of screen absorption and light speed factors depending upon screen phosphor thickness, packing density, grain size, optical properties including reflective layer properties, and also upon film spectral response to screen light output.

In this context, it has been shown that at mammographic beam qualities (less than 2 mm Al half value layer) most screens show similar attenuation properties while at higher beam energies (used in general radiography) a wide variation is observed due to the action of the K edge of the phosphor.

Screens with increased X-ray absorption can produce systems that are faster without being noisier or sharper without being faster. The effective conversion efficiency or light output per unit attenuation is then measured to assess the magnitude of any speed increase coming from an increased light output and leading to an increase in noise. Screens show wide variations of this factor with beam quality. (Moores 1981)

In the context of patient examinations, reasonable exposure values are needed and therefore threshold contrast measurements should play a more important role in assessing the photon utilization of these film-screen combinations. This is important at mammographic beam qualities for which a high proportion of incident photons are absorbed in most screens (Parker-Hodds and Moores, 1979) and therefore any speed variation exhibited by a film-screen unit must reflect variation in the number of exposure quanta per resolution element.

5. X-RAY IMAGE INTENSIFIER TELEVISION SYSTEMS

5.1 Principle of image intensifier

An evacuated tube with a fluorescent screen at one end converts the X-ray pattern to light photons which strike a photocathode that is in intimate contact, giving rise to electrons. These electrons are accelerated by a potential difference of at least 25 kV on the anode and focussed by an electrode system onto an output screen that converts electrons back to light. The gain of the intensifier results from a) acceleration of the electrons and b) reduction, usually to about one tenth in each dimension, of the image size on the output phosphor. The light output passes normally through a tandem lens system and is focussed onto a television camera which produces the video signal. A camera control unit processes this signal into a form suitable to operate a television monitor in which the automatic gain control maintains the brightest part of the image at a constant level. Feedback control of X-ray exposure factors can also be incorporated.

5.2 Optimum performance

The major aim in using image intensifiers in medicine is to reduce patient exposure. In this way, the exposure rate on the intensifier screen lies between 20 and 100 μR.s^{-1}.

5.3 Threshold contrast

The threshold contrast is proportion to noise level, with a proportionality constant between 3 and 5. For a typical X-ray beam quality of 80 kV and a HVL of 7 mm Al, one Roentgen is given by 2×10^8 photon. mm^{-2}; so 50 μR.s^{-1} would correspond to 10^4 photons. mm^{-2}.s^{-1} or 10^6 photons.cm^{-2}.s^{-1}.

The time over which the signal is sampled is determined by the lag of the television camera and/or the eye performance and is about 0.1s. Therefore the signal would correspond to 10^5 photons.cm^{-2} falling on the input screen and the noise

($\sqrt{\text{signal}}$) to 3×10^2 photons or 0.3% of the input signal.

A $ZnCdS_2$ fluorescent screen absorbs about 20% of the X-ray quanta incident on it, so the input signal is really produced by the absorption of 2×10^4 photons. cm^{-2} and the noise is raised therefore to 0.7%.

The performance of the whole system is clearly determined by the properties of the input fluorescent screen and the degree of quantum noise present in the input signal. CsI has now replaced $ZnCdS_2$ since it has a relative quantum detection efficiency 3 to 5 times higher, but its threshold contrast is about half as great.

5.4 Resolution

The resolution of the image intensifier is also determined by the input fluorescent screen properties (thickness, density, structure). For a typical $ZnCdS_2$ intensifier, the maximum resolution is, in practice about 1.2 line pairs.mm^{-1}.

A television system working with a 625 line definition and a monitor image of the same size as the input screen (250 mm diameter) presents a maximum resolution in the vertical direction slightly greater than 2 line pairs.mm^{-1} and in the horizontal direction for a bandwidth of 10 MHz and 25 frames s.$^{-1}$ (10^7/25 x 625 x 250) or 2.5 cycles.mm^{-1}. The television system should therefore be able to handle the maximum resolution of the intensifier and in particular of the input fluorescent screen.

For the CsI intensifier, the bandwidth of the television system should be doubled, since the maximum resolution quoted by the manufacturers is 4 line pairs.mm^{-1}. Since, however, medical interest is usually in the region 0.5 - 1 line pairs.mm^{-1}, this improvement would be reserved for highly specialised examinations.

For further information on image intensifiers, the reader is referred to Conference report series No. 29 of the Hospital Physicists' Association (1978) entitled "Quality Assurance Measurements in Diagnostic Radiology."

6. DIGITAL RADIOLOGY

An exposure of 1 R is obtained at 60 keV by 10^8 photons.mm^{-2}. From a thickness of 20 cm of water, 10^6 emerge unscattered and subject to fluctuations with an uncertainty of $\pm 10^3$ (0.1%). Thus changes of this order should be detectable. However, the contrast range needs to be expanded since the eye can only resolve about 20 levels.

The aim of digital radiology is to improve the statistics of the image and to refine the contrast but where information of clinical interest lies in very small changes, implying a large number of discernible levels, means of range compression are required and this can be most conveniently realised in a digital computer.

Range compression can be approached i) by edge enhancement (rather like a Xerogram) ii) by subtraction between two radiographs before and after contrast infusion.

6.1 Contrast subtraction digital radiography

Imagine two beams of energy E and square cross section of side L traversing a tissue thickness x gm.cm^{-2} of mass attenuation coefficient σ (E). One beam passed through an additional thickness x_a gm.cm^{-2} of material of mass attenuation coefficient σ_a (E).

For N_0 photons entering the tissue, the number N_1 of photons emerging unscattered from tissue alone is

$$N_1 = N_0 \; e^{- \sigma (E).x}$$

and the number N_2 going out similarly from the infused tissue is

$$N_2 = N_0 . e^{- \sigma (E).x} .e^{- \sigma_a (E) . x_a}$$

The probability of detecting a difference in emergent intensity between the two beams and visualising a difference between the two irradiated regions is related to the signal-to-noise ratio by

$$S/N = \frac{N_1 - N_2}{\sqrt{N_1 + N_2}} = \left[\frac{N_o \; e^{- \sigma x}}{2} \right]^{\frac{1}{2}} \sigma_a \, x_a \qquad (1)$$

if $x_a \sigma_a$ is very much less than 1

The exposure, R, is related to N_0 by

$$R = \frac{N_0}{L^2 \, K(E)}$$

where K(E) is the number of photons per Roentgen per cm^2.

Hence

$$(S/N) \ /R = \ \sigma_a \ x_a \ . \ K(E) \ . \ \sqrt{e^{-\sigma(E) \ x}} \ . \ L^2 / \sqrt{2 \ N_0}$$

By analysing expression (1) as a function of energy, it can be shown that for an object thickness of 20 $gm.cm^{-2}$, the addition of 1 $mg.cm^{-2}$ of iodine is best visualised at about 35 keV (Iodine K edge is about 33 keV) and soft tissue changes of 50 $mg.cm^{-2}$, between 60 and 70 keV (effective beam energies of computed tomographic scanners).

6.2 Effect of scattered radiation

Scattered radiation must be taken into account since collimation of the detectors is not perfect and they also respond to scatter. If each detector receives N_S scattered photons in addition to the unscattered photons, the signal-to-noise ratio becomes

$$(S/N)_S = \frac{N_1 - N_2}{\sqrt{N_1 + N_2 + 2N_S}} \simeq \frac{N_1 - N_2}{\sqrt{2} \ \sqrt{N_1 + N_S}} \quad \text{for } N_1 \simeq N_2$$

and after substituting and assuming $\sigma_a \ x_a \ll 1$

$$= (S/N) \ \frac{1}{\sqrt{1 + N_S/N_1}}$$

If $N_S = N_1$

$$(S/N)_S = S/N \ \frac{1}{\sqrt{2}}$$

This relationship shows that scatter degrades the signal-to-noise ratio as a slowly varying function of the ratio of scattered to primary photons.

6.3 Technical feasibility

An array of 300 x 300 detectors each 1 x 1 mm^2 gathers enough data for vessel visualisation in 0.5 second.

The basic requirement for a full digital radiographic system is a low noise, high quantum efficiency two dimensional detector with a resolution power of 2 line pairs. mm^{-1}. Attempts to fulfill this requirement were made using image intensifiers coupled to low noise television cameras. However, whereas image intensifiers have a relatively high quantum efficiency at the energies appropriate for contrast subtraction, even the best TV cameras are too noisy to be used directly and have to be coupled through analogue-to-digital converters into storage devices which sum a number of images so improving the signal-to-noise ratio of the final image.

Charge coupled array devices, with a very wide dynamic range and low noise, allow direct integration of the light from an image intensifier.

In an alternative system, a charge image is produced on aluminised melinex sheet by gathering ions from a high pressure gas under the influence of an electric field (ionography). This image is then read by scanning electrometers. This approach affords the advantage that Xenon filled ionographic chambers have both a high quantum efficiency and high spatial resolution. It is proposed to incorporate an electrode scanning mechanism into a high pressure Xenon chamber and after exposure the mechanism will scan the electrometer head over the charge image stored on the melinex. The output is to be digitised and stored in a computer. Initial experiments are encouraging and for further details the reader is referred to Pullan (1981).

7. X-RAY RESONANCE EFFECTS

7.1 The resonance effect

In general, X-ray attenuation by the photoelectric effect decreases with increasing energy. However, at an absorption edge, the attenuation of monochromatic photons per unit absorbed dose shows an important enhancement. This has been illustrated for example by liberation of halogen atoms from different halides (Garsou 1959, 1976) as well as in the irradiation of 3 5 diiodo-DL-thyronine, 3 5 3'-triiodo-DL-thyronine and thyroxine (Dieffallah et al 1970) and of 5-bromodeoxyuridine and zinc enzyme (Hakpern and Stöcklin 1973 and 1974)

As the basic process involved is the photoelectric effect, heavy atoms in either the mixture or the molecule are the primary targets. This is why halogen atoms, except F, have been studied in organic molecules. An adequate reaction for the detection of the irradiated heavy atoms, halogen atoms for instance, is oxidisation by these halogen atoms of a leucobase and measurement of the dye concentration with a spectrophotometer (Garsou 1959).

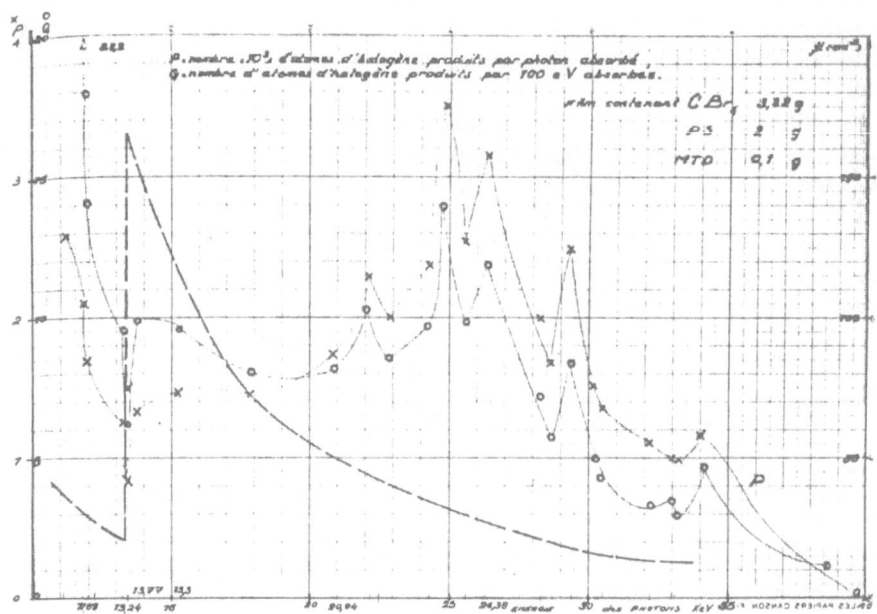

Fig. 4 Release of Br from CBr$_4$ as a function of incident photon energy.

7.2 Evaluation of effect

The effect can be evaluated as a function of the photon energy, by the number of halogen atoms liberated per absorbed photon or by 100 eV absorbed. Experimental results obtained with CBr$_4$ embedded in polystyrene are shown in Fig. 4.

The overall effect can be subdivided into different effects due to all electrons following the photoelectron event and to photoelectrons and Auger electrons as well. Therefore the energy and the number of these electrons must be calculated for every photon energy. The following must be determined:-

C_{Br} contribution of Br to the total sample attenuation

C_{BrK} contribution of K layer of Br to total Br attenuation at photon energies equal to or higher than the K edge energy

$$= \frac{(\mu/\rho)_{up} - (\mu/\rho)_{down}}{(\mu/\rho)_{up}}$$

where $(\mu/\rho)_{up}$ and $(\mu/\rho)_{down}$ are the higher and lower values of (μ/ρ) at the K edge.

C_{BrL} contribution of L layer of Br to the total Br attenuation at photon energies equal to or higher than the K edge energy, taken as a good approximation equal to

$\dfrac{(\mu/\rho)_{down}}{(\mu/\rho)_{up}}$ (all contributions of MN.... layers assumed to be negligible)

With this approximation, C_{Br} can be taken at photon energies lower than the K edge energy as the L contribution of Br to the total Br attenuation.

7.2.1. Number and energy of photoelectrons. At photon energies higher than the K edge energy, since C_{Br} x C_{BrK} is the fraction of photons absorbed in the sample and reacting with K electrons, 1000 x C_{Br} x C_{BrK} = N_{peK} is the number of K photoelectrons ejected for 1000 absorbed photons and similarly 1000 C_{Br} x C_{BrL} = N_{peL} is the number of L photoelectrons ejected for 1000 absorbed photons.

7.2.2 Number and energy of Auger electrons. The number of Auger KLL electrons N_{eAK} is

$$1000\ C_{Br}\ x\ C_{BrK}\ x\ (1 - \omega_K)$$

where ω_K is the fluorescence yield in the K shell, i.e. the fraction of K characteristic photons emitted per primary photon absorbed.

The number of Auger KMM electrons N_{eAL} is the sum of 1000 C_{Br} x C_{BrL} x $(1 - \omega_L)$ coming from the absorption of the primary photons with ω_L, fluorescence yield in the L shell and 1000 C_{Br} x C_{BrK} x ω_K x $\frac{1}{3}$ $(1 - \omega_L)$ coming from the absorption of K characteristic photons, with the assumption that one third are absorbed. The energies of the KLL and LMM Auger electrons are taken equal to 10.085 and 1.272 keV respectively.

362

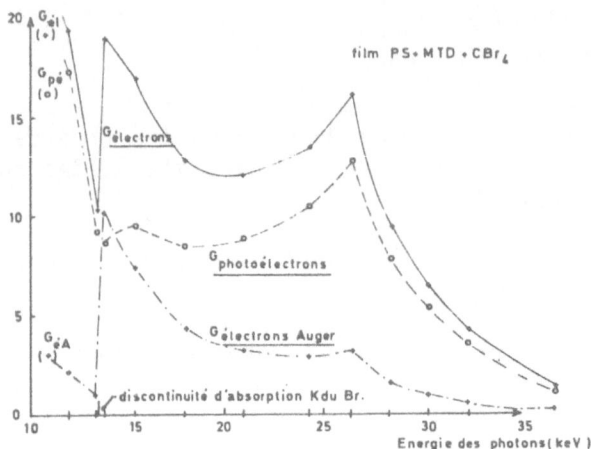

Fig. 5 Variation in halogen atoms liberated per 100 eV absorbed
        as a function of photon energy. (For further details see
        text)

7.2.3 Variation as a function of the primary photon energy of -
    - the number G_{el} of halogen atoms liberated per 100 eV absorbed
      from all electrons, photoelectrons and Auger electrons,
      and its components
    - the number G_{pe} of halogen atoms liberated per 100 eV absorbed
      from photo- electrons and
    - the number G_{eA} of halogen atoms libertated per 100 eV
      absorbed from Auger electrons, is shown in Fig. 5.

    This shows that Auger electrons experience a peak of effective-
ness at primary photon energies slightly higher than the K edge
energy, but photoelectrons also show a peak at about twice the
energy of this K edge.

    Alternatively, if numbers of liberated halogen atoms are
related to the energy of the primary photon which would be entirely
absorbed by all electrons (P_{el}), by photoelectrons (P_{pe}) or by Auger
electrons (P_{eA}) the variation of these quantities
as a function of primary photon energy is illustrated by Fig. 6.

7.2.4 The effectiveness expressed as P_{el} or G_{el} shows peaks for
primary energies just above and at about twice the K
edge energy (13.47 keV). The first of these peaks is mainly due to
Auger electrons, the second to photoelectrons. They are followed by
relatively sharp decreases and these decreases are also observed for
primary photon energies just below the K edge energy.

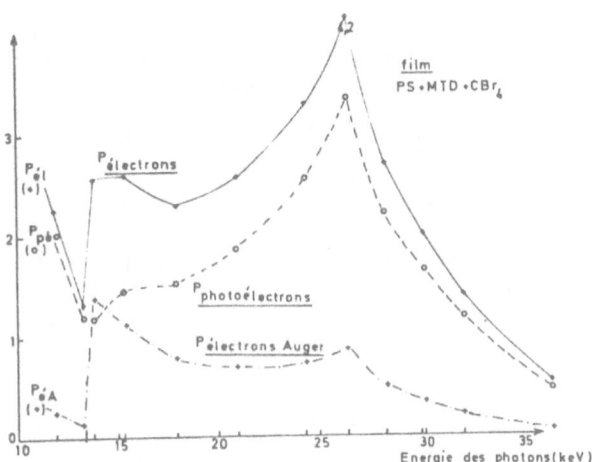

Fig. 6 Variation in halogen atoms liberated per primary photon
absorption event as a function of photon energy.

7.3 Illustration of the resonance effect

The liberation of bromine can be understood not only in terms
of the effect of Auger electrons, but also by the formation of
highly ionized target atoms resulting from the production of Auger
electrons. These target atoms are ejected from the molecule by
Coulomb repulsion (Halpern and Stöcklin 1973).

The model of a Coulombic "explosion" of the molecule due to
the formation of highly ionized target atoms was shown to be
consistent with the recoil energies and the abundances of these
ions (Carlson and White 1966). For example, the X irradiated
products of C_2H_5I are shown in Table I.

The iodine yield from X-irradiated iodoaminoacids is higher
than would be predicted from published values of the K to L shell
Auger yields in iodine. Halper and Stöcklin (1974) suggest that
Auger charging and subsequent molecular fragmentation rather than
electron self-radiolysis are responsible for the observed damage.

It must be emphasised that whenever Auger Cascade occurs,
strong ionisation due to Auger electrons and molecular dislocations
occurs immediately as a short range effect and predominantly at
relatively well-defined photon energies. This phenomenon probably
explains the relatively high radiobiological toxicity of I-125
(Ertl et al 1970).

TABLE I

X-IRRADIATION PRODUCTS OF C_2H_5I

Product	Abundance %	Most probable recoil energy eV
I^+	19%	1.1
I^{2+}	22	3.8
I^{3+}	20	6.1
I^{4+}	18	8.4
I^{5+}	13	15
I^{6+}	4	24
I^{7+}	2	-

(from Carlson and White 1968)

7.4 X-ray resonance systems

The requirements for illustrating the resonance effect are not simply the presence of a heavy atom in the structure, the possibility for it to initiate a molecular explosion, and the availability of a suitable detection reaction of the liberation of the heavy atoms. Thus for example the formation of F centres in X-ray irradiated KBr, RbBr and KCl does not show any resonance effect within the uncertainty of the experiment (Cruz-Vidal et al 1970).

Systems used for the detection of resonance effects are: bromodeoxyuridine (measurement of radical concentration per unit absorbed dose) and carboanhydrase (determination of percentage of Zn liberation and of inactivation) (Halpern and Stöcklin 1973, 1974); 3,5-diiodo-DL-thyronine, 3 5 3'-triiodo-DL-thyronine and thyroxine (measurement of I yield) (Diefallah et al 1970); polystyrene films containing CHI_2CBr_4 and p p' p'' methylidenetris - (N N-dimethyl-aniline) (spectrophotometric measurement of the crystal violet concentration produced by oxidation of the leucobase by I or Br liberated) by the present author. All these groups showed clearly an enhancement of the photon effectiveness per unit of absorbed dose at energies just higher than the K edge. The second peak of photon effectiveness reported here at energies just above twice the K edge has not yet been confirmed by other groups.

7.5 Possible applications of an X-ray resonance dosemeter

Since such a dosemeter highlights resonance effects at definite monochromatic X-ray energies, it could be expected that its anomalous response to photon energies around the absorption edge of target atoms would demonstrate the effect of secondary radiation on depth dose curves. Comparison of depth dose curves obtained with Co-60 and Cs-137 gamma ray beams either by ionization measurements or by plastic films irradiated perpendicularly in a polystyrene block are shown in Figs. 7 and 8.

It is clear that the ionisation chamber and the resonance dosemeter respond differently to the perturbation of the energy spectrum caused by introducing different thicknesses of lead.

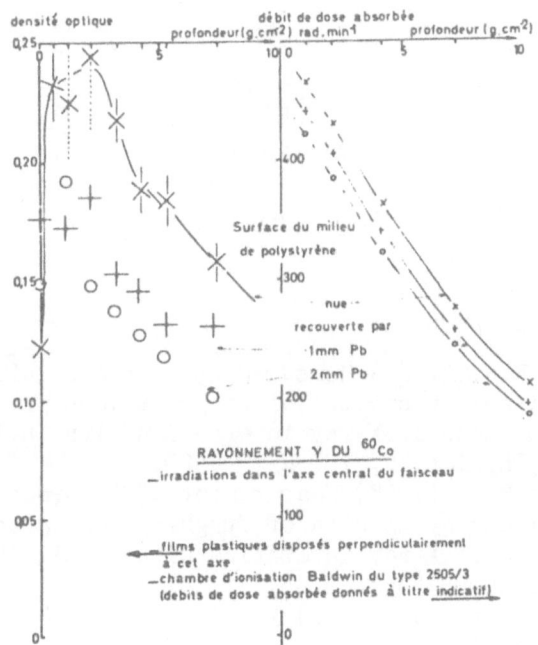

Fig. 7 Depth dose curves for Co-60. Results for the plastic films (resonance dosemeter) are shown on the left and for the ionisation chamber on the right. The introduction of either 1 mm or 2 mm of lead has a much greater effect on the left.

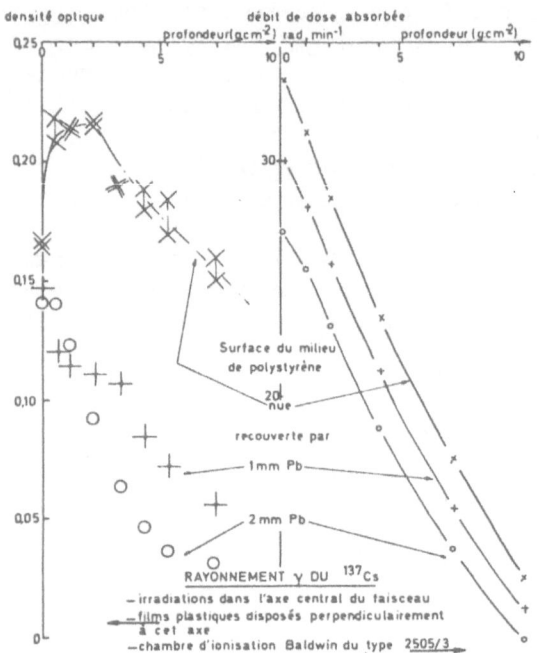

Fig. 8 Depth dose curves for Cs-137. Other details as for Fig. 7.

REFERENCES

Carlson Th.A and White R.M. (1966) Measurement of the relative
abundances and recoil-energy spectra of fragment ions produced
as the initial consequences of X-ray interaction with CH_3I, HI
and DI. J. of Chem. Physics 44, 12, 4510-4520.

Carlson Th.A. and White R.M. (1968) Measurement of the relative
abundances and recoil energy spectra of fragment ions produced as
the initial consequence of X-ray interation with C_2H_5I, CH_3CD_2I
and Pb $(CH_3)_4$. J. of Chem. Physics 48, 11, 5191-5194

Cruz-Vidal B.A. and Gomberg H.J. (1970) The role of K shell
ionisation in the formation of F centres in alkali halides at 78°K
I. KBr. J. Phys. Chem. Solids 31, 1273-1280

Cruz-Vidal, B.A., Gomberg H.J. and Diaz-Hernandez F. (1970) The role
of K shell ionisation in the formation of F centres in alkali
halides at 78°K II. RbBr. J. Phys. Chem. Solids 31, 1281-1285

Cruz-Vidal B.A., Gomberg H.J. and Nino-Rojas L. (1970) The role of
K shell ionisation in the formation of F centres in alkali halides
at 78°K III. KCl. J. Phys. Chem. Solids 31, 1287-1290

Dick C.F., Soares C.G. and Motz J.W. (1978) X-ray scatter data for
diagnostic radiology. Phys. Med. Biol. 23, 1076-1085

Diefallah E., Stelter L. and Diehn B. (1970) Chemical consequences
of the Auger effect: iodine yield from iodoamino acids as a
function of X-ray dose and energy. Rad. Res. 44, 2 273-281

Ertl H.H., Feinedegen I.E., and Heiniger H.J. (1970) I-125, a tracer
in cell biology: physical properties and biological aspects.

Garsou J. (1959) Contribution à l'étude de l'efficacité des rayons
X monochromatiques sur quelques systèmes d'halogenures organiques
solides et liquides. Thèse de doctorat - Universite de Liège,
Belgique.

Garsou J. (1976) Contribution à la caracterisation et à la metrol-
ogie de champs d'électrons et de photons pour la radiothérapie.
Thèse d'agrégation à l'Enseignment Supérieur - Université de
Liège, Belgique

Gomberg H.J. (1956) Suggestion at the basis of Garsou's work.

Halpern A. and Stöcklin G. (1974) A radiation chemical resonance
effect in solid 5 bromodeoxyuridine; chemical consequences of the
Auger effect. Rad. Res. 56, 329-337

Halpern A. and Stöcklin G. (1973) Ein gezielter Strahlenresonanze-
ffekt in Biomolekulen als Folge des Auger-Prozesses. Jahresbericht
1973 der Kernforschungsanlage Jülich Gmbh.

Lassen M. and Bloch P. (1978) Measurements of the effect of X-ray
film-screen characteristics on threshold detectability of small,
low contrast objects. Medical Physics 5, 152-161

Moores B.M. and Brubacker P. (1974) Focal spot studies and electron
focussing in a demountable X-ray tube. Phys. Med. Biol. 19, 605-
618

Moores B.M. (1981) Films and Screens in Physical Aspects of Medical
Imaging Eds. B.M. Moores, R.P. Parker, B.R. Pullan. John Wyley &
Sons Ltd. pp. 253-265

Morlotti R. (1975) X-ray efficiency and modulation transfer function
of fluorescent rare earth screens, determined by the Monte Carlo
method. J. Photographic Sci. 23, 181-189

Parker-Hodds S. and Moores B.M. (1979) Light output and X-ray
attenuation measurements at mammographic beam qualities for nine
commercial intensifying screens. Radiology, 131, 737-742

Pullan B.R. (1981) Digital Radiology in Physical Aspects of Medical
Imaging Eds. B.M. Moores, R.P. Parker and B.R. Pullan. John Wiley
& Sons Ltd. pp. 275-287

Randmer J.A., Holland W.P. and Koller T.J. (1980) X-ray sources and
controls in Radiology of the Skull and Brain. Vol. 5 Technical
Aspects of Computed Tomography. Ed. T.H. Newton, C.S. Mosby,
St Louis.

Rao G.U.V. and Fatouros P. (1978) The relationship between resolution and speed of X-ray intensifying screens. Medical Physics, 5,
205-208

Reece B.L. and Roberts P.J. (1981) Interaction with the patient in
Physical Aspects of Medical Imaging. Eds. B.M. Moores, R.P.Parker,
and B.R. Pullan. John Wiley & Sons, pp. 241-251.

Schade O.H. (1956) Optical and photoelectric analogue of the eye.
J. Optical Soc. Am. 46, 721-739

Stevels A.L.N. (1975) New phosphors for X-ray screens. Medicamundi
20, 12-22

Yaffe M. Taylor K.J. and Johns H.E. (1976) Spectroscopy of
diagnostic X-rays by a Compton scatter method. Medical Physics,
3, 328-334

BIOMEDICAL APPLICATIONS OF ULTRASOUND

C.R. HILL and M. SAMBROOK

Physics Division, Institute of Cancer Research, Royal Marsden Hospital, Sutton, Surrey, UK.

1. ACOUSTIC FIELDS AND THEIR INTERACTIONS WITH TISSUES

Propagation of ultrasound through water-like media, including human soft tissues, is predominantly by longitudinal wave mode. In this situation ultrasound acts as a form of radiation with the following approximate characteristics:

propagation speed: 1500 m s^{-1}
wavelength: $1.5/F \text{ mm}$
attenuation coefficient (tissues): 0.7 F dB cm^{-1}
where F = frequency in MHz

Reflection of the radiation occurs with spatial variations of ρc ("characteristic acoustic impedance": product of density and sound speed) and refraction with variations in c. Both bone and air spaces (e.g. lung) present major barriers to ultrasound propagation.

Proper theoretical treatment of ultrasonic fields involves development of three-dimensional wave theory and solution of the resulting equations under appropriate boundary conditions.

Practical generation of ultrasonic fields is usually by the piezoelectric effect in electro-acoustic transducers having various configurations, such as: plane piston, concave spherical surface and arrays of transducer elements.

Detection and measurement of ultrasound in medical applications is generally either for "diagnostic" or "dosimetric" purposes.

Piezoelectric devices are again very commonly used but for dosimetry they may need to be supplemented by other techniques based on calorimetry, the radiation force phenomenon, and the acousto-optical effect.

2. ACOUSTIC MICROSCOPY AND THE ACOUSTIC FINE STRUCTURE OF TISSUES

For many materials, including human tissues, the frequency dependence of acoustic attenuation coefficient is such that useful acoustic measurements can be made with wavelengths (and corresponding spatial resolution limits) as low as $0.2 \mu m$. Two classes of acoustic microscope have been developed to exploit this situation, working respectively with unfocussed and focussed acoustic beams. In each case both amplitude and phase of the modified beam can be detected and this enables complete fine-scale description of the acoustic properties of the investigated medium.

Measurements of this type demonstrate that human tissues constitute, in mechanical terms, continua that are spatially inhomogeneous in both density (ρ) and adiabatic compressibility (k). This conclusion is supported independently by measurements of the angular scattering behaviour of different tissues.

Such observations provide a scientific basis for the general phenomenon of acoustic scattering by human tissues, which is central to current and developing methods of ultrasonic medical diagnosis.

3. PULSE-ECHO CLINICAL IMAGING: PRINCIPLES AND APPLICATIONS

In the pulse-echo technique a short (1 - 3 wavelength) pulse of ultrasound is propagated into a patient in a directional (usually focussed) manner, and a recording is then made of the resulting train of echoes intercepted by the same (send-receive) transducer. The origin of any particular echo can then be inferred from its return time (and an assumed propagation speed) together with the known propagation vector of the beam. Such information is commonly used to generate two-dimensional tomographic images but specialised techniques are also used to extract data on spatial dimensions and time-variations of position in the interrogated object.

Important limitations to the technique arise from imperfections in directionality of the beam and in pulse length definition (i.e. finite point-spread function) and also in noise, and in the image speckle associated with the coherent nature of the radiation.

With conventional practice it is relatively easy to acquire images at frame rates of up to 25 Hz (and it is feasible to achieve 1000 Hz) with the consequence that cinematic (or "real-time") imaging is readily achievable. This is of very great practical

value in modern diagnostic application, for two reasons: (a) it
enables both normal and abnormal patterns of movement of tissues
to be observed (e.g. resulting from cardiac, pulmonary or foetal
activity) and (b) it extends the tomographic imaging facility from
two to effectively three dimensions.

Applications of pulse-echo investigation are numerous. It can
be estimated that 25 million patient examinations were made worldwide in 1982, thus placing ultrasound an easy second after conventional X-ray in its extent of use as a physical diagnostic technique.
Some of the principal applications are in:

(i)   Obstetrics, for measurement of foetal size, and hence
      maturity; for assessment of placental position; for early
      detection of foetal abnormalities; and for observation of
      foetal movement patterns.

(ii) Cardiology, for diagnosis of heart valve malfunction; for
      detection of heart wall defects; and for investigation of
      myocardial disease.

(iii) Internal Medicine, for detection of both cancer and non-
      malignant lesions in organs such as liver, kidney,
      pancreas, spleen, bladder, prostate, breast and thyroid;
      for assessment of therapeutic response of disease in such
      organs; as a means of "visually" guiding invasive proced-
      ures such as needle biopsy.

4. ULTRASONIC METHODS FOR IN VIVO CHARACTERISATION AND QUANTITATIVE
    ANALYSIS OF TISSUES

One of the major limitations of the conventional pulse-echo
technique is that it is essentially non-quantitative. This impedes
its use, for example in differentiation of some diffuse disease
conditions. There is therefore interest in finding modified, or
different, techniques which will provide supplementary information.

A possible approach is that of reconstruction tomography
(comparable to X-ray CT). With ultrasound it is possible to
reconstruct both attenuation and sound speed values from transmission measurements. It is also possible to use backscatter measurements for reconstructing both attenuation and backscattering coefficients. This approach has the advantage over a transmission
measurement that propagation paths are less often obstructed by
bone and air spaces.

Partly for this reason there is considerable interest in
various other approaches to tissue characterisation by backscattering methods. One such approach is to employ features of the

frequency spectrum of backscattered signals. Another is an
analogue of X-ray diffraction analysis. This uses features of
the diffraction pattern obtained as a result of the dependance of
backscattered signal amplitude on the relative orientation angle
between the ultrasonic beam vector and the structure of an interrogated volume. Finally, and probably most practically, it has
been shown to be possible to characterise tissues from statistical
features of the fine structure (texture) of a corresponding B-scan
image or A-scan record.

5. DOPPLER BLOOD FLOW ANALYSIS AND IMAGING

5.1 General Principles

The first successful attempt to measure blood velocity non-
invasively was in 1959. The instrument used at that time was based
on a theory formulated by Christian Doppler to explain the apparent
change in pitch of moving sources, and verified experimentally by
Ballot, over a century before.

This instrument is a Doppler-shift ultrasonic flow-velocity
meter. Ultrasound incident on the blood cells is scattered and
the scattered wave is received by a second transducer. The transmitted and received signals are mixed and the resulting signal is
the Doppler-shift frequency produced because the scatterers are
moving. This frequency is proportional to the blood cell velocity.

Although, conceptually, the operation of the Doppler flowmeter
is straightforward and described by the Doppler-shift equation,

$$\Delta F = \frac{2f \ v \ \cos a}{c}$$

where f = frequency of excitation
     v = average flow velocity
     a = angle between the sound beam and the direction of
         flow
     c = velocity of sound in blood

since blood cells have a range of velocities (i.e. a velocity
profile) the action of the flowmeter cannot be adequately described
by the above equation. Therefore we have a range of Doppler
frequencies (i.e. a Doppler frequency spectrum) corresponding to
the range of velocities.

There are a variety of ways and means of using this Doppler
frequency spectrum depending on the clinical application, needs of
the user and availability of funds. For example, if an estimation
of the extent of occlusive arterial disease is required a non-
directional continuous wave Doppler system could be used with the
signals processed by simple diode detection and the ear, whereas a

much more elaborate and expensive system is needed for quantitative measurement of blood flow. The accuracy of these measurements is limited by a number of factors depending on the signal processing system used, the direction of the ultrasonic beam in relation to the blood vessel, and on the ratio of the ultrasonic beam size to vessel diameter.

Although combined pulsed-Doppler and imaging systems do solve some of these problems and also the problem of insonation of more than one vessel due to their close proximity, it has self-imposed constraints on its ability to measure range and velocity simultaneously. Such constraints make it possible to design a pulsed Doppler system which will function optimally over a large depth and velocity range.

5.2 Application to measurement of blood flow in tumours

Blood supply is believed to be a very important factor in determining both the pattern of development of a tumour and also its response to therapy. A variety of methods is available for observing and measuring blood flow in tumours but these tend to be invasive and unsuitable for routine and repeated use on patients.

We have used a 10 MHz continuous wave ultrasonic Doppler system to study the blood flow associated with normal and malignant mammary tissue in patients with breast cancer. Some patients were receiving endocrine therapy and were examined regularly over a period of months. This method was used to carry out the study primarily because it is non-invasive and suitable for repeated use.

It was found that the dependence of the average Doppler-shift frequencies on tumour volume is similar to that seen for total volume flow rate in some experimental tumours, including mouse mammary adenocarcinoma. In general blood flow changes appeared to occur in association with and possibly in advance of changes in tumour volume.

The future aim of this work is to study the blood flow of well-established human mammary tumours and in particular to determine whether endocrine therapy can cause early detectable changes in tumour blood flow and if so, whether these changes relate to subsequent clinical response of the tumour. If this did occur it might be possible to identify at an early stage that group of patients (approx. 70%) whose tumours tend not to respond to this form of treatment.

Other applications include a study of breast blood flow in women throughout the menstrual cycle, and the possible use of Doppler for monitoring blood flow changes brought about by hyperthermia.

6. BIOLOGICAL MODIFICATIONS INDUCED BY ULTRASOUND

Under certain conditions (which will generally include acoustic frequency, peak or average acoustic intensity, ambient temperature and pressure) exposure to an ultrasonic beam may induce temporary or permanent modification in human cells and tissues. Such changes can come about through the mediation of several different biophysical mechanisms, including thermal action, cavitation and certain other effects which are neither thermal nor cavitational in nature. The existence of mechanisms in the latter group is now well established although their exact nature is generally not understood.

Such biological action of ultrasound has found practical application in various forms of therapy. A particularly widespread use is in physical medicine but there is now increasing interest in possible uses in cancer therapy.

The demonstration of such biological action of ultrasonic radiation also raises the question of the safety of its use in medical diagnosis. This question has been the subject of a considerable amount of research and discussion in the scientific community, taking in evidence from studies in basic bioacoustics, in specific toxicology, and in epidemiology. At present there seems to be no direct evidence for existence of a hazard but a considerable amount of suggestive but often conflicting indications on the subject.

SUGGESTIONS FOR FURTHER READING

At a reasonably introductory level the subject is dealt with in several text books, including

Biomedical Ultrasonics, P.N.T. Wells, Academic Press (1977)
Ultrasonics in Clinical Diagnosis, 2nd Edition. (P.N.T. Wells Ed.)
  Churchill-Livingstone (1977)
Doppler principles and techniques, D.W. Baker in "Ultrasound: its
  applications in medicine and biology" (F.J. Fry Ed.) Elsevier.
Cardiovascular applications of ultrasound, R. Reneman, N. Holland.

For more advanced treatments see

Physical principles of medical ultrasonics, C.R. Hill, Ellis Horwood
  and J. Wiley. (At present in preparation, scheduled for publication in 1983)

Investigative Ultrasonology
  Vol. 1 - Technical advances, C.R. Hill and C. Alvisi
  Vol. 2 - Clinical advances, C. Alvisi and C.R. Hill - Pitman
  Medical Publications (1980)

Ultrasound in tumour diagnosis. C.R. Hill, V.R. McCready and
   D.O. Cosgrove, Pitman Medical Publications (1977)
Echo cardiology, N. Bom, Martinus Nijhoff (1977)
Acoustical imaging (Vol. 12, E. Ash and C.R. Hill, Eds. 1982)and
   previous volumes, Plenum Press
Ultrasound, Microwave and Radiofrequency radiations: The basis for
   their potential in cancer therapy, C.R. Hill (Ed) British Journal
   of Cancer 45, Supplement 5 (1982)
Medical Ultrasonic Images: Formation, display, recording and
   perception, C.R. Hill and A. Kratochwil, Eds. 1981 Excerpta Medica.

NUCLEAR MAGNETIC RESONANCE PROTON IMAGING

Linda EASTWOOD

Department of Biomedical Physics and Bioengineering,
University of Aberdeen, Foresterhill, Aberdeen AB9 2ZD,
Scotland, U.K.

1. INTRODUCTION

Nuclear Magnetic Resonance (NMR) imaging, first conceived only
ten years ago, has caused growing excitement in medical circles,
especially since good clinical results began to be obtained (Hawkes
et al, 1980; Smith et al 1981a, b; Young et al 1981). The technique
is capable of yielding not only good spatial resolution of anatomy,
but also information on local biochemistry.

Electromagnetic radiation at low radiofrequencies (about 2 to
15 MHz) can easily penetrate human tissues. It may not seem
promising for imaging, since its long wavelength precludes either
simple collimation or the focussing originally suggested by Damadian
(1974), but it can be used if another method of spatial localization
can be found. In NMR radio-frequency (RF) radiation is used to
excite resonant response from selected nuclei in the body under
investigation. In NMR imaging, this resonant response is spatially
localized or coded by using magnetic field gradients (Lauterbur,
1973).

In this chapter the physical and biological background to
NMR proton imaging will first be considered, showing why it should
be valuable clinically. Then the procedure for obtaining spatial
information will be given together with developments of basic techniques. Finally, the safety of NMR imaging, and techniques for
"dosimetry" and image quality control will be discussed. Throughout
the bias is towards techniques used in Aberdeen, but the description
is broadened to cover most NMR imaging methods.

2. PHYSICAL AND BIOLOGICAL BACKGROUND

2.1 Controlling the magnetic nucleus

To understand NMR imaging, look first at the interaction between a single nucleus and a magnetic field. Any nucleus is positively charged. If it also "spins", the spinning charge constitutes a magnet, and interacts with external magnetic fields. Nuclear constituents, protons and neutrons, each possess spin of one half. In some nuclei (Table I), these spins cancel out but in others we observe integral or half-integral spin. Nuclei with spin 0 cannot be studied by NMR. Thus, whilst carbon, because of its biological significance, would seem an interesting candidate for NMR imaging, only the spin 1/2 isotope carbon-13 could be observed and this has a very low natural abundance.

The ratio between magnetic dipole moment, μ and angular momentum, p, or spin I, for a given nucleus is given by γ , the gyromagnetic (or magnetogyric) ratio. Thus

$$\underline{\mu} = \gamma \cdot \underline{p} = \gamma \cdot \hbar \cdot \underline{I}$$

where $\hbar$ is Planck's constant divided by 2π . High γ means high NMR sensitivity. We see from Table I that the hydrogen-1 nucleus, the proton, is the most promising nucleus for NMR imaging for reasons of spin, sensitivity and abundance. Fluorine-19 shows the next highest sensitivity, but its natural abundance is very low. It has been suggested as a contrast agent. Phosphorus-31 is of particular interest, because of its significance in the biological energy cycle, but sensitivity is low.

To control nuclear spins, a static magnetic field, B_0 , which, by convention, is in the +z direction is required. The interaction energy between the field and the magnetic moment of a nucleus is given by the scalar product:

$$E = -\underline{\mu} \cdot \underline{B_0} = -\gamma \cdot \hbar \cdot I_z \cdot B_0$$

By quantum theory, I_z, the z-component of nuclear spin, may take only discrete values. We find corresponding discrete energy states, with energy differences:

$$\Delta E = \gamma \cdot \hbar \cdot B_0$$

For protons, with spin 1/2, there are 2 energy levels, corresponding to approximate alignment with and against the field.

By the Bohr relation, the energy difference between allowed states is associated with a characteristic frequency,

$$\omega = \Delta E / \hbar = \gamma \cdot B_0 \qquad (1)$$

TABLE I

PROPERTIES OF SELECTED NUCLEI

Nucleus	Spin	Elemental Abundance in man * (% wt)	Isotopic Abundance (%)	Relative Atomic Abundance in man	Gyromagnetic ratio rad S^{-1} T^{-1}	Relative NMR sensitivity (const. field equal no. nuclei)	Relative imaging sensitivity (const. field)	NMR frequency at 0.1T (MHz)
^{12}C	0	22.86	98.9	.19	–	0	0	–
^{16}O	0	61.43	99.8	.39	–	0	0	–
^{32}S	0	0.2	95.0	.0006	–	0	0	–
^{40}Ca	0	1.43	96.9	.0035	–	0	0	–
^{1}H	$\frac{1}{2}$	10.0	99.98	1.	2.68×10^{8}	1.0	1.	4.27
^{13}C	$\frac{1}{2}$	22.86	1.11	.0021	0.67×10^{8}	1.59×10^{-2}	3.39×10^{-5}	1.07
^{19}F	$\frac{1}{2}$	.004	100.	.00002	2.52×10^{8}	0.833	1.77×10^{-5}	4.01
^{31}P	$\frac{1}{2}$	1.11	100.	.0036	1.08×10^{8}	6.63×10^{-2}	2.39×10^{-4}	1.72
^{2}H	1	10.0	.02	.0002	4.11×10^{7}	9.65×10^{-3}	1.87×10^{-6}	0.65
^{14}N	1	2.57	99.64	.018	1.93×10^{7}	1.01×10^{-3}	1.86×10^{-5}	0.31

We thus come to the second part of control of nuclear spins. Electromagnetic radiation at the characteristic frequency stimulates transitions between energy states. Further, the value of this characteristic frequency may be controlled via the static magnetic field B_O. As an example, the earth's field is about $50\mu T$ (0.5 Gauss) and gives protons a characteristic frequency about 2kHz. For NMR imaging, we might use a field of 0.1T (1kG), in which protons resonate at 4.27 MHz. The next nearest resonance (Table I) is that of fluorine-19, nearly 300kHz away, so there is no problem in separating the responses of different nuclei.

2.2 Choice of static magnetic field

When RF radiation is used to excite nuclei in a body, only a small proportion of spins contributes to the observed response, which depends on macroscopic magnetization, the product of dipole moment and the number of "active" spins. In NMR imaging, the requirement is for acceptable signal-to-noise ratio, and hence the highest possible macroscopic magnetization.

If we consider, for instance, nuclei of spin 1/2 in a magnetic field, a number $N\downarrow$ of spins is aligned against the field, in the high energy state, and $N\uparrow$ spins are aligned with the field. At equilibrium, the ratio between numbers in the two states is given by the Boltzmann distribution:

$$N\downarrow / N\uparrow = \exp(-\hbar . \omega / k.T),$$

where T is temperature, and k is Boltzmann's constant. In response to excitation, only unmatched spins are observed; the number of active spins is the difference between $N\uparrow$ and $N\downarrow$.

$$N_A = N\uparrow - N\downarrow \approx N.\hbar.\omega / 2.k.T$$

where N is the total number of spins. At low MHz frequencies used in NMR imaging, and at room temperature, we see only about one in a million of the selected nuclei.

Macroscopic magnetization is given, for nuclei of spin 1/2, by

$$\underline{M}_o = N_A . \underline{\mu} \simeq N. \mu^2 .B_o /k/T. \qquad (2)$$

To maximise M_o , then we would like a low sample temperature, but that is hardly feasible in clinical imaging! We also want a large number of nuclei, and high magnetic moment, demonstrating again the advantage of protons for clinical NMR imaging.

From equation 2, macroscopic magnetization is proportional to the field, suggesting that the field should be as high as possible.

However, a high field means a more expensive magnet. Further, by
equation 1, a high magnetic field is associated with high frequency
signals, and attenuation in the body starts to be significant at
about 10MHz. This suggests an upper frequency limit for whole-
body imaging of about 10-15MHz, corresponding, for protons, to a
maximum field of 0.35T. There are other limitations to signal-to-
noise ratio at high frequency because of problems of detector
design, so the cost-benefit curve of field versus signal-to-noise
ratio tends to flatten off at high field.

In practice, a wide range of field strengths is used. A
0.35T magnet (proton resonance 15MHz) has been used with good
results in San Francisco (Crooks et al 1982). It is a large superconducting solenoid, requiring a supply of liquid nitrogen and
helium for cooling. Capital cost for the magnet would probably now
be about half to three-quarters of a million dollars, before an
imager was built around it. At the low field end of the range is
the 0.04T magnet (proton resonance 1.7MHz) used for the Aberdeen
Mk.I imager (Hutchison et al 1980). The magnet consists of four
resistive coils, with capital cost now about 60,000 dollars,
roughly ten times less than the previous example. It requires about
16 kW electrical power and cooling water. This magnet, with the
newer "spin-warp" imaging method described by Edelstein et al 1980,
was used to produce the clinical images shown later in this paper.

2.3 Observation of nuclear spins

Signals observed from nuclear spins in a body depend on macroscopic magnetization, $\underline{M}$, the behaviour of which is more simply
described by classical, rather than quantum mechanics. In a
magnetic field, $\underline{B}_0$, $\underline{M}$ precesses around the field direction, just
as the axis of a spinning top precesses about the earth's gravitat-
ional field. The rate of change of angular momentum is given by the
torque, $\underline{M} \times \underline{B}_0$, and $\underline{M}$ is related to angular momentum by the gyromagnetic ratio γ , so: $\quad d\underline{M}/dt = \gamma . \underline{M} \times \underline{B}_0$

However, if the precession (or "Larmor") frequency is ω_L, then:

$$d\underline{M}/dt = \underline{\omega}_L \times \underline{M}$$

so, equating these, we have:

$$\underline{\omega}_L = -\gamma . \underline{B}_0$$

The precession frequency in this classical model, then, is the same
as our earlier characteristic frequency (equation 1) for transitions
between energy levels.

382

At equilibrium, $\underline{M}$ is parallel to, and generally much less than, $\underline{B}_o$. To observe $\underline{M}$, we must tip it away from $\underline{B}_o$. This can be achieved by applying a field $\underline{B}_1$, perpendicular to $\underline{B}_o$ and rotating at ω. Then $\underline{M}$, while precessing about $\underline{B}_o$, experiences a second field at fixed orientation to itself, and precesses around that, too. The result is a spiral motion shown in Fig. 1. The larger the RF field, $\underline{B}_1$, or the longer the application time, the greater the precession angle.

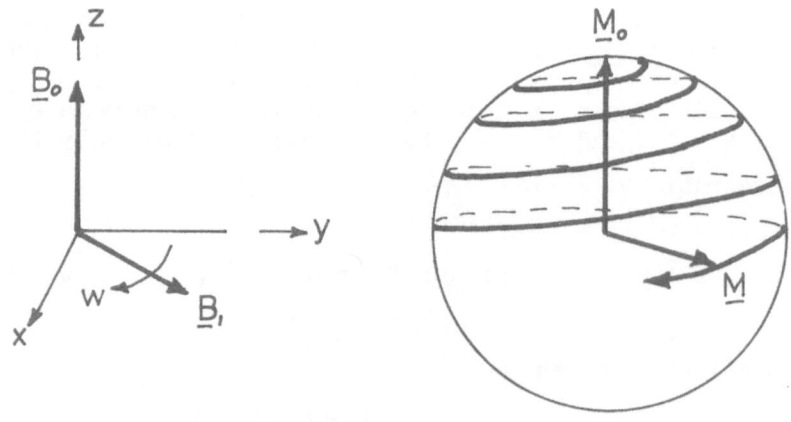

Fig. 1   Manipulation of macroscopic magnetization. a) Application of radiofrequency field $\underline{B}_1$ b) Precession of macroscopic magnetization $\underline{M}$.

If an RF pulse is just sufficient to tip $\underline{M}$ through 90°, it is termed a 90° pulse. This is our read-out pulse because $\underline{M}$ is now perpendicular to $\underline{B}_o$, so we can measure it. $\underline{M}$ still precesses about $\underline{B}_o$, and will induce a current in a coil wrapped around the body under investigation. A larger, or longer, RF pulse can tip $\underline{M}$ until it opposes $\underline{B}_o$. This is the 180 degrees, or spin inversion, pulse. While useful as part of a sequence of pulses, on its own it provides no signal, since $\underline{M}$, antiparallel to $\underline{B}_o$, does not precess. If $\underline{M}$ is at any other angle to $\underline{B}_o$, a signal smaller than that following a 90° pulse is seen. Only that component of $\underline{M}$ which is perpendicular to $\underline{B}_o$ contributes to the observed signal.

This point demonstrates that there are fundamental limitations to the signal-to-noise ratio in NMR imaging. An increase in RF power input beyond that required for a 90 degree pulse will yield a decrease, not an increase, in resultant signal.

The 180° pulse demonstrates another point of great practical importance, namely, that a spin-system which yields no observable signal is not necessarily in equilibrium.

To observe nuclear spins, then, we need to generate, and respond to, RF magnetic fields perpendicular to the main $\underline{B}_o$ field. In the Aberdeen whole-body imagers, with $\underline{B}_o$ perpendicular to the patient axis, this can be achieved with a solenoidal coil around the patient (Fig. 2a, b). When $\underline{B}_o$ is along the patient (the more common arrangement), more complex, inherently less efficient, RF coil designs are required (Hoult 1982)

A second requirement is for radio-frequency shielding. Signals from the body are small (typically tens of microvolts), so other radiowaves must be screened out. Some commercial imagers require a shielded room. The "local" shielding used on the Aberdeen Mk.II imager is shown in Fig. 2c.

2.4 Nuclear spin relaxation mechanisms

After excitation as described above, nuclear spins return to equilibrium. This relaxation, when it occurs in living tissues, yields interesting biological information.

The first of two closely related relaxation mechanisms is spin-lattice relaxation. If $\underline{M}$ is tipped through any angle, spins are in an excited state, and lose energy to their environment, the "lattice" by stimulated emission. For protons in soft tissues, the relaxation time T_1 for this process is about 0.1 to 1 second. By

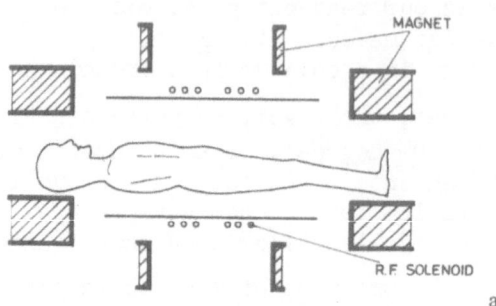

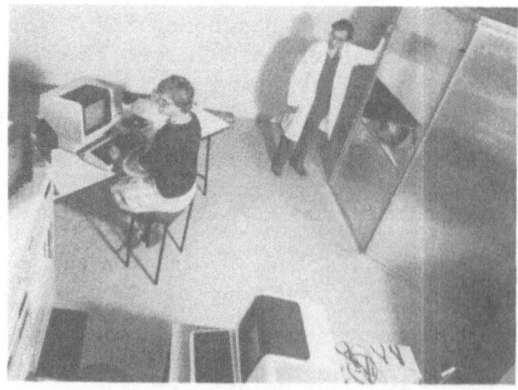

Fig. 2   The Aberdeen Mk.II NMR imager. a)Schematic b) The "patient
tube" and RF solenoid during construction c) During construction, showing local RF shielding.

this mechanism, M_z returns to its equilibrium value M_o. To a first-order approximation,

$$\partial M_z / \partial t = (M_0 - M_1)/T_1$$

Since M_z is "hidden" by $\underline{B}_o$, this process cannot be observed directly. Measurement of T_1 is discussed in section 2.6.

Relaxation of transverse magnetization M_{xy}, with relaxation time T_2, is brought about by the additional phenomenon of "spin-spin" relaxation, the exchange of energy between spins. Spins interact by creating local fluctuations in magnetic field, so that some spin presessions become slower, some faster, and they lose coherence. Without loss of energy, there is, in effect, a loss of entropy, and transverse magnetisation disappears.

$$\partial M_{xy} / \partial t = - M_{xy}/T_2$$

The transverse relaxation time T_2, like T_1 cannot be measured directly. In practice, transverse magnetisation, and so observed signal, is strongly affected by inhomogeneity in $\underline{B}_o$. This, too, leads to de-phasing of spin precessions, and gives an effective transverse relaxation time T_2^*, dependent on both fundamental mechanisms and field inhomogeneity. In NMR imaging, magnetic field gradients are introduced deliberately, leading to very rapid decay of transverse magnetization.

Thus, after application of a 90° RF pulse to a body, an RF signal with approximately exponential decay is observed. The initial size of the signal indicates the number of resonant nuclei present, but its decay indicates chiefly field inhomogeneity. More ingenuity is therefore required to measure fundamental relaxation times. Only spin-lattice relaxation is considered in detail in this paper.

2.5 Proton density; biological significance

Using proton NMR, we could try to image proton density (proton concentration), or relaxation times, or some combination. Most techniques image an inextricable combination but we shall consider here methods used in Aberdeen to extract proton density and T_1 values and examine their limitations.

The first limitation is that not all protons in the body are seen. In general, the more solid a material, and the more fixed its structure, the shorter its relaxation time. In cortical bone, for instance, excited proton spins revert to equilibrium so fast that, in most imaging techniques, resultant signals are never seen.

Similarly, protons in proteins, which are highly structured, are not seen directly, although they do influence observed relaxation times as discussed in section 2.6. Membrane lipids are too structured to give observable signals, but protons in free lipids are seen. In general, the strongest contribution to the signal is that of water.

The expected range of proton density values in soft tissues is not large. If the value for water is taken as 100, then most soft tissues would give values of about 70 to 80. We would not expect good contrast in an image. Cortical bone would give a low value, as would gas, so clear discrimination of bone, lung or gas in the gut would be expected.

To image proton density, excitation by $90°$ RF pulses is used. The signal amplitude then indicates the size of $\underline{M}$, the macroscopic magnetization, and therefore the proton density. In imaging we need to repeat the readout many times, allowing some recovery time, τ_R, between RF pulses for the spin system to return to equilibrium. Ideally τ_R would be long compared with any spin-lattice relaxation time, T_1 in the body. In practice, this would make imaging time unacceptably long, so a shorter time is used, leading to signal size S_1 given to a first approximation by:

$$S_1 = kM_0[1 - \exp(-\tau_R/T_1)] \quad \ldots \quad (3)$$

where k depends on the sensitivity of the instrumentation.

The "proton density" images produced in Aberdeen use a recovery-time of one second, long compared with most soft-tissue T_1 values, but leading to saturation of magnetization in body fluids. Where T_1 is long, signal size is reduced. (By measuring T_1 it would be possible to correct proton density images for this effect, but this correction has not been applied. For long T_1 values and hence small S_1 values it would yield noisy results.)

An example of a proton density image, a 24-week foetus made just before termination of pregnancy, is shown in Fig. 3. Amniotic fluid, with long T_1, gives a false appearance of low proton density. Soft tissue contrast is, as expected, not good, although fat is seen to have a high mobile-proton density. Pelvic bones, with low proton density, contrast well.

If an anatomical image is required, without information fundamental NMR parameters, then recovery time can be made much shorter, giving signals in which proton density and T_1 are inextricably mixed. This method is commonly used to provide so-called "T_1-enhanced" images (e.g. Crooks et al 1982).

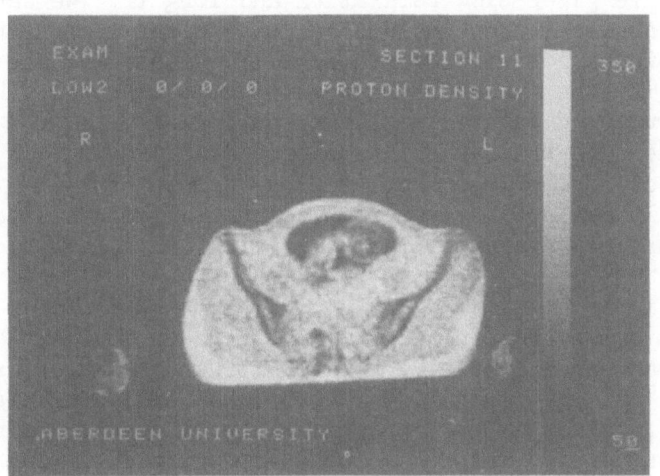

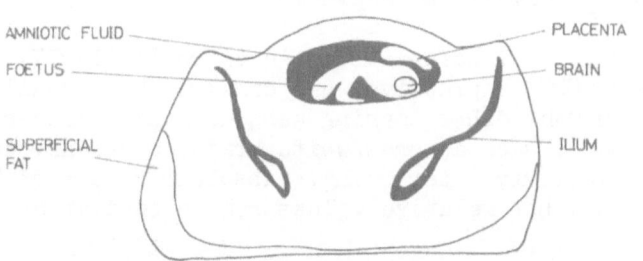

Fig. 3 Proton density image of 24-week foetus (just before termination). Image at Aberdeen Mk.I imager, .04T, 1.7MHz. Field
of view 450 mm square, slice 16 mm thick. Four minutes to
produce proton density and T1 images on 128 x 128 matrix
(displayed interpolated to 256 x 256).

2.6 Spin-lattice relaxation: biological significance

Since, as mentioned in section 2.4, spin-lattic relaxation occurs predominantly by stimulated emission, its rate depends on interaction between a spinning proton and its environment. A close, rigid structure gives fast relaxation and short T_1. A "wider" more random, structure gives slow relaxation and long T_1. We can relate this to, for instance, protein in a cell. Protons within the highly-structured proteins possess a very short T_1. Those in free water show long T_1. At the interface between the two there is a hydration sheath of "structured water", with a gradation of T_1 values. Diffusion is fast enough that, within a single measurement time of several milliseconds, a water molecule may sample many environments. Thus we normally measure an averaged T_1 for all intracellular water. In clinical imaging, extracellular water also contributes to observed T_1.

T_1 values observed clinically will depend on several factors. First, the total water content and water structure may interact in a complex way, but, in general, the more free water, the higher T_1. The range of T_1 values is much greater than the corresponding range of proton density values. Second, free lipid protons have much lower T_1 than water protons. Two tissues with very different water/fat ratios may have similar proton density, but will be clearly differentiated on a T_1 image. Third, there may also be effects due to paramagnetic materials, and maybe due to pH, although these will normally be negligible.

T_1 values for normal soft-tissues in vitro (Table II) show a range of around 400%, impressive when compared with small contrasts available in many other imaging methods. Differences between neighbouring tissues, such as grey/white brain, or kidney cortex/medulla, are particularly interesting. Absolute values of T_1 are frequency-dependent, but relative values remain similar at other frequencies.

True T_1 images require at least two measurements to be made for each point in the image, even if single-exponential decay is assumed. The most common procedure for T_1 measurement (Farrar and Becker 1971) uses the $180°$, τ, $90°$ pulse sequence. The $180°$ pulse completely inverts bulk magnetization $\underline{M}$ from $\underline{M}_o$ to $-\underline{M}_o$, yielding no transverse magnetization and thus no signal. M_z then relaxes back towards $+M_o$ at a rate given by:

$$M_z = M_0[1 - 2\exp(-t/T_1)]$$

TABLE II

NMR T_1 VALUES FOR NORMAL RABBIT AT 2.5MHz, 20°C
IN VITRO (from Ling et al 1980)

Liver	141 ± 16
Red Femur Marrow (fatty)	175 ± 50
Intercostal Muscle	191 ± 14
Kidney Cortex	206 ± 27
Stomach Wall	227 ± 20
Spleen	258 ± 4
White Brain	264 ± 11
Grey Brain	332 ± 22
Whole Blood (heparinized)	372 ± 34
Kidney Medulla	426 ± 61
Testis	463 ± 47
Blood Serum	820 ± 12
Bile	888 ± 388

If a 90° pulse is applied at time $t = \tau$, the resultant signal gives an instantaneous readout of the value M_z has reached at time τ. The pulse sequence may be repeated with different τ to give a complete plot of M_z relaxation.

If single exponential decay occurs, then T_1 can be calculated from the results of two pulse sequences, one a 90 degree pulse yielding signal S_1, the other a 180 degree, τ, 90 degree sequence yielding S_2. If we assume complete equilibrium of magnetization before each sequence, then:

$$S_2 = S_1[1 - 2\exp(-\tau/T_1)]$$
$$T_1 = \tau/\ln[2S_1/(S_1 - S_2)] \quad \dots (4)$$

T_1 images produced in Aberdeen use this basic method, although adiabatic fast passage (AFP), rather than a 180° pulse, is used for good spin inversion (Johnson et al 1982). A look-up table is used to implement a T_1 algorithm based on equation 4, but modified to allow for incomplete inversion etc. (Redpath 1982). Since the calculation assumes single-exponential decay, some "average" T_1 for each pixel is displayed. The calculation must be abandoned for very small values of S_1 (and hence $S_1 - S_2$), which yield spurious results.

Another method which can be used to produce T_1 images is based on equation 3, varying the recovery time, t^1, between 90° pulses.

Good T_1 contrast has, as expected, been found in practice in images created at 1.7MHz on the first Aberdeen imager, with occasional exceptions such as reduction of contrast between kidney cortex and medulla, due perhaps to tissue perfusion by blood and urine in life, or to movement of the kidney. In Fig. 4, a normal image (of the author's abdomen) shows the superiority of T_1 images over proton density images for soft-tissue discrimination. Note that large blood vessels are clearly seen without the use of contrast media.

The original impetus for developing clinical NMR imaging came from results of in vitro T_1 measurements by Damadian and co-workers (1973), who showed consistently elevated T_1 values in malignant tissues when compared with their normal counterparts. Damadian was indeed the first to propose NMR imaging (Damadian, 1974), although the first efficient implementation was that of Lauterbur (1973). Fig. 5 shows an example of such elevated T_1 values in vivo. Metastatic deposits are clearly seen in the liver, with T_1 values above 250 ms, as compared with normal liver values of about 150 ms (at 1.7MHz). The ascitic fluid surrounding liver and spleen also has a long relaxation-time, in this case about 900 ms. More

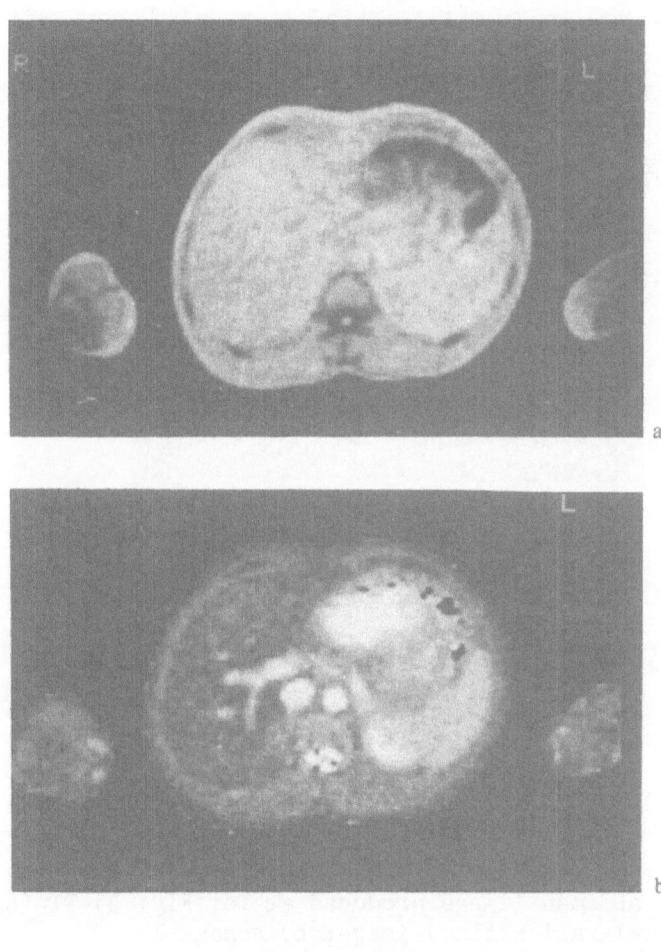

Fig. 4 Normal abdomen (images produced as for Fig. 3) a) Proton
    density image b) T1 (spin-lattice relaxation time) image
    c) schematic

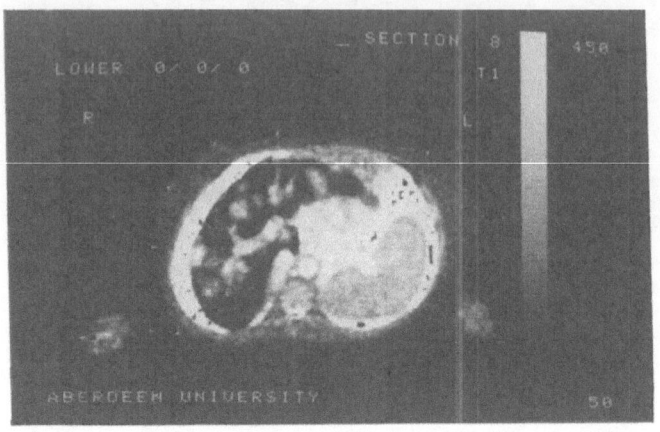

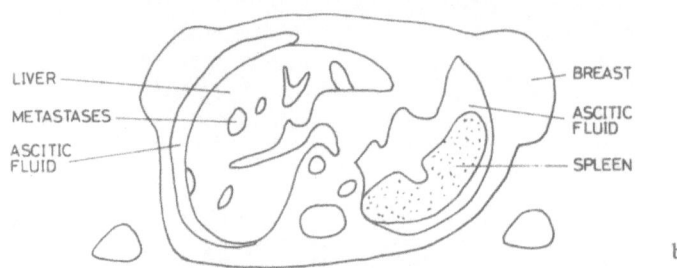

Fig. 5 Abnormal abdomen (image produced as for Fig. 3) T1 (spin-
lattice relaxation time) image; b) schematic

generally, observed T_1 values of ascitic fluid seem to vary between
about 500 and over 1000 ms, presumably depending on protein
content.

Moving up to the thorax, Fig. 6 shows a subject with right
pulmonary carcinoma, after earlier left mastectomy. Also seen
clearly is the upper heart, with blood-filled chambers (T_1 about
400 ms) clearly distinguished from heart wall (measured T_1 about
220 ms), even though this is not a cardiac-gated image. Such
differentiation is not normally possible on proton density images.

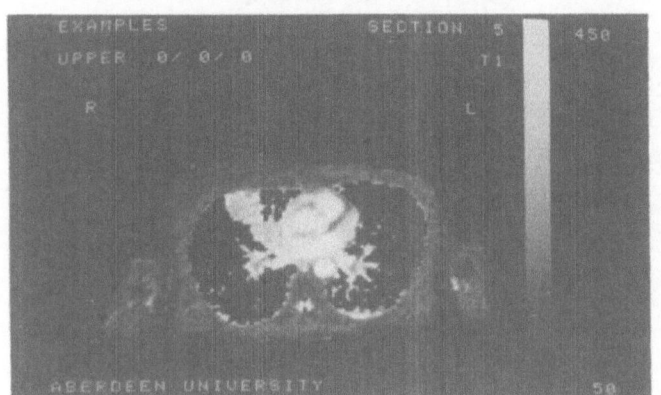

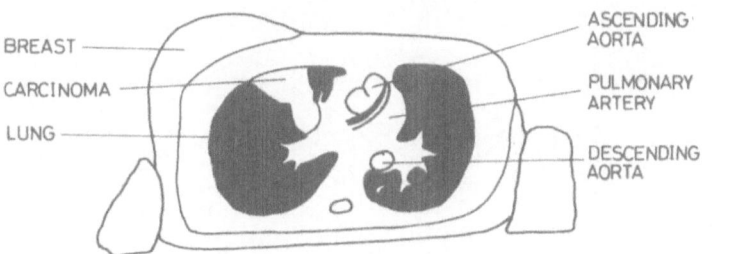

Fig. 6 Abnormal thorax (image produced as for Fig.3) a) T1 image
b) schematic

T_1 of the carcinoma in this case is about 300 ms. This lesion
would also be clearly seen on the proton density image, because
its density is much greater than that of the surrounding lung.
Measurement of T_1 can, however, yield further information. This
was illustrated for us by two patients with mediastinal masses
which appeared similar on X-rays. In the first case, with T_1 about
250 ms (greater than muscle, less than blood), the mass was
correctly thought, after NMR, to be carcinoma of the bronchus. The
second mass showed a T_1 value the same as aortic blood, and was
correctly deduced to be an aortic aneurysm.

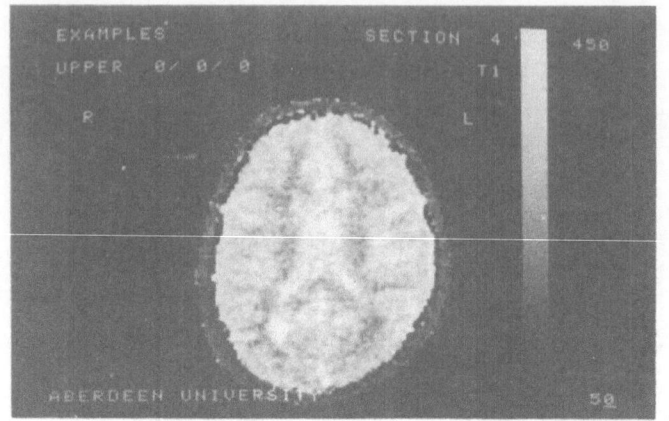

a

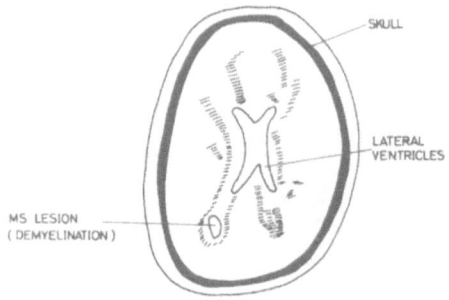

b

Fig. 7 Brain, with multiple sclerosis (image produced as for Fig. 3)
    a) T1 image b) schematic. Note that brain white matter
    appears darker than grey.

Figure 7 shows a T_1 image through the brain of a subject
with multiple sclerosis, and clearly shows discrimination between
normal grey and white brain (not normally distinguished on proton
density images), as well as high T_1 at the demyelination characteristic of multiple sclerosis.

Care will have to be taken in distinguishing pathological
changes in T_1 from physiological changes, as illustrated for
example by increases in T_1 of thigh muscles after exercise.
Work by Foster and co- workers (personal communication)

suggests that the increase in T_1 is related to changing pH with lactic acid build-up, as well as possibly to increased water content.

3. IMAGING TECHNIQUES

3.1 Classification of techniques

So far, only the background to NMR measurements has been discussed, without reference to image-formation. There are many variants of NMR imaging, which are not equivalent in either what they image or the artefacts to which they are prone. Most methods, however, are based on just four techniques of spatial discrimination, an understanding of which will clarify the whole subject.

NMR imaging may also be classified by two other criteria. One is the RF pulse sequence used. This affects the contribution of proton density, relaxation times or flow to the final image. In most cases (with the exception of "selective sensitivity" techniques) it may be chosen independently of the methods of spatial discrimination. Several possible RF pulse sequences were discussed in section 2.5 and 2.6; they will not be covered further here.

Another classification is by "interrogated volume". The region from which signals are received can be defined in a much more flexible way than for, say, X-ray or photon-emission tomography. Point, line, plane and volume selection are all possible. In the first, only one "voxel" (volume element) is excited at a time. This voxel is scanned through the object to build up an image. In "line" imaging, a single line is studied, and the emitted signals must be coded according to position along it. The line can be scanned through the object. "Plane" imaging requires two-dimensional and "volume" imaging three-dimensional, coding. Volume imaging is most efficient in terms of signal-to-noise ratio per unit imaging time and resolution, but it tends to result, at present, in long imaging times, and in data-handling problems. "Plane" imaging, currently the most common, will be assumed for most of this chapter.

3.2 Field gradients for spatial discrimination

Referring back to equation 1 ($\omega = \gamma \cdot B_0$), the characteristic frequency of selected nuclei is controlled by the static magnetic field, B_0. This is the basis for all NMR imaging showing that magnetic field gradients can be used to map space to frequency (Lauterbur 1973).

In practice, field gradients are generated by sets of current-carrying windings either inside or outside the main magnet. In the Aberdeen Mk.II imager, the three gradient windings all lie on a cylindrical former surrounding the patient (Hutchison et al 1980; Sutherland 1980). Each winding generates a vertical magnetic field with magnitude increasing in one direction. Gradients across the patient are generated by straight wires along the cylinder, with winding densities proportional to cos. 2ϑ and sin.2ϑ, giving quadrupolar fields. (These windings also act as a Faraday shield between RF coil and patient). The gradient along the patient is generated by a set of hyperbolic windings on the same cylinder. This gradient set has high efficiency and low inductance, but·is suitable only for a "side-access" magnet. Many other configurations are possible.

When field-gradients are generated, they can be used to give spatial discrimination by four different methods, classified here as i) frequency coding, ii) phase encoding, iii) selective excitation, iv) selective sensitivity. NMR imaging techniques may apply these methods to the various dimensions in almost any combination.

3.3 Selective sensitivity

The method referred to as "selective sensitivity" is the least widely used, and is also the most restrictive in that it alone uses a special RF pulse sequence. It has, however, been used to good effect at Nottingham (Moore and Holland 1980; Hawkes et al 1980), and so should be considered.

The RF pulse-sequence consists of a long series of a° pulses, alternating in phase, and separated by time t much shorter than T_2. This sets up an equilibrium condition known as steady-state free precession (SSFP), in which the signal never decays to zero between successive pulses (Carr 1958). SSFP produces relatively large and continuous resonance signals, and is thus very efficient, but it is not easy to separate contributions of T_1, T_2 etc. to the resulting signals.

The equilibrium condition described above relies on everything except the RF field remaining constant. Selective sensitivity (the "sensitive point" method of Hinshaw 1976) uses an oscillating field gradient so that the magnetic field is time-independent only in one plane. Signal-averaging removes contributions from all but the "sensitive plane". Two orthogonal gradients can be used to set up a rotating gradient, giving a "sensitive line". Three varying gradients can give a "sensitive point". If line or plane sensitivity is used, another method is required to give spatial coding of the resultant signal.

The method has yielded good images, but the use of SSFP is restrictive, and patient tissue movement leads to signal loss when SSFP equilibrium is destroyed.

3.4 Selective excitation: choosing the image plane

The other three methods of spatial discrimination (selective excitation, frequency coding and phase encoding) may be illustrated by the so-called "spin-warp" method developed in Aberdeen (Edelstein et al 1980), which uses all of them. The basic sequence of RF and field-gradient pulses is shown in Fig. 8. To produce an image, the entire sequence is repeated many times, with, typically, a one second repetition period. The 90 degree pulse is a "readout" pulse, so we might expect a decaying free-induction signal immediately following it. Instead, because of the applied magnetic field gradients, we obtain a later "spin-echo", modified by each gradient to yield spatial information.

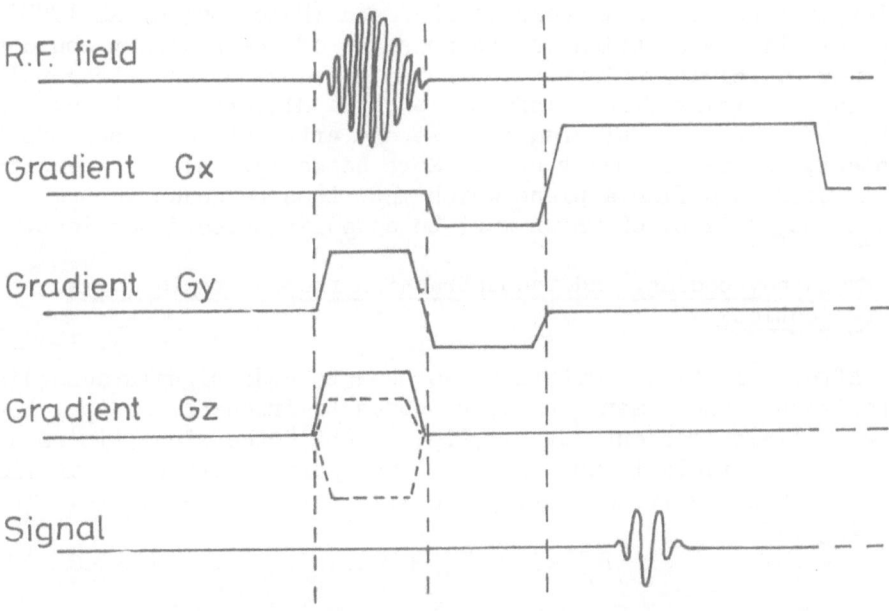

Fig. 8 Basic pulse-sequence for "spin-warp" imaging.

Consider first the gradient G_y along the patient. This is the only gradient applied during the 90 degree RF pulse, and gives slice selection by "selective excitation", used in many NMR imaging techniques. While the gradient is on, proton resonance frequency varies linearly along the patient. The RF pulse is shaped so that it contains only a narrow band, $\Delta\omega$, of frequences. It excites only protons in planes with the corresponding resonant frequencies in the patient. The resultant slice width is given by:

$$\Delta y = \Delta\omega \cdot G_y$$

and so may be controlled via either the RF spectrum or the y gradient. Slice profile is approximately (bot not exactly - rotation is not a linear operator!) equivalent to the shape of the RF spectrum, and may also be controlled.

During excitation, proton precession frequency varies across the slice width, so spins are out of phase after excitation. This dephasing, if left uncorrected, would result in signal loss. Referring again to Fig. 8, we see that the positive G_y during the RF pulse is followed by a pulse of negative G_y. This changes precession frequencies across the slice for just long enough to re-phase the spins.

Selective excitation may be extended to more than one dimension, as for instance in early work in Aberdeen (Hutchison et al 1980) when selective excitation in the presence of one gradient could be followed by selective inversion in the presence of an orthogonal gradient, yielding "line" information by a difference method. This technique, however, suffered from severe artefacts due to patient movement, and is no longer used. More commonly, selective excitation is used to define a plane which must then be coded in two dimensions, or a block which must be coded in three dimensions.

3.5 Frequency coding, and reconstruction from multiple angle projections

After selective excitation, we need to code signals according to position. One dimension may be coded to frequency by applying a magnetic-field gradient (G_x in Fig. 8) while the signal is read out. G_x is a field in the z direction (parallel to the main field, $\underline{B}_0$), but with magnitude varying linearly with x co-ordinate. Thus during readout,

$$B_z(x) = B_0 + x \cdot G_x$$

and signal frequencies correspond to the x co-ordinate by

$$\omega(x) = \gamma \cdot (B_0 + x \cdot G_x)$$

A Fourier transform (the frequency spectrum) of the signal thus gives us a projection of the selected plane onto the x axis, analagous to one of the many "shadow" projections obtained in X-ray computed tomography.

The variation in frequency along x will, as discussed in the previous section, lead to de-phasing of spins. This may be compensated, as shown in Fig. 8 by applying a negative G_x for a short time deliberately to de-phase spins, so that the positive read-out gradient will bring them back into phase as a "spin-echo" at the required time.

The "shadow" analogy to frequency projections, as given above, suggests one possible, and indeed widely used, method of creating NMR images (e.g. Lauterbur 1973; Young et al 1981). This is the technique of reconstruction from multiple angle projections (MAP), used in X-ray CT. Given a set of projections in different directions, the image plane could be reconstructed (and indeed the method has been extended to reconstruction from projections of a body in all three dimensions). The different projections may be obtained either by rotating a single set of gradient windings or, more elegantly, by combining two or three orthogonal gradients in varying proportions.

The method, although successful, places stringent requirements on the homogeneity of both static and gradient magnetic fields. Field errors create frequency errors, and thus distorted projections. If, for example, the static field B_o were high at one point in the image plane, then that point would be shifted sideways along every projection, and would be reconstructed as a disc or ring, rather than a point. Generally B_o varies smoothly across the field of view, resulting in overall blurring of the image. If the field errors were known, and were such that applied gradients still resulted in a monotonic increase of field in the given direction, then each projection could, in principle, be corrected before reconstruction. The procedure is however, computationally demanding, and would not allow for changing field homogeneity, as occurs during "warm-up" of a resistive magnet. The effect could also be reduced by increasing the field gradients, so that a given frequency error corresponded to a smaller spatial shift. This would, however, increase the bandwidth required to cover the width of a slice, and would thus increase thermal noise, reducing signal-to-noise ratio. Blurring due to field errors could also be reduced by improving field homogeneity, but this can be both difficult and expensive.

3.6 Phase encoding, and Fourier transform imaging

The effect of field errors would be reduced if frequency coding were used for only one dimension, so that a field error caused a lateral shift of position on the image, but no blurring, and hence

400

no loss of information. If we use selective excitation to define a slice, and frequency coding to code one dimension, we need another method for the third dimension. We can use phase-encoding. This is achieved by applying a variable pulse of one field gradient (G_z in Fig. 8) before, rather than during, readout. The original implementation (Kumar et al 1975) varied the time for which the gradient was applied, with amplitude constant. The approach illustrated in Fig. 8 (Edelstein et al 1980) applies the gradient for a constant time, but varies the pulse amplitude. (This reduces errors due to uncontrolled phase variations at B_o inhomogeneities. Such dephasing is the same for every pulse sequence, and the effects cancel out.)

Gradient G_z, not present during readout, does not affect signal frequencies, but it does influence phase. After a positive gradient pulse, spins at the top of a "column" (in the z direction) have, for a short time, precessed faster than those at the bottom, and are relatively advanced in phase. A phase-twist (or "spin-warp") has been applied to each column. At readout, phase depends on z according to:

$$\phi_z = \gamma \cdot z \int_0^T G_z \, dt \qquad (6)$$

Where T is the time for which the G_z pulse was applied. Each signal recorded is effectively multiplied by $\exp(- i \cdot \Phi_z)$. Decoding requires a set of signals with different Φ_z. These are established by varying the "warp" gradient, G_z, linearly with sequence number, n, according to the formula

$$\int_0^T G_z \, dt = G_0[n - (N+1)/2]$$

where G_0 is constant, and n steps from 1 to N. At each application of G_z, the phase twist between the two ends of a column increases by 2π.

If M samples are taken of each signal, and the signal is recorded N times, as described above, then the proton density matrix P (k,l) is calculated from the discrete Fourier transform of the sampled signal S(m,n).

$$P(k,\ell) = C \left| \sum_{m=1}^M \sum_{n=1}^N S(m,n)\exp(-2\pi jkm)\exp(-2\pi j\ell n) \right|$$

We thus see that the two dimensions of our data array are equivalent. Each is decoded via a Fourier transform. One dimension is time, the other may be regarded as "pseudo-time". The change in phase-twist along a column in "pseudo-time", as imposed by varying G_z, exactly simulates the change in phase-twist along a row with time, due to frequency-dispersion imposed by G_x.

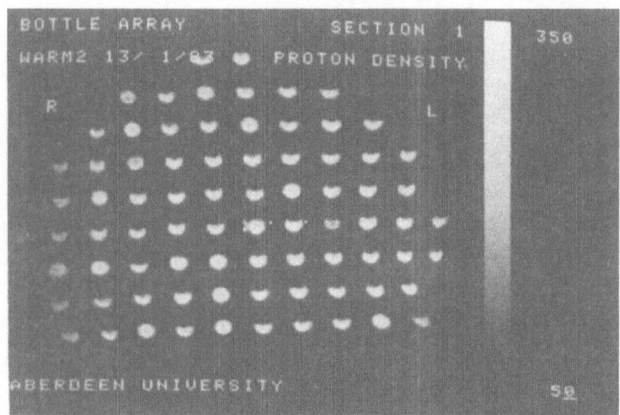

Fig. 9 Proton density image of an array of small bottles of $CuSO_4$
solution, each lying perpendicular to the image plane.
Fourier transform imaging used.

In the final image, the number of independent pixels in each
dimension is given by M (the number of time-samples per signal) and
N (the number of G_z steps). The height and the width of the field
of view can be controlled independently.

$$Width = 2\pi/\gamma G_x^+ \, dt$$

where Δt is the sampling period, and

$$Height = 2\pi/\gamma G_0$$

where G_0 is the gradient "step" of equation 5. The field of view
is thus not restricted to a square. A rectangle, more suitable
for most human frames, is equally feasible.

If, as suggested at the start of this section, field errors
are to cause no vertical shift on the image, then phase-encoding
must be unaffected by field error. This is true because the
"pseudo-frequency" (the phase-shift between samples) depends only
on G_0, the change in G_z amplitude, and not on absolute field
strength. This is illustrated by Fig. 9, which is an image
of a rectangular array of liquid-filled (or part-filled!) bottles,

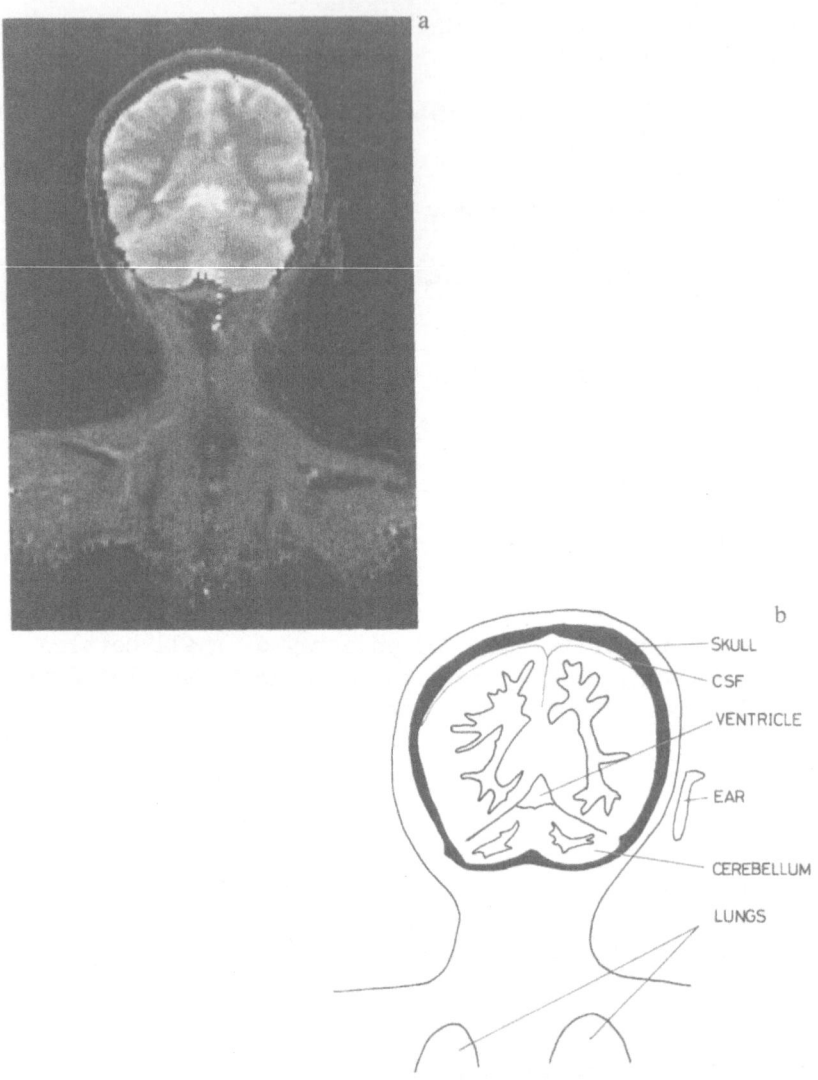

Fig. 10   Coronal T1 section through normal brain (image at Aberdeen
         Mk II imager .08T, 3.4MHz). The patient lies in the same
         position as for a transaxial image. Field gradient pulses
         are interchanged (slice selection G_z, frequency coding G_y,
         phase-encoding G_x)

403

taken just before a recent shim of our .04T magnet. There is a
slight anti-clockwise rotation of the whole image, due to rotation
of the gradient coils from their correct position. Effects of B_0
inhomogeneity are most marked at the right-hand side of the image,
where significant lateral shifts are seen, but with no blurring.
If field errors were known (or checked using an array "phantom"),
then distortions could be corrected at the final image.

3.7 Development of the basic techniques

The discussion has so far tended to imply that NMR imaging is
restricted to single transaxial slices across the patient. This is
by no means true. If, for instance, gradient directions are interchanged, then sagittal and coronal sections may very easily be
obtained (Fig. 10). Indeed, sections of any orientation are
feasible by combining gradients in suitable proportions.

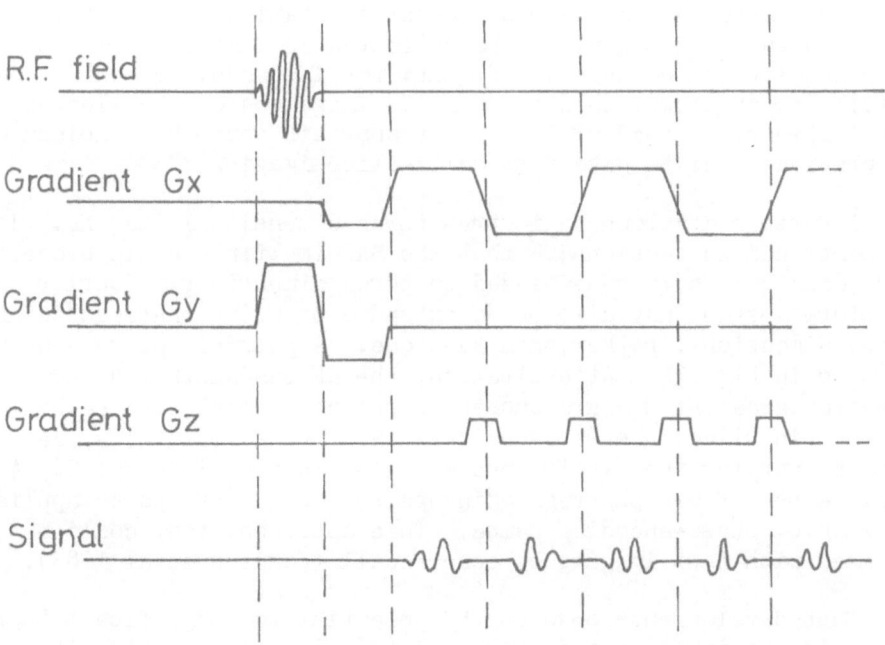

Fig. 11 A "fast-imaging pulse sequence.

NMR imaging is being developed in many ways, generally with the aim of making more efficient use of imaging time. Consider again the pulse sequence shown in Fig. 8. After each 90 degree (readout) pulse, a signal is recorded for, say, one second before the next excitation. This gives a useful "duty cycle" of only about 1 in 100, which it would clearly be desirable to improve.

One approach ("multi-slice" imaging) starts by noting that, because selective excitation is used, only one slice in the object is affected by the basic pulse sequence. During the recovery time for slice one, a second slice could be excited by using a 90 degree RF pulse with a different centre frequency in the presence of the same "selection" gradient, and the resulting signal recorded. By extension, imaging of several slices may be time-multiplexed.

Another approach notes that it is possible to create more than one "echo" signal after each excitation. This is because the dominant de-phasing (T_2*) mechanism acting is the field inhomogeneity due to applied field gradients, and this is reversible. Irreversible processes are occurring on a longer time scale. If, in Fig. 8, the frequency-coding gradient G_x were reversed after the first signal had been recorded, a second echo would be obtained. This approach is illustrated in Fig. 11, showing a series of echoes, alternately time-reversed. Also shown is the application of the phase-encoding gradient, G_z , between echoes, so that the phase-twist on each is that given by equation 6 , where $\int G_z$ now applies to all G_z pulses since the most recent excitation. The total time for which the echo signal may be collected is limited by T_2 and by inhomogeneity in the main field. If, for instance, 16 echoes could usefully be obtained, then a 128 x 128 image could be collected from 8 excitations, instead of 128. This approach could be combined with the previous one to give fast multi-slice imaging.

A further development is true three-dimensional imaging. It was mentioned in section 3.5 that the MAP (multiple angle projection) technique could be extended to three dimensions. Fourier transform imaging may also be so extended by using phase-encoding in two dimensions, rather than just one. A possible pulse sequence is shown in Fig. 12. At excitation, the RF bandwidth and the selection gradient G_y are chosen to select a thick slab rather than a thin slice. Phase-encoding is applied across the slice-width by varying the G_y "re-phase" pulse in time interval 2. A complete set of G_y phase-encoding pulses would have to be applied for each G_z phase-encoding pulse. This approach, too, could be combined with fast imaging by echo recall (Johnson et al 1983).

Such developments open up the potential for significant improvements in imaging time and/or signal-to-noise ratio (when signal averaging is used) in NMR imaging.

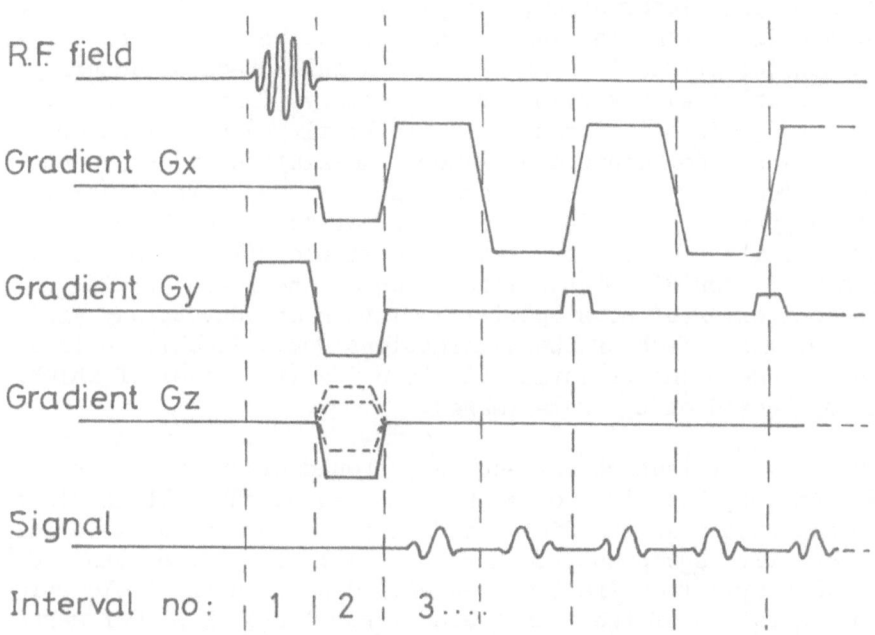

Fig. 12 A pulse-sequence for 3D Fourier transform imaging.

4. IMAGE QUALITY CONTROL

NMR imaging offers the possibility of acquiring information of both anatomical and biochemical significance. If, however, it is to be used to the full, the accuracy of this information should be monitored regularly.

Anatomical information requires that blurring (most likely in "MAP" techniques - section 3.5) is minimal, and that geometric distortion is either minimal or is itself accurately known. Blur and distortion are both caused by local variations in magnetic field, and might be introduced unexpectedly if, for instance, a ferro-magnetic object (such as an oxygen cylinder) were inadvertently placed too close to the magnet. A regular check could be kept by imaging an "array" phantom: either a very simple one, as shown in Fig. 9, or a more complex design with various "object" sizes to yield information on resolution as well as distortion. It is worth noting that such a phantom does not show variations in the flatness or thickness of the image slice. Work is currently in progress on simple phantoms, based on "edges" oblique to the image slice, to monitor these defects.

More important perhaps, are the accuracy and reproducibility of the numerical information, especially relaxation-time measurements, acquired from the images. Errors can be introduced via any of the static, gradient or RF fields. A full check of accuracy requires samples with varying proton density, T_1 and, perhaps, T_2 to be imaged. The first can be varied by mixing normal water with, for instance, deuterated water. Relaxation times can be varied by "doping" water with paramagnetic ions, perhaps using copper sulphate. More significant for day-to-day use, however, is reproducibility of results. Imaging a simple "flood" phantom (a container of "doped" water filling much of the field of view) allows monitoring of both spatial variation and day-to-day variation of results. Each of these variations would ideally be less than measurement uncertainties due to noise (the level of which could be checked on the same image).

Concern for measurement accuracy should extend not only to the NMR imager, but also to neighbouring equipment. It should be remembered that magnetic fields affect the paths of mobile electrons. Sensitive equipment, such as photomultiplier tubes, should be kept well away. The affected area will depend on the field strength and configuration of the magnet used. For Aberdeen's .08T magnet, for instance, the field is down to 0.1 mT (twice the earth's field) at a horizontal distance of 7 m from the centre of the magnet.

5. NMR IMAGING: SAFETY AND "DOSIMETRY"

5.1 Possible source of hazard

In this section, the most likely sources, and mechanisms, of biological effect will be indicated briefly. No attempt will be made to review all the evidence, much of which is discussed by Saunders (1982).

The three possible "direct" sources of biological effect or hazard in NMR imaging are static magnetic field, "slow" varying magnetic fields (the switched gradients) and radio-frequency fields. Evidence on these is encouraging, but still sparse. In Britain, the National Radiological Protection Board has drawn up a set of guidelines for safety in NMR imaging (NRPB, 1981), suggesting upper

limits for the various fields, and giving, as contra-indications
for imaging, a history of epilepsy or cardiac disease, use of a
cardiac pacemaker, or pregnancy. Such ideas are not based on known
hazards, but on prudence until such hazards can be discounted or
proved. The guidelines are likely to change as more evidence
becomes available.

When assessing these "direct" hazards, it is important to
keep them in the context of the overall hazard to the patient.
Other possible sources (all of which should be considered in imager
design) are electric shock, suffocation by gas from a quenching
superconducting magnet, and mechanical injury by magnetic drag on
small tools, mechanical collapse of a magnet, or even falling off
the couch!

5.2 Static field

Static magnetic fields used in NMR imaging vary from about
.04T to 0.5T, with the current trend being towards higher fields.
The applied field is generally very accurately known, but could
also be checked by using a Hall effect probe or (more accurately)
a proton resonance magnetometer. There remains the unsolved problem
of knowing local field variations at, for instance, ferromagnetic
implants in the patient.

The NRPB guidelines currently suggest an upper limit for the
static field of 2.5T. Published work to date suggests no lasting
effect of exposure to such fields. The most likely mechanism for
biological effect would appear to be the creation of electric
potential at flowing material. If, for instance, blood is flowing
at velocity v, perpendicular to the main field B, in a cylindrical
vessel of diameter d, then the potential evoked across the vessel
will be $E=B.d.v$. In, for instance, human aorta, peak velocity is
about 0.6 m/s, diameter about .025m, so E is about 15 mV/T across
the whole vessel. The observation by Beischer (1969) of an
electric field of about 0.07 mV/T superimposed on the ECG from
squirrel monkeys in static magnetic fields was considered to be
due to this effect. The electrical potential is across the whole
vessel; that across any single cell will be much less than the cell
depolarization threshold of about 40 mV. Thus there would appear
to be little hazard arising directly from the static field.

5. 3 Switched fields

Two possible sources of switched fields are the field-gradient
coils at gradient pulsing, and the main magnet at switch-on or
switch-off. In the latter case, the rate of change should be
limited by the time-constant of the magnet inductance and resistance. In the two Aberdeen resistive imagers, for instance, time

constants of 0.3s (.04T magnet) and 1s (.08T magnet) limit rates of field-change to .12T/s and .08T/s respectively. This may be compared with an NRPB-suggested upper limit of 20T/s if the "pulse" is more than 10 ms long. (Superconducting magnets, in which eddy currents will be set up in metal dewars and in the copper matrix supporting the superconductor, will have very long time-constant at "quench").

The maximum rate of change of field due to gradient coils generally occurs well outside the imaging plane, for instance at the patient's head and feet during selection of a slice across the abdomen. Typical values would be in the order of a few T/s or less. Actual values may easily be checked using a search coil. (The most practical approach is probably to use the search coil, and a known sinusoidal current through the gradient coil, to plot the magnetic field per unit current, and then to monitor the actual current through the gradient coil.)

Any effects of switched fields are likely to be due to eddy currents in loops of soft tissue, which is highly conductive. Consider, for example, a toroid of radius r, thickness dr, conductivity σ, through the middle of which passes a field B. The evoked current density is given by:

$$i/dr^2 = (\sigma.r/2).(dB/dt).$$

If a "loop" in a large organ had radius of 10 cm. and typical soft-tissue conductivity of 0.2 siemens/m, then a field changing at 1T/s would induce a current of $1 \mu A/cm^2$, negligible compared with normal ion currents of about $1mA/cm^2$. It is thus difficult to envisage any effects of eddy currents in clinical NMR imaging, and indeed none has been observed. Published work on effects (visual phosphenes, ventricular fibrillation etc.) of electrical currents applied via external electrodes also suggest that currents induced in NMR imaging would be well below effect thresholds. It should, however, be borne in mind that little is known about where current loops would occur in-vivo, or whether loops could overlap to give "hot spots". Further work is required on this.

5.4 Radio-frequency fields

The main effect of exposure to RF fields is thought to be temperature rise. NRPB recommendations suggest that a rise of less than 1 degree C is acceptable. In the Aberdeen imagers, the peak RF field is currently $40 \mu T$, applied for 10 ms every 2 seconds, and the time-averaged power deposition is small compared with the basal metabolic rate of about 1W/kg. It should, however, be noted that some systems, and some RF pulse sequences, apply significantly more RF power, and also that power absorption in the body increases with frequency. Each NMR technique should be considered separately.

409

Particular care should be taken when applying RF power to organs such as testes or the lens of the eye that may have little or no blood supply to carry heat away.

The applied RF field in any imager may easily be checked using a search coil. Total power absorption in a patient may be found by monitoring either the change in Q-factor of the RF coil, or the change in applied power needed to maintain the same field, when the patient enters the coil. Local variations in power absorption are not, however, amenable to measurement.

ACKNOWLEDGEMENTS

The Aberdeen work described in this paper comes from a team, with imager-development led by Dr J.M.S. Hutchison, biological work by Dr M.A. Foster, clinical work by Dr F.W. Smith, and the whole under the guidance of Professor J.R. Mallard. The author would like to acknowledge help from the whole team and financial support from an MRC (Medical Research Council) Research Fellowship. M & D Technology Limited sponsored the author's attendance at this meeting; their overall collaboration with Aberdeen University in the development of NMR imaging is also valued.

REFERENCES

Carr H.Y. (1958) Steady-state free precession in nuclear magnetic resonance. Phys. Rev., 112, 1693-1701
Crooks L.E., Mills C.M., David P.L. et al (1982) Visualization of cerebral and vascular abnormalities by NMR imaging. The effects of image parameters on contrast. Radiology, 144, 843-852
Damadian R., Zaner K., Hor D. et al (1973) Nuclear magnetic resonance as a new tool in cancer research: human tumours by NMR. Ann. N.Y. Acad. Sci., 222, 1048-1076
Damadian R. (1974 - but filed Mar.1972) Apparatus and method for detecting cancer in tissue. U.S. Patent 3, 789, 832.
Edelstein W.A., Hutchison J.M.S., Johnson G. and Redpath R.(1980) Spin-warp NMR imaging and application to human whole-body imaging. Phys. Med. Biol., 25, 751-756
Farrar T.C., and Becker E.D. (1971) Pulse and Fourier Transform NMR. Academic Press
Hawkes R.C., Holland G.N., Moore W.S., Worthington B.S. (1980) NMR tomography of the brain: a preliminary clinical assessment with demonstration of pathology. J. Comput. Assist. Tomogr., 4, 577-586
Hinshaw W.S. (1976) Image formation by nuclear magnetic resonance: the sensitive-point method. J. Appl. Phys. 47, 3709-3721.

Hoult D.I., (1982) Radio-frequency coil technology in NMR scanning. pp. 33-39 in "NMR Imaging" Proc. International Symposium on NMR imaging. Winton-Salem. October 1981. Ed. Witcofski R.L. Karstaedt N., Partain C.L.

Hutchison J.M.S., Edelstein W.A., and Johnson G. (1980) A whole-body NMR imaging machine. J. Phys. E: Sci. Instrum., 13, 947-955.

Johnson G., Hutchison J.M.S., and Eastwood L.M. (1982) Instrumentation for NMR spin-warp imaging. J. Phys. E.: Sci. Instrum. 15, 74-79

Johnson G., Hutchison J.M.S., Redpath T.W., Eastwood L.M. (1983) Improvements in performance time for simultaneous three-dimensional NMR imaging. Submitted to J. Mag. Resonance

Kumar A., Welti D., Ernst R.R. (1975) NMR Fourier Zeugmatography J. Mag. Resonance, 18, 69-83

Lauterbur P.C. (1973) Image formation by induced local interaction: examples employing nuclear magnetic resonance. Nature 243, 190-191.

Moore W.S., Holland G.N. (1980) Experimental considerations in implementing a whole-body multiple sensitive point nuclear magnetic resonance imaging system. Phil. Trans. Roy. Soc. Lond.B, 289, 511-518

NRPB (1981) Exposure to nuclear magnetic resonance clinical imaging. Radiography 47, 258-260

Redpath T.W. (1982) Calibration of the Aberdeen NMR imager for proton spin-lattice relaxation time measurements in vivo. Phys. Med. Biol. 27, 1057-1065

Saunders R.D. (1982) Biological hazards of NMR. pp. 65-71 in "NMR Imaging". Proc. International Symposium on NMR imaging, Winston-Salem, Oct. 1981, Ed. Witcofski R.L., Karstaedt N., and Partain C.L.

Smith F.W., Mallard J.R., Hutchison J.M.S., et al (1981a) Clinical applications of nuclear magnetic resonance. Lancet, Jan. 10, 78 - 79.

Smith F.W., Mallard J.R., Reid A. and Hutchison J.M.S. (1981) Nuclear magnetic resonance tomographic imaging in liver disease. Lancet, May 2, 963-966

Sutherland R.J. (1980) Selective excitation in NMR and considerations for its use in three-dimensional imaging. Ph.D. thesis, University of Aberdeen

Young I.R., Hall A.S., Pallis C.A. et al (1981) Nuclear magnetic resonance imaging of the brain in umltiple sclerosis. Lancet, Nov. 14, 1063-1066.

INDEX

418